Organic Light Emitting Diode

edited by
Marco Mazzeo

Organic Light Emitting Diode
Edited by Marco Mazzeo

Published by InTech
Janeza Trdine 9, 51000 Rijeka, Croatia

Publishing Process Manager Iva Lipovic

Technical Editor Teodora Smiljanic

Cover Designer InTech Design Team

Additional hard copies can be obtained from orders@intechopen.com

Organic Light Emitting Diode, Edited by Marco Mazzeo
p. cm.
ISBN 978-953-307-140-4

For EU product safety concerns:
IN TECH d.o.o., Prolaz Marije Krucifikse Kozulić 3, 51000 Rijeka, Croatia,
info@intechopen.com or visit our website at intechopen.com.

Meet the editor

Marco Mazzeo graduated in Physics with honors in 2001 and the Ph.D. in Physics in 2005 at the University of Salento. In 2006 he was in the group of Prof. Karl Leo at the Institute of Advanced Photophysics (IAPP) in Dresden (Germany) as a postdoc working on the development and physics of blue light emitting OLEDs for artificial lighting applications realized through the technique of "electric doping". Scientific coordinator of the "electroluminescent molecular devices" section of the National Laboratory of Nanotechnology (NNL) of the CNR-Nano unit of Lecce, he was and is technical-scientific responsible for several industrial projects with Italian and foreign industries (FIAMM, Leuci).

Contents

Preface

Organic Light Emitting Diodes have made great progress since their first presentation based on small molecule organic materials by Tang and Van Slyke in 1987. After more than two decades of research, the OLEDs emerged as an important and low-cost way to replace liquid crystal displays and recently lighting sources. Indeed organic semiconductors combine novel semiconducting optoelectronic properties with the scope for much simpler processing than their inorganic counterparts. The purpose of this book is to present an introduction to the subject of OLEDs and their applications. Although it is not possible to fully do justice to the vast amount of published information concerning these devices, we have selected those areas in materials, fabrication and applications that we feel are most relevant to practical devices. Some aspects of the field have reached a reasonable level of maturity, while others are in the process of rapid development. The volume begins with a few contributions dealing with materials for high efficiency OLEDs. Several materials are explored such as oligothiophenes (chapter 1) and iridium(III) complexes (chapter 2). The aim of these chapters is to show how new emitting compounds (fluorescent and phosphorescent) can be used to improve the efficiency of the devices by chemical functionalization. In addition, the possibility to tune the emission wavelength in a very wide range, from blue to near infrared, makes the devices made of these classes of molecules strongly competitive with respect to inorganic ones. Nevertheless, the synthesis of new emitting materials is not the only way to improve the efficiency. Transporting Materials are also important. In chapter 3 new transporting materials for foldable and flexible OLEDs have been reported, paying also attention to materials for fabricating high efficiency transparent displays. Another strategy to improve the efficiency of the devices is the use of inorganic nanoparticles.

The chapter 4 gives an overview of the recent works on nanocomposites used in OLEDs. Adding metallic, semiconducting and dielectric nanocrystals into polymer matrices enables to enhance the efficiency and duration of the devices because they can positively influence the mechanical, electrical and optical properties of the polymer in which they are embedded. The section devoted to materials ends with chapter 5, where the transport properties of disordered organic materials are analyzed, such as the dependences of carrier transport behavior and luminescence mechanism on dopant concentration of OLEDs. In the second section new technological structures have been reported, such as single-layered ionic p-i-n PHOLED (chapter 6), where the balance in the charge injection due to the ionic p-i-n structure was improved significantly by controlled adsorption of ions at the interfaces. This can simplify the conventional structure of the OLEDs, showing new perspectives for displays and lighting applications.

Chapters 7-9 report new strategies to improve the characteristics of organic display, such as the contrast and the mechanical flexibility. Indeed high contrast and mechanical flexibility are the real factors which make these devices strongly competitive with those based on liquid crystals. In conclusion, chapter 10 shows the technology to fabricate efficient white light OLEDs for lighting applications. In particular, the various techniques to improve the efficiency and the color quality of these devices are discussed. We are confident that such range of contributions gathered in this volume should constitute an adequate survey of present research on these new kinds of devices, which are a revolution in standard technology for information and lighting.

Editor

Marco Mazzeo
National Nanotechnology Laboratory (NNL) of INFM-CNR and Dip. Ingegneria Innovazione, Università del Salento, Via Arnesano Km. 5, I-73100 Lecce Italy

1

Organic light emitting diodes based on functionalized oligothiophenes for display and lighting applications

Marco Mazzeo[a], Fabrizio Mariano[a], Giuseppe Gigli[a] and Giovanna Barbarella[b]
[a]*National Nanotechnology Laboratory (NNL) of INFM-CNR and Dip. Ingegneria Innovazione, Università del Salento, Via Arnesano Km. 5, I-73100 Lecce (Italy)*
[b]*Consiglio Nazionale Ricerche (ISOF), Mediteknology srl, Area Ricerca CNR, Via Gobetti 101, I-40129 Bologna (Italy)*

1. Introduction

The electroluminescence properties of oligothiophenes are here reviewed. It is shown that thanks to joint molecular engineering and device improvement remarkable results have been achieved in recent years in terms of device operational stability and lifetime. These results open new perspectives in the search for tailor-made oligothiophenes with improved EL properties. Since the first report on the phenomenon of organic electroluminescence by M. Pope et al. in 1963 (Pope et al., 1963) and the description of the first organic light-emitting diode based on 8-hydroxyquinoline aluminum (Alq_3) as emissive and electron-transporting material by C. W. Tang et al. in 1987 (Tang et al., 1987), astonishing progress has been made in the field of Organic Light Emitting Diodes (OLEDs) owing to improved materials and device design (Burroughes et al., 1990; Greenham et al., 1993; Kraft et al., 1998; Friend et al., 1999; Pei & Yang, 1996; Yu et al., 2000; Scherf & List, 2002; Hung et al., 2005; Müllen & Scherf, 2006; Kalinowski, 2005; Shinar, 2004; D'Andrade, 2007; Misra et al., 2006; Baldo et al., 1998; Baldo et al., 2000; D'Andrade & Forrest, 2004; Kawamura et al., 2005; Yang et al., 2006; Chou & Chi, 2007). The promise of low-power consumption and excellent emissive quality with a wide viewing angle has prompted the interest for application to flat panel displays. High-efficiency OLEDs in various colours have been demonstrated and a few commercial products are already in the market, like displays for cell phones and digital cameras. Today much research is being carried out on white OLEDs for lighting applications, in order to attain lifetimes and brightness that would allow replacing current indoor and outdoor light sources at costs competitive with those of existing lighting technologies (D'Andrade, 2007; Misra et al., 2006).

One of the key developments in the advance of organic LED technology was the discovery of electrophosphorescence which lifts the upper limit of the internal quantum efficiency of devices from 25% to nearly 100% (Kawamura et al., 2005). Indeed, one of the factors contributing to device efficiency is the ratio of the radiatively recombining excitons (from injected holes and electrons) to the total number of excitons formed. With fluorescent emitters, statistically (parallel spin pairs will recombine to triplet excitons while antiparallel spin pairs will recombine to singlet and triplet excitons) only 25% of the generated excitons can recombine through a radiative pathway, causing an intrinsic limitation on the external quantum efficiency of the OLED. In phosphorescent materials - complexes containing heavy metals - strong spin-orbit coupling leads to singlet-triplet state mixing which removes the spin-forbidden nature of the radiative relaxation from the triplet state. Thus, when phosphorescent emitters are used, an internal quantum efficiency up to 100% can in principle be achieved since in phosphorescent emitters both singlet and triplet excitons can radiatevely recombine. The synthesis of phosphorescent triplet emitting materials (phosphors) has lead to remarkable improvements in EL quantum efficiencies and brightness (D'Andrade, 2007; Misra et al., 2006; Baldo et al., 1998; Baldo et al., 2000; D'Andrade & Forrest, 2004; Kawamura et al., 2005; Yang et al., 2006; Chou & Chi, 2007). Nevertheless, although much research is focused today on the synthesis of new phosphorescent emitters, a great number of laboratories are still working on fluorescent compounds. The reason for this lies in the higher chemical and electrical stability shown by many of these compounds. Another advantage is that most fluorescent materials can be deposited without dispersing them in a matrix. While indeed the phosphors need to be deposited into a wide gap material to avoid self quenching, there are numerous fluorescent compounds, including thiophene oligomers, which do not suffer this problem. Moreover, the problem of self-quenching together with the wide absorption band of phosphors implies that the host material must have a gap wider than those of the emitters, so the minimum voltage that it is possible to apply to the device is high compared to the voltage of devices based on fluorescent compounds.

So far, thiophene materials have played a little role in the development of organic LEDs compared to other materials such as polyphenylenevinylenes (Burroughes et al., 1990; Greenham et al., 1993; Kraft et al., 1998; Friend et al., 1999), polyfluorenes (Pei & Yang, 1996; Yu et al., 2000; Scherf & List, 2002; Hung et al., 2005), or phosphorescent complexes (D'Andrade, 2007; Misra et al., 2006; Baldo et al., 1998; Baldo et al., 2000; D'Andrade & Forrest, 2004; Kawamura et al., 2005; Yang et al., 2006; Chou & Chi, 2007) and the research in this field has mainly been confined to the understanding of basic properties. The electroluminescence of thiophene materials is a poorly investigated field, probably due to the fact that in the early days of OLEDs the most investigated thiophene materials displayed low electron affinities and photoluminescence quantum yields in the solid state and were believed to be mainly suited for application in field-effect transistors (Garnier, 1999). Moreover, the few investigations carried out later on phosphorescence in thiophene materials afforded rather disappointing results (Wang et al., 2004). Nevertheless, the finding that appropriate functionalization of thiophene oligomers and polymers may increase both electron affinity (Barbarella et al., 1998 a) and photoluminescence efficiency in the solid state (Barbarella et al., 2000), allows to achieve high *p*- and *n*-type charge carrier mobilities (Yoon et al., 2006), may lead to white electroluminescence via spontaneous self-assembly of a single oligomer (Mazzeo et al., 2005), may allow the realization of optically pumped lasers (Zavelani-Rossi et al. 2001) and very bright electroluminescent diodes (Mazzeo et al., 2003 a), has risen again the interest on the potentialities of these compounds, also in view of the next generations of organic devices like light-emitting transistors or diode-pumped lasers. This paper reviews the various approaches used to obtain electroluminescence from oligomeric thiophene materials and recent progress with various device designs and synthetic products. In section 2, electroluminescence from linear oligothiophenes is discussed focusing on bilayer device structures realized by spin coating. Section 3 presents the results obtained using V-shaped thiophene derivatives and section 4 describes the different approaches employed to achieve white electroluminescence with oligothiophenes. Section 5 reports new results obtained in heterostucture devices using a thermally evaporated compound.

The choice to focus on the eloctroluminescence of oligomeric thiophene materials is due to the fact that there has been little progress in polythiophenes as electroluminescent materials from earlier studies (Braun et al., 1992; Berggren et al., 1994; Barta et al., 1998) to more recent investigations (Charas et al., 2001; Pasini et al., 2003; Cheylan et al., 2007; Melucci et al., 2007).

2. Linear thiophene oligomers

The first attempt to get electroluminescence from thiophene oligomers dates back to 1994 (Horowitz et al., 1994). A detailed study was reported three years later based on an end capped sexithiophene (EC6T) used as emissive and hole transporting layer in a single layer device (Väterlein et al., 1997). The molecular structure and the photoluminescence and electroluminescence spectra of ECT6 at various temperatures are shown in Figure 1. The *I-V* and *EL-V* curves measured for an ITO/EC6T-/Ca-OLED at forward bias for temperatures in the range 30-270 K (thickness 65 nm) are also reported in the figure. The photoluminescence and electroluminescence spectra were virtually the same, indicating that the radiative recombination of excitons proceeded from the same excited states in both cases. The current-voltage (*I -V*) curves exhibited strong temperature and thickness dependence. External quantum efficiencies in the range 1-8x10^{-5} at room temperature were measured. The orange electroluminesce generated by the device could be observed with the naked eye but lasted only for a few seconds.

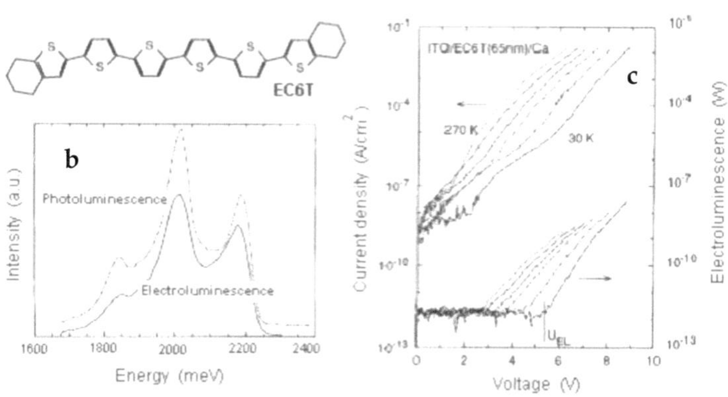

Fig. 1. a) Molecular structure of EC6T; b) photoluminescence and electroluminescence spectra at 4 and 20 K, respectively; c) *I - V* curves (top, left-hand scale) and EL-*V* curves (bottom, right-hand scale) of a ITO/EC6T/Ca OLED (thickness 65 nm) as a function of temperature (30, 90, 120, 150, 210, and 270 K from right- to left).

Two of the main drawbacks of conventional thiophene oligomers such as EC6T for applications in OLEDs are the low electron affinity (EA) and the non-radiative phenomena induced by packing causing the quenching of photoluminescence (PL) in the solid state.
Conventional oligothiophenes are easy to oxidize but difficult to reduce, as demonstrated by cyclovoltammetry (CV) measurements (Meerholiz & Heinze, 1996; Barbarella et al., 1998 a). In light emitting devices the low electron affinity generates a huge energy barrier between the cathode and the organic layer. In consequence, only a small number of electrons are injected, resulting in poor electron-hole balancing.
The low PL efficiency in the solid state is largely determined by the intermolecular interactions governing the supramolecular organization. The analysis of conformation and packing modalities in oligothiophene single crystals has pointed out the existence of numerous intra- and intermolecular interactions, such as van der Waals, π-π stacking, weak CH...S and CH...π hydrogen bondings and S...S contacts (Marseglia et al., 2000). Intermolecular interactions induce additional non-radiative channels so that the photoluminescence efficiency of oligothiophenes is lower in the solid state than in solution. For example, a photoluminescence quantum yield of about 40% was reported by several authors for solutions of quinquethiophene, which, however, dropped by several orders of magnitude in thin films of the same compound (Oelkrug et al., 1996; Kanemitsu et al., 1996; Ziegler, 1997). Owing to low photoluminescence efficiency caused by non-radiative phenomena induced by packing and to intrinsically low electron affinities, conventional oligothiophenes display very poor electroluminescence characteristics.

2.1 Achieving the simultaneous increase of electron affinity and solid-state PL via chemical modification

There are several ways to modify the molecular structure of oligothiophenes via chemical synthesis, as indicated in Scheme 1:

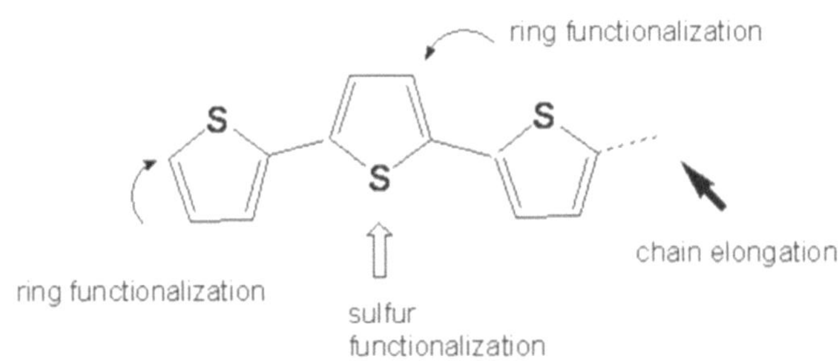

Scheme 1. Functionalization modalities in oligothiophenes

A considerable improvement in solid-state PL and electron affinity values was achieved via chemical modification of the thiophene rings through functionalization of the sulphur atom with oxygen (Barbarella et al., 1998 a). Indeed, in thiophene, sulphur has unshared lone-pair electrons which can be exploited to form chemical bonds with oxygen. In this way, a new class of oligomeric thiophene materials - namely oligothiophene-S,S dioxides, in which two oxygen atoms are linked to one or more thienyl sulphurs - was synthesized (Barbarella et al., 1998 b). By using this approach, compounds with greater electron affinities and

photoluminescence efficiencies in the solid state became available and proved to be useful for applications in electroluminescent diodes (Barbarella et al., 1999) and photovoltaic devices (Camaioni et al., 2004).
Scheme 2 shows the molecular structure of a few quinquethiophenes with one or more oxidized thienyl units and the variation of the corresponding redox peak potentials measured by cyclic voltammetry (Barbarella et al., 1998 a). Oxidation and reduction potentials are related to HOMO and LUMO orbital energies hence to ionization energies and electron affinities, respectively. The oxygen atoms cause the de-aromatization of the thiophene ring and allow the frontier orbitals to shift towards lower energies, causing in particular a sizeable increase in electron affinities (Barbarella et al., 1998 a). The scheme shows that the reduction and oxidation potentials of quinquethiophene can be tuned by changing the number and the position of the oxigenated units. In particular, it can be seen that the oligomer with alternating oxidized and non-oxidized thiophene rings is easier to reduce than to oxidize, opposite to the precursor quinquethiophene.

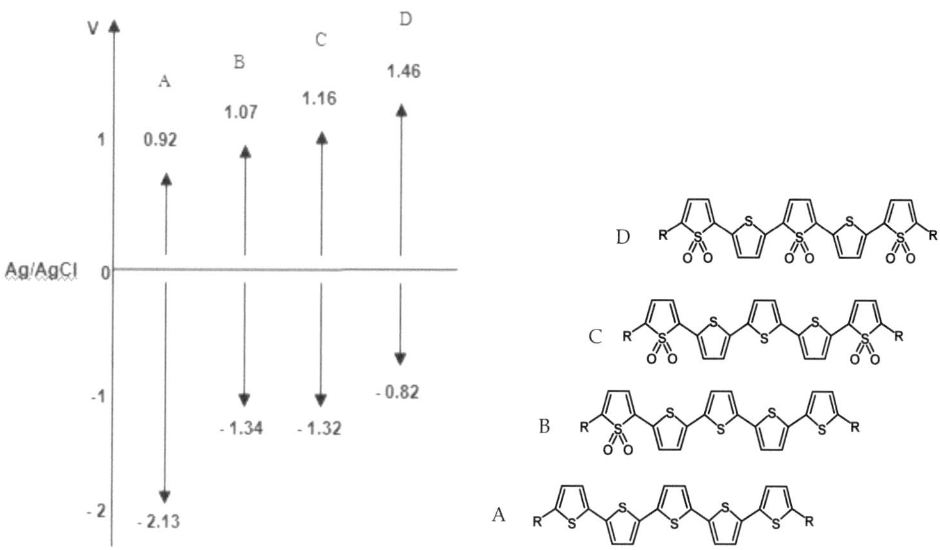

Scheme 2. Reduction and oxidation potentials of selected quinquethiophene-S,S-dioxides. R= $Si(CH_3)_3C(CH3)_3$.

Figure 2 shows an example of cyclic voltammogram of a quinquethiophene-S,S-dioxide, namely compound B in Scheme 2. The CV in the oxidation region shows two reversible waves with $E_{p,a1}$ = 1.00 V and $E_{p,a2}$ = 1.30 V. The CV in the reduction region shows two reversible waves with $E_{p,c1}$ = -1.28 V and $E_{p,c2}$ = -1.63 V, the first of which probably corresponds to the formation of the radical anion. It should be noted that the first oxidation potential is 0.15 V larger than that of the parent unmodified quinquethiophene, while the first reduction potential is shifted by 0.79 V towards less negative values, indicating a remarkable increase of the electron affinity of the molecule.
Since the first report in 1998 (Barbarella et al., 1998 a), the increase in molecular electron affinity upon inclusion of a thienyl-S,S-dioxide unit into the aromatic skeleton of conjugated oligomers and polymers has been observed by several authors (Hughes & Bryce, 2005;

Beaupré & Leclerc, 2002; Berlin et al., 2003; Perepichka et al., 2005; Casado et al., 2006; Liu et al., 2008).

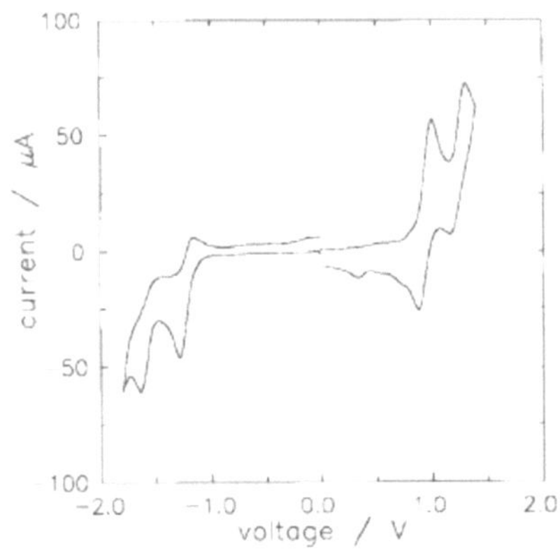

Fig. 2. Cyclic voltammogram (200 mV/s, 1 mM in CH_2Cl_2/Et_4NBF_4 0.2 M. V vs.Ag/AgCl) of quinquethiophene-S,S-dioxide B depicted in Scheme 2.

Recently, the redox potentials of fully oxidized (conjugated but no more aromatic) bi- and terthiophene have also been reported, showing that complete oxidation of the thienyl rings has a dramatic effect on both redox potentials and may cause a marked energy gap increase (Amir & Rozen, 2005).

Time-Dependent Density Functional Theory (TD-DFT) simulations in adiabatic approximation, carried out on a prototype terthiophene oxidized in the inner position (Raganato et al., 2004), indicated that the oxidation of the thiophene ring leads to the formation of new interactions in the LUMO orbital. The kinetic energy of the electrons in this orbital is lowered, while the energy of the electrons in the HOMO orbital is almost unchanged. As a consequence, the electron affinity of the whole molecule is increased.

The functionalization with oxygen obliges the molecules to pack far apart from each other in the solid state thus preventing the photoluminescence quenching caused by the close packing of non oxidized thiophene oligomers (Antolini et al., 2000). Consequently, the PL quantum yield (η, the number of photons re-emitted radiatively as a percentage of the number of photons absorbed) is increased, by an amount up to one order of magnitude (Barbarella et al., 1999). Scheme 3 illustrates the trend of variation of photoluminescence quantum yield (PLQY) in solution and in the solid state of α-quinquethiophene upon introduction of different chemical modifications in the aromatic skeleton (Barbarella et al., 1999). Upon insertion of alkyl chains, there is a dramatic decrease in PLQY in solution, from 40% to 9%, while in the solid state the PLQY value remains unchanged. Modification of the inner ring with oxygen atoms causes the PLQY in solution to drop to less than 1% but increases remarkably the PLQY value in the solid state, from 2% to 11%. Further functionalization of the thienyl groups with methyl substituents in the head-to-head orientation does not modify the PLQY value in solution, but further enhances that in the solid state, which reaches a very significant 37%.

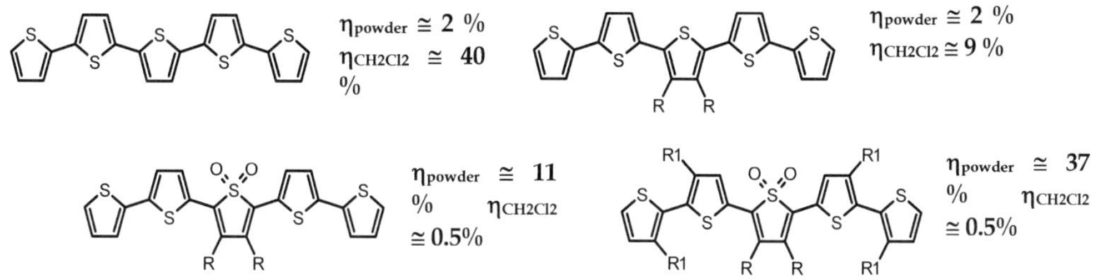

Scheme 3. Trend of variation of the photoluminescence quantum yield (η%) of α-quinquethiophene in solution and in the solid state following different chemical modifications. R= Hexyl, R1 = Methyl.

One of the highest PLQY values, measured in the solid state for an oligothiophene-S,S-dioxide was shown by the 'rigid-core' oligomer 3,5-dimethyl-2,3′-bis(3-methylthiophene)-dithieno[3,2-b;2′,3′-d]thiophene-4,4-dioxide, DTTOMe4:

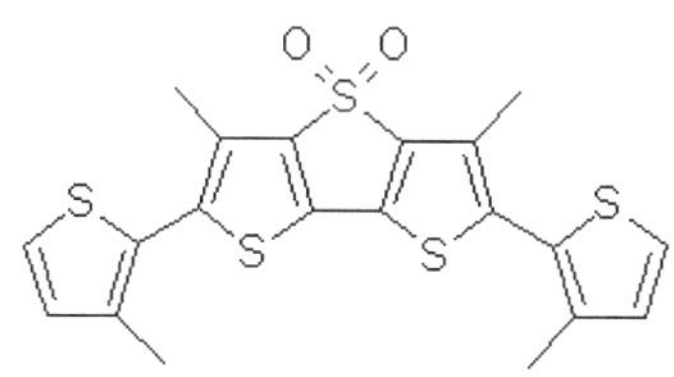

3,5-Dimethyl-2,3′-bis(3-methylthiophene)-dithieno[3,2-b;2′,3′-d]thiophene-4,4-dioxide (DTTOMe4)

This 'rigid core' compound displayed a PLQY value in the microcrystalline powder of η = 48% (Barbarella et al., 2001). DTTOMe4 belongs to a class of oligothiophene-S,S-dioxides characterized by high photoluminescence efficiency both in solution and in the solid state, contrary to conventional thiophene oligomers and conformationally flexible oligothiophene-S,S-dioxides (Barbarella et al., 2001). The conformation and the crystal-packing modalities of DTTOMe4 could be established directly from microcrystalline powder diffraction data (Tedesco et al., 2003). Semiempirical Intermediate Neglect of Differential Overlap with Single Configuration Interaction (INDO/SCI) theoretical investigations then allowed to obtain semiquantitative correlations between the structural characteristics of the compound and the main intermolecular factors that are known to affect the solid-state photoluminescence of organic molecules (Tedesco et al., 2003). The energy of the first singlet excited state and the oscillator strength of the optical transition, the exciton resonance interactions and the electron and hole transfer integrals for each pair of molecules were calculated using molecular geometries from the crystal structure. The calculations showed that there was very good agreement between the singlet excitation energies and the experimental maximum energy values of the absorption spectra in the solid state and in solution, indicating that the INDO/SCI approximation reproduces well the optical properties of this type of molecules. This result also suggested that exciton resonance interactions in the solid state (i.e., the interactions between the neutral excited states generated by photoexcitation) were weak. In agreement with this, the calculated exciton resonance interactions, which are proportional to the rate of excitation transfer between molecules, were small. According to the calculations, the intermolecular interactions play a major role in determining the solid-state photoluminescence efficiency, which correlates well with the rate of formation of non radiatively decaying charge-transfer pairs upon photoexcitation. The theoretical results obtained with DTTOMe4 indicated that a similar mechanism may also explain the very different photoluminescence quantum yields measured for ter-, quinque-, and heptathiophene-*S,S*-dioxides in the solid state (45, 12, and 2%, respectively) (Antolini et al., 2000). The trend in the PLQY values of these conformationally flexible molecules was first temptatively ascribed to the different orientations of the long molecular axes in single crystal structures: markedly tilted in the trimer, strictly parallel in the heptamer, with the pentamer in an intermediate situation. However, the theoretical and experimental data obtained for DTTOMe4 - in which the molecules pack with their long molecular axes parallel - show that, even when oligothiophene-*S,S*-dioxide molecules pack with their long molecular axes parallel, the photoluminescence efficiency can be quite high if the molecules are sufficiently distant from each other.

It is worth noting that first-principles Time-Dependent Density-Functional Theory (TD-DFT) calculations on terthiophene-S,S-dioxide have also shown that another important result of the functionalization of the thienyl ring with oxygen atoms is that the separation between the triplet state T2 and the singlet state S1 is enhanced with respect to the parent unmodified terthiophene (Della Sala et al., 2003; Anni et al., 2005). In this way, the probability of intersystem crossing from singlet states to optically forbidden triplet states is reduced, advantaging further the PL efficiency.

2.2 Electroluminescence in linear oligothiophene-S,S-dioxides

The increased electron affinities and PL quantum efficiencies of oligothiophene-S,S-dioxides in the solid state, combined with the optical and chemical stability of these compounds, allowed the fabrication of electroluminescent devices with much better characteristics than those obtained with conventional thiophene oligomers. The lifetime of the devices went from seconds to days and the characterization could be carried out in ambient atmosphere. The characteristics of the devices were well retained after a few hours of operation in air atmosphere, showing a good stability of the device.

Scheme 4 shows the molecular structure of a series of oligomers (compounds 1-6) containing a central thiophene-S,S-dioxide that were used as active materials in light emitting diodes (Gigli et al., 2001). The LEDs were prepared by spin coating onto indium-tin-oxide (ITO) coated glass substrates an hole transporting material, namely poly(3,4-ethylenedioxythiophene), PEDOT, doped with poly(styrene sulphonate), PSS, and then a dicloromethane (CH_2Cl_2) solution of the oligothiophene-S,S-dioxide. The deposition of the PEDOT-PSS layer was aimed at increasing the hole injection from the ITO anode into the oligothiophene-S,S-dioxide layer. The cathode was Ca capped with Al and prepared by thermal evaporation. The devices were characterized in air.

The electrooptical characteristics of 1-6 are reported in Table 1, while the electroluminescence spectra of all compounds, together with the current-voltage (*I-V*), luminance-voltage (*L-V*) characteristics and EL efficiency of the device obtained with compounds 3 are shown in Figure 3.

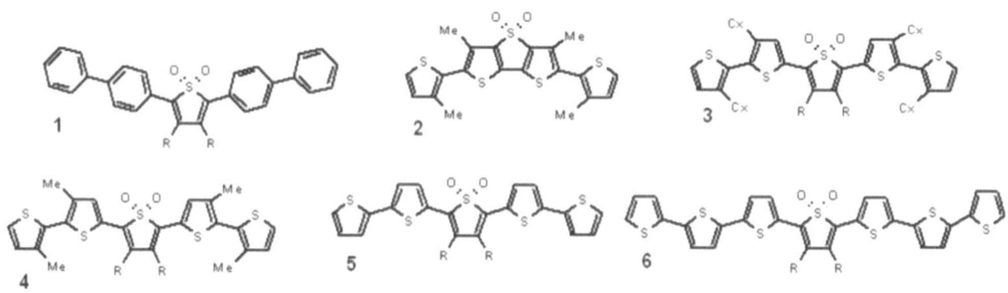

Scheme 4. Molecular structure of linear oligothiophene-S,S-dioxides 1-6. R = Hexyl; Me = Methyl; Cx = Cyclohexyl.

Changing oligomer size and substituents from 1 to 6 allowed to tune the electroluminescence from green to near-infrared (Gigli et al., 2001). Pentamers 1-5 emit in the green-red region, the colour tuning being obtained either by replacing the thienyls with phenyl groups or by distorsion of the oligomer chain length via β-functionalization with methyls or cyclohexyl groups. Light emission in the NIR region was obtained by using compound 6 as the active material. This long oligomer, a derivative of heptathiophene, adopts a fully planar conformation in the solid state (Antolini et al., 2000), which determines a large electron delocalization and a strong decrease of the optical gap.

	η %	EA (eV)	V (Volts)	Lum_M (cd/m^2)	φ %
1	70	3	3.2	100	0.03
2	48	2.9	2.3	110	0.004
3	22	3.1	4.8	400	0.2
4	37	3	1.9	110	0.08
5	13	3.1	2	105	0.03
6	2	3	4.9	80	0.002

Table 1. Electro-optical characteristics of componds 1-6[a]
a) η: PL efficiency; EA: electron affinity, extrapolated from CV data; V: turn-on voltage; Lum_M : maximum luminance; φ: EL efficiency.

As shown in Table 1, the turn-on voltages for luminance at 0.01 cd/m^2 were all between 2 and 5 V, strongly reduced as compared to the values reported for poly(alkylthiophenes)-based devices (Barta et al., 1998). This was the result of the increased electron affinity of compounds 1-6 induced by the S,S-dioxide functionalizaty and to the consequent reduction of the electron injection barrier.

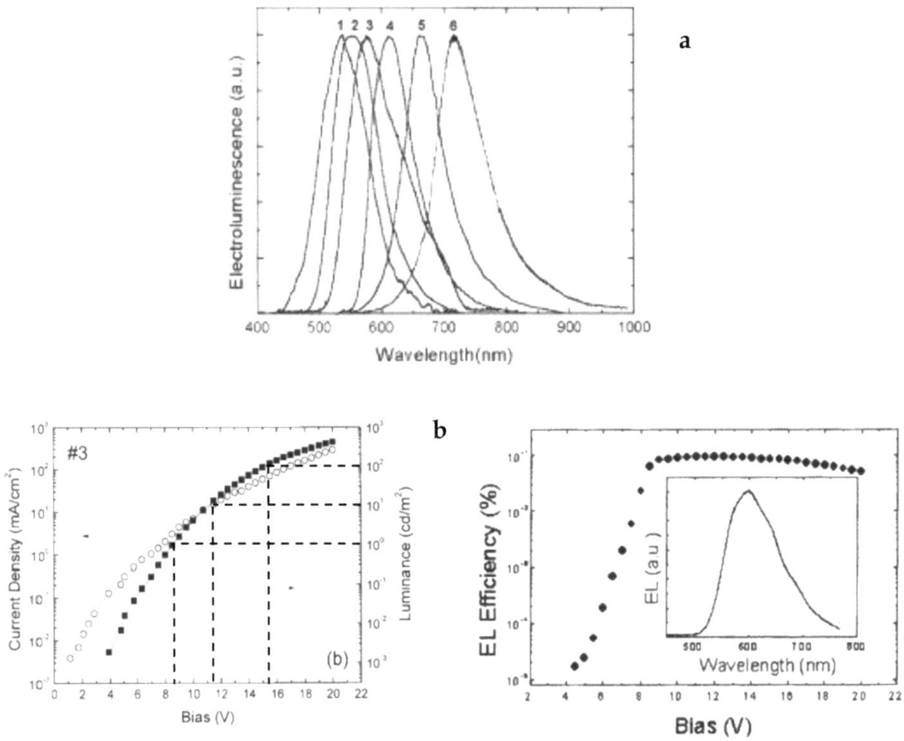

Fig. 3. a) Electroluminescence spectra of compounds 1-6 spanning from green to near IR; b) Current-voltage (*I -V*), luminance-voltage (*L -V*) characteristics (left) and EL efficiency (right) of the device obtained with compounds 3 as the active layer (120-nm-thick with a 70-nm-thick PEDOT layer).

The maximum luminance reached using compound 3 (Figure 3b) was 400 cd/m^2 at 20 V, a value which was already good enough for display applications. The device with compound 3 displayed also the highest EL efficiency, 0.2%, which was at least one order of magnitude larger rather than those already reported for the best oligo- and polythiophene based devices. The devices obtained with compounds 1-6 showed that it was possible to obtain multicolor electroluminescence from oligomeric thiophene materials and greatly improve the electroluminesce characteristics compared to conventional oligomers.

3. V-shaped oligothiophene-S,S-dioxides with high photo and electroluminescence performance

A remarkable improvement was achieved in 2003 with a new approach based on the replacement of the conventional linear structure of oligothiophenes and oligothiophene-S,S-dioxides with branched benzo[b]thiophene based structures (Mazzeo et al., 2003 a; Barbarella et al. 2005). These compounds (V-shaped oligothiophenes), in combination with the oxygen functionalization of the core thienyl sulphur and the cyclohexyl substitution of the lateral thienyl rings, allowed to achieve a remarkable luminance value of 10500 cd/m^2, which was the highest value obtained for LEDs based on oligothiophenes. The rationale

behind the synthesis of V-shaped compounds was the need to replace crystalline by amorphous thin films in order to avoid strong intermolecular interactions and then reduce the contribution of non radiative intermolecular deactivation pathways.
The molecular structure of selected V-shaped oligothiophenes is shown in Scheme 5, while the corresponding electro-optical characteristics are reported in Table 2. The luminance vs. voltage plots and the electroluminescence spectra of the devices fabricated with 9, 10 and 11 as the active materials, are shown in Figure 4.

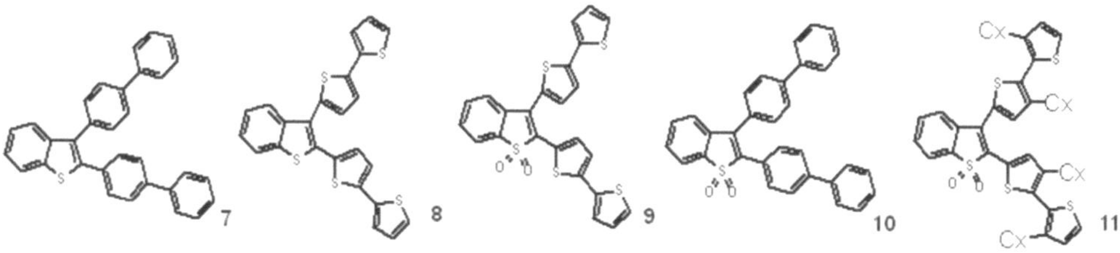

Scheme 5. Molecular structure of V-shaped oligothiophene-S,S-dioxides 7-11. Cx = Cyclohexyl.

	η %	Epc	Epa	Lum_M (cd/m^2)	φ %
7	4	<-2	1.60	35	0.001
8	2	<-2	1.35	1250	0.02
9	4	-1.26	1.43	2500	0.14
10	50	-1.45	>2	500	0.06
11	21	-1.36	1.48	10500	0.45

Table 2. Electro-optical characteristics of componds 7-11[a]
a) η: PL efficiency; Epc, Epa: reduction and oxidation peak potentials (vs calomel electrode) measured by cyclovoltammetry; LumM : luminance max; φ: EL efficiency.

Table 2 shows that the functionalization of the benzothienyl moiety with oxygen affects slightly the oxidation potentials but causes a relevant displacement of the reduction potentials towards less negative values (by an amount up to 0.74 eV), indicating a marked increase in the electron affinity of the compounds, in line with what was observed for linear oligothiophene-S,S-dioxides. All compounds were employed as active layers in OLEDs in which ITO/PEDOT:PSS and calcium/aluminum were used as the anode and the cathode, respectively, i.e. the same conditions employed with linear oligothiophene-S,S-dioxides. Most devices showed much better performance and operational stability than those achieved using the linear oligomers.

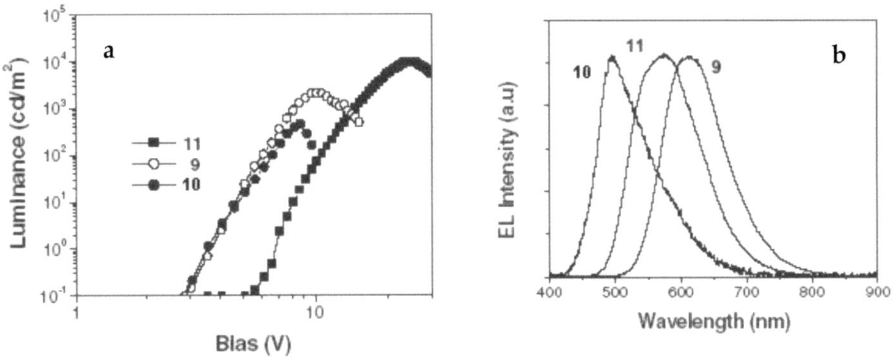

Fig. 4. a) Luminance vs. voltage and b) electroluminescence spectra of devices fabricated using compounds 9, 10 and 11.

Comparison of the the luminance values reported in tables 1 and 2, shows that the V-shaped structure was crucial to improve the brightness of the devices. For example, the non oxidized branched compound 8, shows a maximum brightness value of 1250 cd/m^2, three times higher than the best functionalized linear oligothiophene-S,S-dioxide reported in Table 1. While, in contrast to linear oligothiophene-S,S-dioxides, the functionalization with oxygen does not result in a substantial enhancement in photoluminescence efficiency, the EL efficiency is significantly improved. This is shown, for example, by comparison of the efficiency and luminance of the devices fabricated with compound 8 (0.02% and 1250 cd/m^2, respectively) and with the corresponding oxigenated derivative 9 (0.14% and 2400 cd/m^2). This result is due to the fact that the oxygen atoms induce a strong reduction in the energetic barrier between the cathode and the emissive layer, as in linear oligothiophene-S,S-dioxides. The maximum luminance (10500 cd/m^2) for the LED fabricated with compound 11 is more than 20 times larger than the maximum luminance displayed by the LED fabricated with the corresponding linear oligothiophene-S,S-dioxide, i.e. compound 3 (400 cd/m^2). As shown by comparison of the data reported in tables 1-2, also the maximum luminance of the devices based on compounds 9 (2500 cd/m^2) and 10 (500 cd/m^2) are much higher than those obtained with the devices based on the corresponding linear compounds 5 (105 cd/m^2) and 1 (100 cd/m^2). Since theoretical calculations, optical and CV data indicate that V-shaped oligothiophene-S,S-dioxides have electronic and optical features very similar to those of the corresponding linear compounds, the reason for the improved performances was ascribed to the much better film-forming properties of V-shaped compared to linear compounds and to changes in morphology from crystalline to amorphous films. There are several studies in the literature indicating that amorphous thin films, obtained either by vapor deposition or spin coating, enhance the electroluminesce properties (Robinson et al., 2001; Su et al., 2002; Doi et al., 2003).

The good film-forming properties and the amorphous morphology of V-shaped oligomers are due to their branched structure and asymmetric molecular conformation. TD-DFT calculations showed indeed that the molecular geometry of V-shaped oligothiophenes was not planar due to the large dihedral angle (>60°) between the branch in the β-position and the rigid core (Mazzeo et al., 2003).

The best performance of OLEDs based on V-shaped oligomers was obtained with the oxigenated compound functionalized with β-cyclohexyl substituents, namely compound 11,

in which electronic de-excitation via intermolecular interactions and internal conversion processes - which are the most important non radiative relaxation channels in oligothiophene-S,S-dioxides (Lanzani et al., 2001; Della Sala et al., 2003; Anni et al., 2005) - are strongly reduced. Functionalization with the bulky cyclohexyl groups has several effects. First, the large intermolecular distances due to the bulky substituents reduce the intermolecular interactions. Second, as shown by DFT (ground state) and TD-DFT (excited state) molecular geometry optimizations (Mazzeo et al., 2003 a), the molecular distortion is increased both in the ground and in the excited state. In the first singlet excited state the thiophene branches lie in two different planes, making the formation of non radiative aggregates unlikely. Third, the flexibility of the branches is strongly reduced. The calculations show, for example, that while compounds 8 and 9 are very flexible and can exist in different conformations of similar energy, for compound 11 only one ground-state energy minimum is found. Thus, the cyclohexyl substituents stabilize the conformation and make the molecule more rigid. All factors lead to enhanced photoluminescence in the solid state. The luminance of 10500 cd/m^2 reached with the device based on compound 11 is one of the highest values reported so far in the literature for devices with spin coated active layers.
Recently, OLEDs using as emitting layers nicely engineered branched oligomers (compounds 12-13) containing a dibenzothiophene-S,S-dioxide core and triarylamine branches, have been reported (Huang et al., 2006). The thiophene-S,S-dioxide group was introduced for its beneficial effect on the electron affinity of the molecules, while the triaryl amino groups were introduced because of their beneficial effect on charge (holes) transport and film forming properties. The molecular structure of compounds 12 and 13 is reported in Scheme 6.

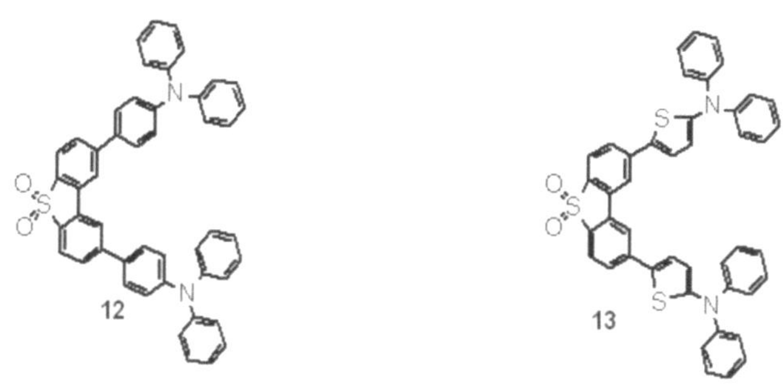

Scheme 6. Molecular structure of branched oligomers 12 and 13.

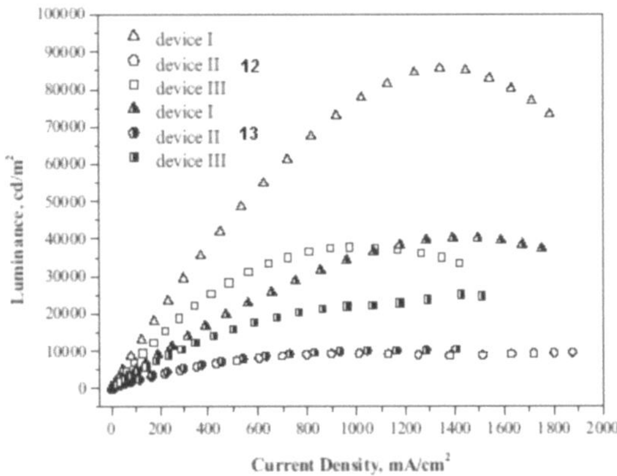

Fig. 5. Luminance versus current density characteristics for single layer devices of compounds 12 and 13. Structure of the devices: I) ITO/12 (or 13) (40 nm)/TPBI (40 nm)/LiF (1 nm)/Al (150 nm); II) ITO/NPB (40 nm)/12 (or 13) (40 nm)/LiF (1 nm)/Al (150 nm); III) ITO/12 (or 13) (80 nm)/LiF (1 nm)/Al (150 nm).

Using compounds 12 and 13, excellent luminances were obtained with single layer (spin coated) devices, as shown in Figure 5. The optical characteristics of compounds 12-13 and the relevant parametrs of the devices based on these compounds are reported in Table 3.

	12	13
V_{on} [V]	2.5, 2.3, 2.2	2.2, 2.5, 2.0
L_{max} [cdm^{-2}]	85475 (12.5), 9537 (15.0)	40140 (13.0) 10521 (11.5)
(V at L_{max},[V])	37699 (12.5)	25159 (14.5)
λ_{em} [nm]	492, 492, 496	540, 536, 542
$\eta_{ext,max}$ [%]	4.9, 1.3, 3.1	1.4, 0.87, 1.3
$\eta_{p,max}$ [lmW^{-1}]	9.7, 3.3, 7.2	4.9, 3.3, 5.0
$\eta_{c,max}$ [cdA^{-1}]	11, 3.1, 7.7	5.1, 3.1, 4.7
L [cd m^{-2}][a]	10778, 2107, 7529	4904, 2272, 4245
η_{ext} [%][a	4.7, 0.94, 3.1	1.4, 0.65, 1.2
η_p [lmW^{-1}][a]	6.5, 1.3, 3.9	2.8, 1.7, 2.1
η_c [cdA^{-1}][a]	10.8, 2.1, 7.5	4.9, 2.3, 4.2

Table 3. Optical characteristics of compounds 12-13 and performance of the corresponding devices[a]

a) *V*on: turn-on voltage; *L*max: maximum luminance; ηext,max: maximum external quantum efficiency; ηp,max: maximum power efficiency; ηc,max: maximum current efficiency.
[a] Measured at a current density of 100 $mAcm^{-2}$. V_{on} was obtained from the x-intercept of a plot of log(luminance) vs applied voltage

A maximum luminance value of about 90000 cd/m^2 at 1300 mA/cm^2 was reached for 13 and of about 90000 cd/m^2 at a similar current density for 2. The good performance of the devices was likely to be related to a much better balance of electron- and hole-transport properties than that achieved with linear or V-shaped oligothiophene-S,S-dioxides.
These results underline the potential impact that molecules containing thiophene-S,S-dioxide moieties could have on light emitting devices if more sophisticated device structures were realized with these materials.

4. Oligothiophenes for white OLEDs applications

Application in displays is only one among the several possible technological developments of oligomeric thiophene materials. Another important application is in the lighting sector where the replacement of standard white sources with flat organic devices is currently a matter of intense research.
One of the first approaches to realize white OLEDs (WOLEDs) using thiophene materials consisted in exploiting the high electron affinity of 2,5-bis-trimethylsilyl-thiophene-1,1-dioxide (STO) used as acceptor to generate exciplex states in combination with a very low electron affinity material (triphenyldiamine, TPD) used a donor (Mazzeo et al., 2003 b). Figure 6 shows the molecular structure of both TPD and STO. While the electron affinity of TPD is around 2,3 eV, that of STO is around 3,0 eV, i.e. close to that of longer linear oligothiophene-S,S-dioxides. The reason for this is in the fact that the LUMO state of the longer compounds is almost entirely localized in the central oxidized ring (Della Sala et al., 2003; Anni et al., 2005). Once the exciton is formed on the TPD molecule, the electron can move to a near STO molecule, with higher electron affinity. In consequence, two radiative transitions become allowed, one from the TPD molecules and the other from the transition between the LUMO level of STO and the HOMO level of TPD. As a result, a peak at 420 nm and a band at 570 nm are obtained, the two transitions resulting in white emission.
In Figure 6, PL spectra and images of blended films with different relative donor/acceptor concentrations, spin-coated on quartz substrates, are reported. It is seen that enhancing the concentration of STO a broad red-shifted emission due to exciplex states appears, in addition to the blue emission due to TPD, which is responsible for the white emission within a concentration range 17-53% of STO in TPD. The normalized EL spectra were similar to the PL spectra for the concentration used (20%), showing that the shape of the low-energy exciplex spectrum is almost independent of the applied voltage. The CIE coordinates of the EL spectra indicated a balanced white emission (0.39, 0.40) (Mazzeo et al., 2003 b).

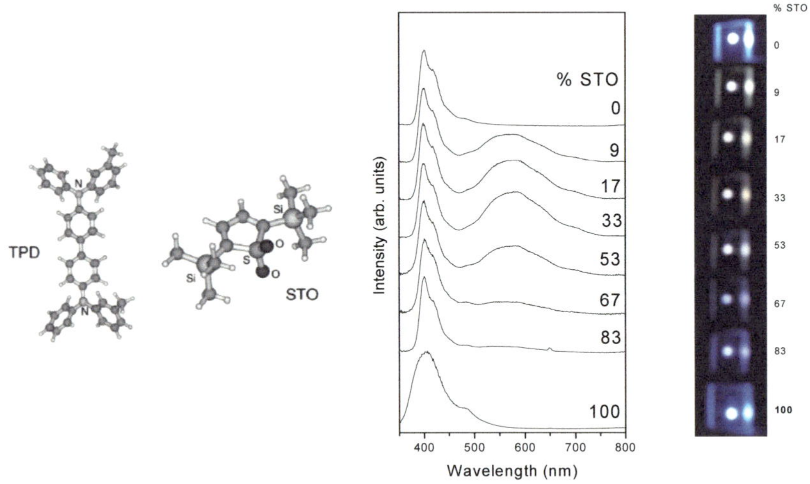

Fig. 6. Molecular structure of TPD and STO (left); PL spectra of the blends realized through TPD and STO (middle) and images of the blends in solid state films (right).

Although these results were promising for the generation of a new class of devices, their luminance was not very high. Much better results were obtained with a different approach, i.e. using a single thiophene material emitting in the white by virtue of its supramolecular organization (Mazzeo et al., 2005). The material in question is 3,5-dimethyl-2,6-bis(dimesitylboryl)-dithieno[3,2-b:2′,3′-d]thiophene, whose molecular structure is reported in Figure 7.

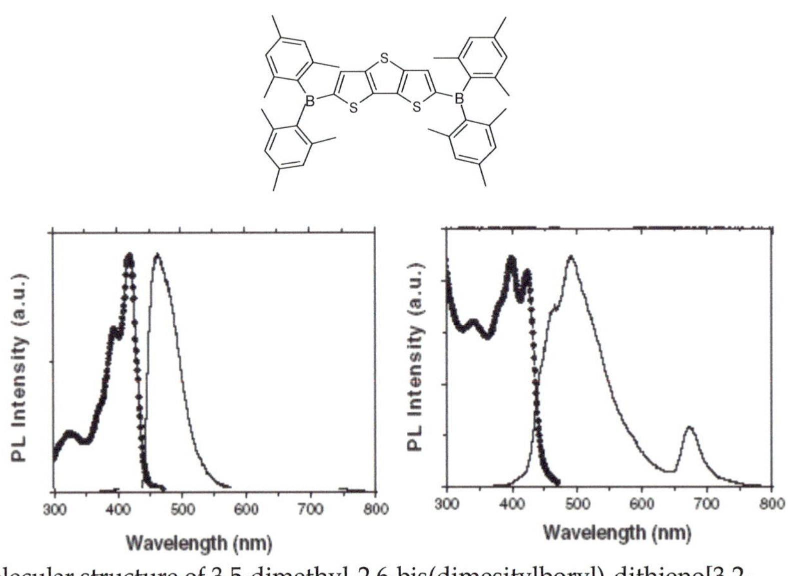

Fig. 7. Molecular structure of 3,5-dimethyl-2,6-bis(dimesitylboryl)-dithieno[3,2-b:2′,3′d]thiophene and PL spectrum in solution (left) and in the solid state (right).

Figure 7 shows that while in solution only a blue-green emission is observed, in the solid state an additional narrow red-shifted emission at 680 nm is also present in the PL spectrum. This red shifted absorption was peculiar to the solid state and could not be observed in solution in the concentration range 10^{-5} - 10^{-2} M. The appearance of similar red shifted absorption peaks had already been reported for several organic compounds and were assigned to triplets activated in the solid state or particular aggregation states (Lupton et al., 2003). By the aid of time-resolved photoluminescence (TR-PL) the red-shifted emission could be ascribed to the formation of aggregates or excimers. However, contrary to what it is generally observed with aggregates and excimers that are characterised by broad PL spectra (Lupton et al., 2003), the linewidth of the peak at 680 nm was narrow. In order to elucidate this point, INDO/SCI calculations were carried out. The calculations suggested that the narrow line was the result of the very peculiar supramolecular arrangement assumed by the compound in the aggregated state (Mazzeo et al., 2005). Indeed, due to the planar and rigid conformation of the inner dithienothiophene core and the presence of the bulky mesityl substituents, the molecules tend to fit together in a *cross-like* configuration as shown in Figure 8. Only very small movements of the molecules through translations along the x-y axis, or small angular (α) deviations are allowed.

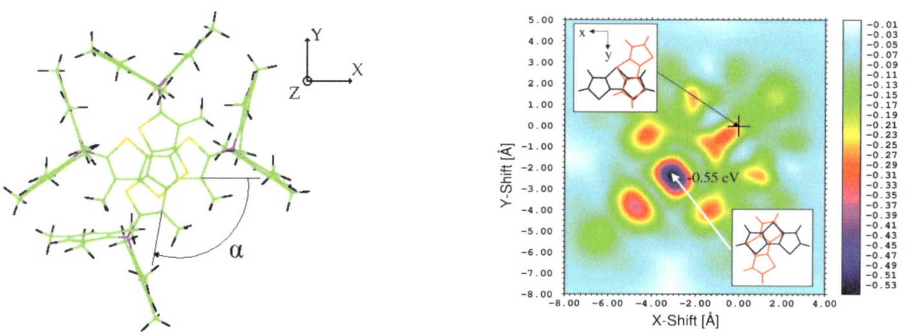

Fig. 8. Molecular structure of two interacting molecules forming a cross-like dimer (left) and (right) Intermediate neglect of differential overlap/single configuration interaction (INDO/SCI) excitation-energy shifts due to intermolecular interactions. The scale on the right corresponds to calculated excitation energy shifts.

Commonly observed *HB*-type dimeric aggregates (i.e. with α=180°) are completely forbidden for this rigid compound, due to the repulsion of the mesityl substituents. In such fixed *cross-like* configuration the peak broadening induced by supramolecular conformational dispersion is strongly reduced. The plot reported in Figure 8 shows that the excitation energy shift is indeed dominated by only one deep minimum. This means that only one single arrangement is responsible for the additional red emission observed in the solid state, leading to a very narrow emission. The calculations also showed that this kind of intermolecular arrangement induces a red-shift as high as 0.55 eV, a value which is in good agreement with the experimental result (0.7 eV).
The white emitting dithienothiophene derivative displayed good film forming properties and could be used as active material in light emitting diodes. The emissive layer was spin-coated between ITO/PEDOT:PSS and LiF/Al, used as anode and cathode, respectively. The

LiF layer was employed in order to enhance the carrier injection in the emissive layer (Hung et al., 1997). The EL spectrum at a LiF thickness of ≈ 5 nm and the device performance are shown in Figure 9.

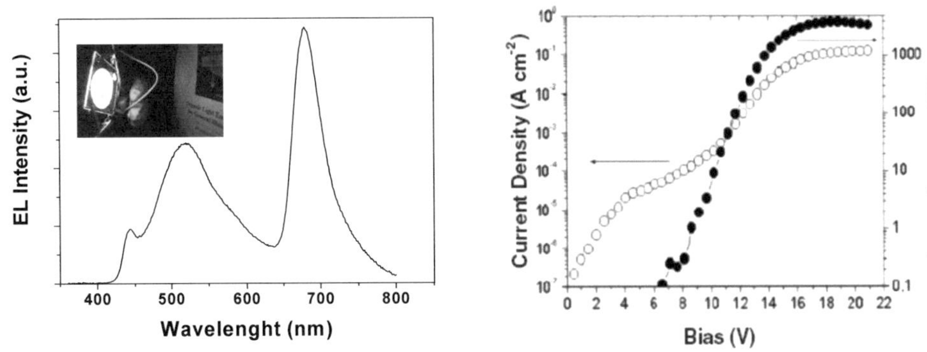

Fig. 9. EL spectrum of device with d_{LiF} ≈ 5nm (left) (inset: image of a large area device); Luminance-current density-voltage characteristics (right).

For 4.8 nm of LiF the performances are 50 times higher than the device in which only the Aluminium was used as cathode. In particular, a brightness of 3800 cd/m^2 at 18 V (Figure 9) and a maximum QE of 0.35% could be achieved. It is worth noting that the luminance of this device overcomes the minimum value of 1000 cd/m^2 required for lighting systems. The white electroluminescence was achieved by the superposition of the broad blue-green emission originating from the single molecule and the red-shifted narrow peak assigned to the formation of cross-like dimers in the solid-state. This was one of the first examples in the literature of white emission from a single molecular material in the solid state. The good performance of the device was due to an unusual mixing of favourable factors, i.e. the very peculiar self-organization properties of the dithienothiophene derivative, the well known electron-acceptor properties of the boron atom and the good film forming properties of the material. Nevertheless, the results obtained, indicate that the fabrication of a new class of white emitting devices combining the simplicity and low-cost of single layer spin-coated devices is achievable through appropriate molecular engineering.

5. Very low voltage and stable oligothiophene OLEDs.

As shown in the previous section, thiophene oligomeric materials have great potential for application in displays and lighting. All the devices described in the previous sections have been realized in a single-layer or bilayer configuration by depositing the active material by spin coating. This is a strong limitation for oligomeric materials since, even if the material is highly performant in terms of PL, the devices are not efficient enough due to the limits of wet deposition processes like spin coating. Polymeric materials have the same type of problems. However, while polymeric materials cannot be evaporated, this is possible for oligomeric materials owing to their small molecular weight. Thus, a possible improvement in OLEDs based on thiophene oligomeric materials can be realized if these compounds are deposited in a heterostructure system (Walzer et al., 2007; Zhou et al., 2001; Huang et al.,

2002). So far, no attempts have been made in this direction and the data shown below are the first reported to date.
We fabricated a much more sophisticated device using compound 3 (Mariano et al., 2009) and, to check the limit in brightness and stability of the compound, we realized an OLED based on electrically doped transport layers, i.e. in the so-called *p-i-n* (p-type-intrinsic-n-type) configuration (Walzer et al., 2007).
In order to obtain low driving voltages, low ohmic losses at the interface between the metal and the transport layers are an important factor. Organic light-emitting diodes are usually realized with un-doped thin organic films, requiring high operating voltages to overcome the energy barriers between the contacts and the transport layers and to drive the opposite charges into the emissive layer. Contrary to inorganic LEDs, the typical driving voltage is much higher than the thermodynamic limit, which is given by the energy gap of the active layer. Recently, controlled electrical doping in transport layers of the OLEDs has been introduced (Walzer et al., 2007; Zhou et al., 2001; Huang et al., 2002). The typical dopants explored have been the 2,3,5,6-tetrafluoro-7,7,8,8 tetracyanoquinodimethane (F4TCNQ) as donor of holes in an hole transport layer and alkali metals such as Cs or Li as donors of electrons in an electron transport layer. The doping of the transport layers leads to the formation of thin space charge layers which are formed at the interface with the metal contact layer, allowing for a good injection (ohmic) of the carriers by tunneling despite the barriers. This effect removes completely the energy barrier between the metal layers ad the transport layers, thus reducing the voltage. Moreover, the high electrical conductivity of the doped layers reduces also the drop in voltage caused by the usually high resistance of the undoped organic films. The *p* and *n*-doping of the transporting layers permits to reach a conductivity of 10^{-5} S/cm, which is enough in order to have a negligible drop in the voltage across these layers. Due to the incorporation of these very conductive transporting layers, which form Ohmic contacts with the electrodes, *p-i-n* architectures supply more current density than conventional OLEDs, under the same driving voltage (Zhou et al., 2001; Huang et al., 2002). Therefore, higher brightness can be obtained at low bias.
This type of OLED has an electrically intrinsic emission layer (EML), a hole transport layer (HTL) and an electron transport layer (ETL). Additional blocking layers between the charge transport layers and the emissive layer are generally also introduced to prevent problems related to the lack of charge balance, exciton quenching by excess of charge carriers, and exciplexes formation at the interface. All these layers complicate the structure of the device but increase its efficiency and stability. The aim of *p-i-n* technology is to reduce the applied voltage in order to have a given luminance and improve power efficiency giving more stability to the device and less power consumption.
The structure of the device realized with the linear oligothiophene-S,S-dioxide 3 (whose molecular structure is shown in Scheme 4) is shown in Figure 10. It consists of the following layers: ITO transparent anode; a 35 nm thick layer of *N,N,N′,N′* tetrakis(4-methoxyphenyl)-benzidine (MeO-TPD) doped with 2.7 wt % of 2,3,5,6-tetrafluoro-7,7,8,8 tetracyanoquinodimethane (F4-TCNQ) that was evaporated as *p*-doped hole injection and transport layer; a 7 nm thick film of 2,2′,7,7′-tetrakis-(diphenylamino)-9,9′-spirobifluorene (Spiro-TAD) which acts as electron blocking layer; an emitting layer consisting of 30 nm of compound 3 of Figure 5; 10 nm of 4, 7-diphenyl-1,10-phenanthroline (Bphen) as hole blocking layer; 35 nm of Bphen doped with Cs as electrons injecting and transporting layer;

200 nm Al deposited as cathode. The optically thick metallic film acts as a reflector and thereby aids the output coupling of light from the device.
All films were deposited by thermal evaporation in a base pressure of about 10^{-8} mbar, at a rate in the range 0.5-1.0 Å/s. Before the deposition of the organic compounds, ITO substrates were cleaned in acetone, isopropanol and deionized water for 10 min at 60 °C in an ultrasonic bath. No plasma oxygen was performed because of the electrical doping of the transport layers.
Figure 10a shows the luminance and the current density of the device as a function of the voltage. The turn-on voltage was around 2.1 V, while the luminance reached the remarkable value of 11000 cd/m^2 at only 9 V. The device showed a maximum EQE of 0.55%. It is worth noting that the same material deposited by spincoating in a bilayer configuration shows a maximum luminance of about 400 cd/m^2 obtained at the very high voltage of 19 V (Table 1). This result obtained with the LED in *p-i-n* configuration underlines how all the properties of the device can be strongly improved thanks to the possibility of vacuum evaporating oligomeric materials.
The device reported in Figure 10 was encapsulated using a lid attached to the sample by an epoxy resin in order to carry out aging measurements and to check the stability of both the device and the material. We recall that the lifetime of OLEDs is defined as the time taken to reach half of the starting luminance.

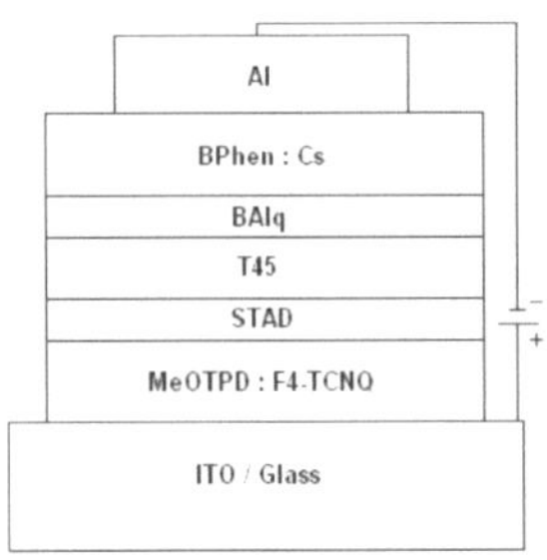

Bphen: 4, 7-diphenyl-1,10-phenanthroline
STAD: 2,2′,7,7′-tetrakis-(diphenylamino)--9,9′-spirobifluorene
BAlq: bis-(2-methyl-8-quinolinolato)-4--(phenyl-phenolato) aluminum-III
MeO-TPD: *N,N,N′,N′* tetrakis (4-methoxyphenyl)-benzidine (MeO-TPD)
F4-TCNQ: 2,3,5,6-tetrafluoro-7,7,8,8 tetracyano--quinodimethane

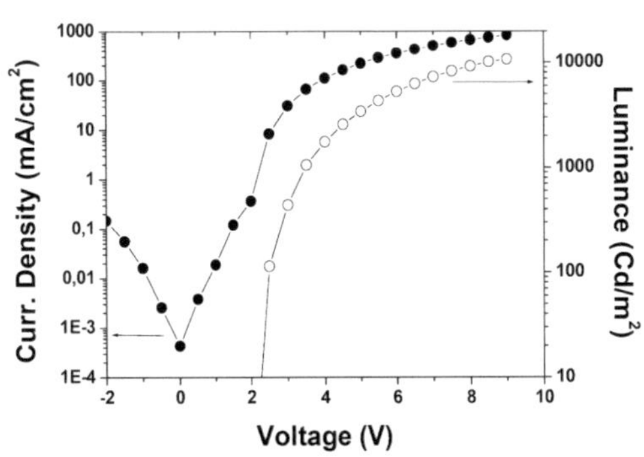

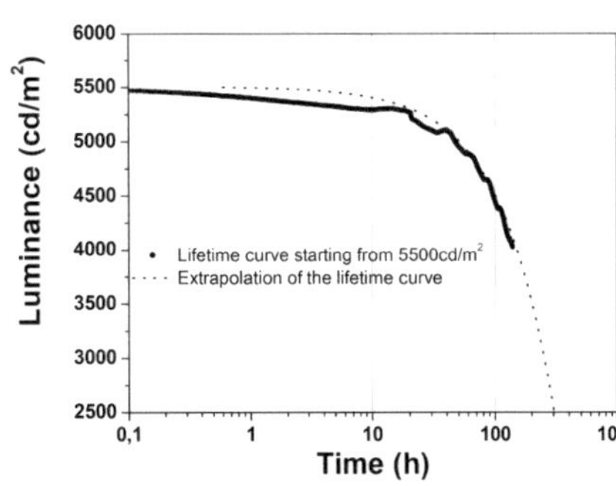

Fig. 10. Top: device structure. Bottom: (A) Luminance-Current Density vs. Voltage (left); Luminance decay time for a starting value of 5500Cd/m^2 at fixed current density (right).

Figure 10b shows the plot of the luminance as a function of time for a starting value of 5500 cd/m^2 at a fixed current density of 320 mA/cm^2. The black curve represents the experimental data while the dotted curve represents the extrapolated behavior. The figure shows that a remarkable lifetime of about 270 hours was reached. This result demonstrates that heterostructure devices are the tools where thiophene oligomeric materials should be tested to reveal all their potential as emissive compounds. Moreover, the demonstration that oligthiophene-S,S-dioxides show very high stability is an important step forward that allows to classify these materials among the best so far available for electroluminescence (Mariano et al., 2009).

6. Conclusions and Outlook

Today, the field of electroluminescence of organic semiconductors is dominated by two kinds of materials: phosphorescent and fluorescent small molecules, in particular for applications where high emission power is needed, like lighting. Although phosphorescent compounds seem to be the most promising in terms of external quantum efficiency and low power consumption, fluorescent compounds show high stability and the possibility to be deposited avoiding codeposition with a host. This is particularly true for oligothiophenes which show high stability and the possibility to tune the emission wavelength in a very wide range, from green-bluish to near infrared without the need of coevaporation. The possibility to functionalize these compounds in a very flexible way and finely tailor their properties, make this class of molecules strongly competitive with respect to standard ones (also phosphorescent), although much research must still be carried out to further improve the stability and the efficiency of devices based on these materials. We are currently pushing up this research field trying to mix the best technology for OLEDs (*p-i-n* technology) with the best thiophene oligomeric materials with the aim to generate new kinds of electroluminescent devices for different pourposes: from display to lighting and automotive.

7. References

Amir, E. & Rozen, S. (2005). *Angew. Chem. Int. Ed.*, 44, p. 7374.

Anni, M.; Della Sala, F.; Raganato, M. F.; Fabiano, E.; Lattante, S.; Cingolani, R.; Gigli, G.; Barbarella, G.; Favaretto, L. & Görling, A. (2005). *J. Phys. Chem. B*, 109, p. 6004.

Antolini, L.; Tedesco, E.; Barbarella, G.; Favaretto, L.; Sotgiu, G.; Zambianchi, M.; Casarini, M. D.; Gigli, G. Cingolani, R. (2000). *J. Am. Chem. Soc.*, 122, p. 9006.

Baldo, M. A.; O'Brien, D. F.; You, Y.; Shoustikov, A.; Sibley, S.; Thompson, M. E. & Forrest, S. R. (1998). *Nature*, 395, p. 151.

Baldo, M. A.; Thompson, M. E. & S. R. Forrest. (2000). *Nature*, 403, p. 750.

a) Barbarella, G.; Favaretto, L.; Zambianchi, M.; Pudova, O.; Arbizzani, C.; Bongini, A. & Mastragostino, M. (1998). *Adv.Mater.*, 10, p. 551.

b) Barbarella, G.; Favaretto, L.; Sotgiu, G.; Zambianchi, G.; Antolini, L.; Pudova,O. & Bongini, A. (1998). *J.Org.Chem.*, 63, p. 5497.

Barbarella, G.; Favaretto, L.; Sotgiu, G.; Zambianchi, M.; Fattori, V.; Cocchi, M.; Cacialli, F.; Gigli, G. & Cingolani, R. (1999). *Adv.Mater.*, 11, p. 1375.

Barbarella, G.; Favaretto, L.; Sotgiu, G.; Zambianchi, M.; Bongini, A.; Arbizzani, C.; Mastragostino, M.; Anni, M.; Gigli, G. & Cingolani, R. (2000). J. Am. Chem. Soc., 122, p. 11971.

Barbarella, G.; Favaretto, L.; Sotgiu, G.; Antolini, L.; Gigli, G.; Cingolani, R. & Bongini, A. (2001). *Chem.Mater.*, 13, p. 4112.

Barbarella, G.; Favaretto, L.; Zanelli, A.; Gigli, G.; Mazzeo, M.; Anni, M. & Bongini, A. (2005). *Adv. Funct. Mater.*, 15, p. 664.

Barta, P.; Cacialli, F.; Friend, R. H. & Zagórska, M. (1998). *J. Appl. Phys.*, 84, p. 6279.

Beaupré, S. & Leclerc, M. (2002). *Adv. Funct. Mater.*, 12, p. 192.

Berggren, M. Inganäs, O. Gustafsson, G. Rasmusson, J. Andersson, M. R. Hjertberg, T. Wennersträm, O. (1994). *Nature*, 372, p. 444.

Berlin, A.; Zotti, G.; Zecchin, S.; Schiavon, G.; Cocchi, M.; Virgili, D. & Sabatini, C. (2003). *J. Mater. Chem.*, 13 , p. 27.

Braun, D.; Gustafsson, G.; McBranch, D. & Heeger, A. J. (1992). *J. Appl. Phys.*, 72, p. 564- 568

Burroughes, J. H.; Bradley, D. D. C.; Brown, A. R.; Marks, R. N.; Friend, R. H.; Burn, P. L. & Holmes, A. B. (1990). *Nature*, 347, p. 539.

Camaioni, N.; Ridolfi, G.; Fattori, V.; Favaretto, L. & Barbarella G. (2004). *Appl. Phys. Lett.*, 84, p. 1901.

Casado, J.; Zgierski, M. Z.; Ewbank, P. C.; Burand, M. W.; Janzen, D. E.; Mann, K. R.; Pappenfus, T. M.; Berlin, A.; Pérez-Inestrosa, E.; Ortiz, R. P. & López Navarrete, J. T. (2006). *J. Am. Chem. Soc.*, 128, p. 10134.

Charas, A.; Morgado, J.; Martinho, J. M. G.; Alcácer, L. & Cacialli, F. (2001). *Chem. Commun.*, p. 1216.

Cheylan, S.; Bolink, H.J.; Fraleoni-Morgera, A.; Puigdollers, J.; Voz, C.; Mencarelli, I.; Setti, L.; Alcubilla, R. & Badenes, G. (2007). *Organic Electronics*, 8, p. 641.

Chou, P. T. & Chi, Y. (2007). *Chem. Eur. J.*, 13, p. 380.

D'Andrade, B. W. & Forrest, S. R. (2004). *Adv. Mater.*, 16, p. 1585.

D'Andrade, B. (2007). *Nature Photonics* , 1, p. 33.

Della Sala, F.; Raganato, M.F.; Anni, M.; Cingolani, R.; Weimer, M.; Görling, A.; Favaretto, L.; Barbarella, G. & Gigli, G. (2003). *Synth. Met.*, 139, p. 897.

Doi, H.; Kinoshita, M.; Okumoto, K. & Shirota, Y. (2003). *Chem. Mater.*, 15, p. 1080.

F. Garnier (1999). Acc. Chem. Res., 32, p. 209.

Friend, R. H.; Gymer, R. W.; Holmes, A. B.; Burroughes, J. H.; Marks, R. N.; Taliani, C.; Bradley, D. D. C.; Dos Santos, D. A.; Brédas, J. L.; Lögdlund, M. & Salaneck, W. R. (1999). *Nature*, 397, p. 121.

Gigli, G.; Inganas, O.; Anni, M.; De Vittorio, M.; Cingolani, R.; Barbarella, G. & Favaretto, L. (2001). *Appl. Phys. Lett.*, 78, p. 1493.

Greenham, N. C.; Moratti, S. C.; Bradley, D. D. C.; Friend, R. H. & Holmes, A. B. (1993). *Nature*, 365, p. 628.

Horowitz, G. Delannoy, P. Bouchriha, H. Deloffre, F. Fave, J.-L. Garnier, F. Hajlaoui, R. Heyman, M. Kouki, F. Valat, P. Wittgens, V. & Yassar A. (1994). *Adv. Mater.*, 6, p. 752.

Huang, J.; Pfeiffer, M.; Werner, A.; Blochwitz, J.; Leo, K. & Liu, S. (2002). *Appl. Phys. Lett.*, 80, p. 139.

Huang, T. H.; Lin, J. T.; Chen, L. Y.; Lin, Y. T. & Wu, C. C. (2006). *Adv. Mater.*, 18, p. 602.

Hughes G. & Bryce, M. R. (2005). *J. Mater. Chem.*, 15, p. 94.

Hung, L. S.; Tang, C. W. & Mason, M. G. (1997). *Appl. Phys. Lett.*, 70, p. 152.

Hung, M. C.; Liao, J.L.; Chen, S. A.; Chen, S. H. & Su, A. C. (2005). *J. Am. Chem. Soc.*, 127, p. 14576.

Kalinowski, J. (2005). Organic *Light-Emitting Diodes: Principles, Characteristics and Processes*, Marcel Dekker, New York.

Kanemitsu, Y. Shimizu, N. Suzuki, K. Shiraishi, Y. Kuroda, M. (1996). *Phys. Rev. B* , 54, p. 2198.
Kawamura, Y.; Goushi, K.; Brooks, J.; Brown, J. J.; Sasabe, H. & Adachi, C. (2005). *Appl. Phys. Lett.*, 86, p. 071104.
Kraft, A.; Grimsdale, A. C. & Holmes, A. B. (1998). *Angew. Chem. Int. Ed.*, 37, p. 402.
Lanzani, G.; Cerullo, C.; De Silvestri, S.; Barbarella, G. & Sotgiu, G. (2001). *J.Chem.Phys.*, 115, p. 1623.
Liu, Y. L.; Feng, J. K. & Ren, A. M. (2008). *J. Phys. Chem.* A, 112, p. 3157.
Lupton, J. M.; Craig, M. R. & Meijer, E. W. (2002). *Appl. Phys. Lett.*, 80, p. 4489.
Mariano, F.; Mazzeo, M.; Duan, Y.; Barbarella, G.; Favaretto, L.; Carallo, S.; Cingolani, R. & Gigli, G. (2009). *Appl. Phys. Lett.*, 94, p. 063510.
Marseglia, E. A.; Grepioni, F.; Tedesco, E. & Braga, D. (2000). *Mol. Cryst. Liq. Cryst.*, 348, p. 137.
a) Mazzeo, M.; Vitale, V.; Della Sala, F.; Pisignano, D.; Anni, M.; Barbarella, G.; Favaretto, L.; Zanelli, A.; Cingolani, R. & Gigli, G. (2003). *Adv. Mater.*, 15, p. 2060.
a) Mazzeo, M. Pisignano, D. Della Sala, F. Thompson, J. Blyth, R. I. R. Gigli, G. Cingolani, R. Sotgiu, G. Barbarella, G. (2003). *Appl. Phys. Lett.*, 82, p. 334.
Mazzeo, M.; Vitale, V.; Della Sala, F.; Anni, M.; Barbarella, G.; Favaretto, L.; Sotgiu, G.; Cingolani, R. & Gigli, G. (2005). *Adv. Mater.*, 17, p. 34.
Meerholiz, K. & Heinze, J. (1996). *Electrochim. Acta*, 41, p. 1839.
Melucci, M.; Favaretto, L.; Barbarella, G.; Zanelli, A.; Camaioni, N.; Mazzeo, M. & Gigli, G. (2007). *Tetrahedron*, 63, p. 11386.
Misra, A.; Kumar, P.; Kamalasanan, M. N. & Chandra, S. (2006). *Semicond. Sci. Technol.*, 21, R35-R47.
Müllen, K. & Scherf, U. (2006). *Organic Light-Emitting Devices: Synthesis, Properties and Applications*, Wiley-VCH, Weinheim, Germany.
Oelkrug, D.; Egelhaaf, H. J.; Gierschner, J. & Tompert, A. (1996). *Synth. Met.*, 76, p. 249.
Pasini, M.; Destri, S.; Porzio, W.; Botta, C. & Giovanella, U. (2003). *J. Mater. Chem.*, 13, p. 807.
Pei, Q. & Yang, Y. (1996). *J. Am. Chem. Soc.*, 118, p. 7416.
Perepichka, I. I.; Perepichka, I. F.; Bryce, M. R. & Pålsson, L. O. (2005). *Chem. Commun.*, p. 3397.
Pope, M.; Kallmann, H.P. & Magnante, P. (1963). *J. Chem. Phys.*, 38, p. 2042.
Raganato, M.F.; Vitale, V.; Della Sala, F.; Anni, M.; Cingolani, R.; Gigli, G.; Favaretto, L.; Barbarella, G.; Weimer, M. & Gorling, A. (2004). *J. Chem. Phys.*, 121, p. 3784.
Robinson, M. R.; Wang, S. J.; Heeger, A. J.; Bazan, G. C. (2001). *Adv. Funct. Mater.*, 11, p. 413.
Scherf, U. & List, E. (2002)., 14, 477.
Shinar J. (2004). *Organic Light-Emitting Devices: A Survey, Springer-Verlag*, New York.
Su, Y. Z. Lin, J. T. Tao, Y. T. Ko, C. W. Lin, S. C. Sun, S. S. (2002). *Chem Mater.*, 14, p. 1884.
Tang, C. W. &Van Slyke, S. A. (1987). *Appl. Phys. Lett.*, 51, p. 913.
Tedesco, E.; Della Sala, F.; Favaretto, L.; Barbarella, G.; Albesa-Jové, D.; Pisignano, D.; Gigli, G.; Cingolani, R. & Harris, K. D. M. (2003). *J. Am. Chem. Soc.*, 125, p. 12277.
Väterlein, C.; Neureiter, H.; Gebauer, W.; Ziegler, B.; Sokolowski, M.; Bäuerle, P. & Umbach E. (1997). *J. Appl. Phys.*, 82, p. 3003.
Walzer, K. Maennig, B. Pfeiffer, M. Leo K. (2007). *Chem. Rev.*, 107, p. 1233.
Wang, X.; Andersson, M. R.; Thompson M. E. & Inganäs, O. (2004). *Thin Solid Films*, 468, p. 226.

Yang, X.; Müller, D. C.; Neher, D. & Meerholz, K. (2006). *Adv. Mater.*, 18, p. 948.
Yoon, M. H.; Kim, C.; Facchetti, A. & Marks, T. J. (2006). *J. Am. Chem. Soc.*, 128, p. 12851.
Yu, W.-L.; Pei, J.; Huang, W.; Heeger, A. J. (2000). *Adv. Mater.*, 12, p. 828.
Zavelani-Rossi, M.; Lanzani, G.; De Silvestri, S.; Anni, M.; Gigli, G.; Cingolani, R.; Barbarella, G. & Favaretto, L. (2001). *Appl. Phys. Lett.*, 79, p. 4082.
Zhou, M.; Pfeiffer, M.; Blochwitz, J. & Leo, K. (2001). *Appl. Phys. Lett.*, 78, p. 401.
Ziegler C. (1997). *Handbook of Organic Conductive Molecules and Polymers*, Vol. 3, Ch. 13, (Ed. H. S. Nalwa) Wiley, New York.

2

The efficient green emitting iridium(III) complexes and phosphorescent organic light emitting diode characteristics

Kwon Soon-Ki[a], Thangaraju Kuppusamy[a],
Kim Seul-Ong[a], Youngjin Kang[c] and Kim Yun-Hi[b*]
[a]School of Material Science and Engineering & Engineering Research Institute (ERI),
[b]Department of Chemistry and Research Institute of Natural Science (RINS),
Gyeongsang National University, Jinju – 660 701,South Korea
[c]Division of Science Education, Kangwon National University, Chuncheon, 200-701, South Korea

1. Introduction

Organic light emitting diodes (OLEDs) have attracted enormous attention in the recent years due to their potential application in the solid state lighting and full color flat panel displays because of their high efficiency, low fabrication cost, ease of fabricating large area devices, and wide range of emission colors (Shen et. al, 1997; Tang & VanSlyke 1987). When compared to fluorescent OLEDs where only singlet state excitons can emit the light and luminescence is reduced due to triplet formation, the phosphorescent organic light emitting diodes (PHOLEDs) are efficient as both singlet and triplet excitons can be harvested for the light emission (Baldo et al. 1998; Baldo et al. 1999; Adachi et al. 2001; Ikai et al. 2001; Lo et al. 2002). The various phosphorescent light emitting materials have been synthesized and intensively studied worldwide. A promising phosphorescent metal complex should have a short lifetime in the excited state and high phosphorescent efficiency for the highly efficient OLEDs. Among the various phosphorescent emitters, iridium complexes remain the most effective material due to their ability to achieve maximum internal quantum efficiency, nearly 100%, and high external quantum efficiency in the devices (O'Brien et al. 2003; Duan et al. 2003; Xie et al. 2001; Noh et al. 2003; Wang et al. 2001; Neve et al. 2001). The most widely used homoleptic green emitting iridium(III) complex, *fac*-tris(2-phenylpyridine)iridium [$Ir(ppy)_3$], derivatives have shown a number of advantages such as ease of tuning emission energy by functionalizing the 'ppy' ligand with electron donating and electron withdrawing substituents and high phosphorescent quantum efficiency at room temperature (Grushin et al. 2001; Jung et al. 2004; Lee et al. 2009). The structural modifications of ppy ligand with various substistuents have been carried out for the fine tuning of emission colors in the green region and to improve the device efficiencies for the lighting and display applications.

During our investigations the following substitutions have been examined. The influence of substitutions of (i) methyl groups, (ii) bulky trimethylsilyl groups, (iii) trimethylsilyl groups, and (iv) rigid and bulky cycloalkene units in the homoleptic green emitting iridium(III) complex [Ir(ppy)$_3$] on the electrochemical, photo-physical, and electroluminescence properties has been investigated (Jung et al. 2004; Jung et al. 2006; Jung et al. 2009; Kang et al. 2008).

2. Substitution of Methyl groups on photophysical and electrophosphorescence characteristics of Iridium(III) complex

We attached the methyl groups as substituents on the ppy ligand of Ir(ppy)$_3$ to prepare a series of *fac*-[Ir(dmppy)$_3$] complexes [Hdmppy = n-methyl-2-(n′methylphenyl)pyridine and H = n,n′] and studied their electrochemical, photophysical and electroluminescence properties in devices (Jung et al. 2004). The methyl moiety was chosen due to (i) the phosphorescence emission of Ir(ppy)$_3$ derivatives is strongly affected by the triplet energy of the ortho-chelating C^N ligands. Hence the methyl substituents on the ppy rings could play a key role in alteration of the triplet energy and (ii) although a methyl group is a weaker electron-donating group than other functional groups (e.g. OMe, NR$_2$ etc), it can influence on the highest occupied molecular orbital (HOMO) and/or the lowest unoccupied molecular orbital (LUMO) level as well as ^{3}MLCT state of the iridium(III) complex. The molecular structures of Ir(dmppy)$_3$ derivatives are shown in Fig.1.

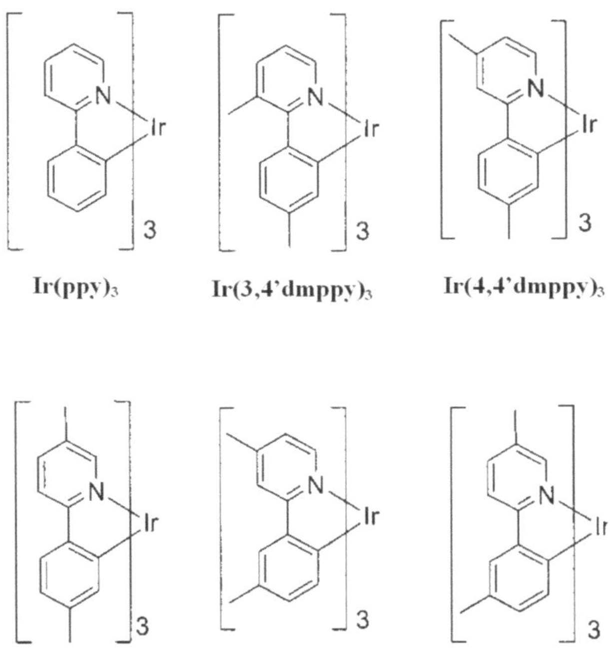

Ir(ppy)$_3$ Ir(3,4'dmppy)$_3$ Ir(4,4'dmppy)$_3$

Ir(5,4'dmppy)$_3$ Ir(4,5'dmppy)$_3$ Ir(5,5'dmppy)$_3$

Fig. 1. The molecular structures of methyl substituted Ir(dmppy)$_3$ derivatives.

2.1 Structural, Electrochemical, Photophysical properties

The methyl substituted Ir(dmppy)$_3$ derivatives were synthesized by the Suzuki coupling reaction of 2-bromo-n-methylpyridine with the corresponding n′-methylphenylboronic acid in the presence of K_2CO_3 and [Pd(PPh$_3$)$_4$] catalyst as reported by our research group (Jung et al. 2004). We found that the *fac-* isomer was formed as the major component, although some *mer-* isomer was formed with low yields and can be isolated. All Ir(dmppy)$_3$ derivatives are very stable up to 300°C without degradation in air. All the Ir(dmppy)$_3$ complexes were characterized by mass spectrometry and elemental analysis. The crystal structures of Ir(4,4′dmppy)$_3$ and Ir(4,5′dmppy)$_3$ complexes exhibit only the *fac-* configuration, with a distorted octahedral geometry around the Ir atom as shown in Fig. 2 and Fig. 3, respectively.

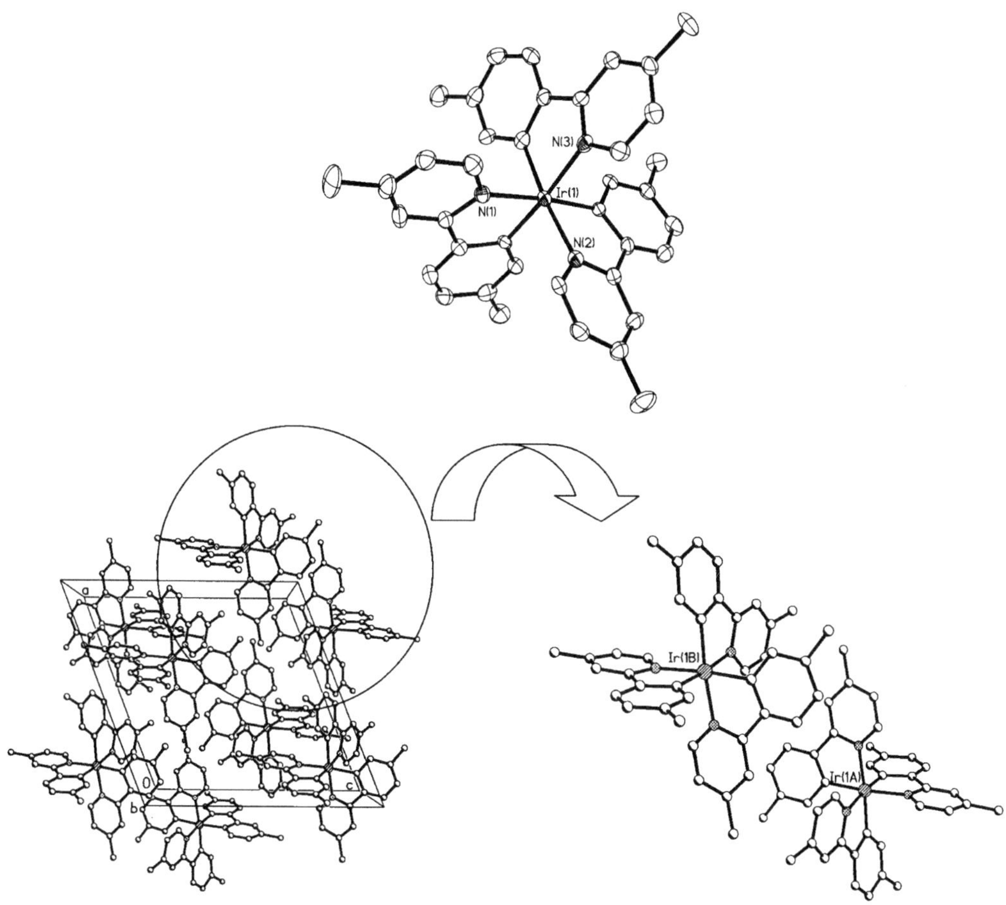

Fig. 2. Top: Molecular structure of Ir(4,4′dmppy)$_3$ with atom labeling schemes and 50% thermal ellipsoids; all hydrogen atoms have been omitted for clarity; bottom: crystal packing diagram between two adjacent molecules of Ir(4,4′dmppy)$_3$ showing the lack of a π-π stacking interaction in the solid state.

The dihedral angle between the phenyl and pyridine rings in Ir(4,4′dmppy)$_3$ complex (8.81°) is approximately three times larger than that of Ir(4,4′dmppy)$_3$ (3.4°), indicating a decreased conjugation of 4,4′-ppy ligands. The UV-visible absorption and emission spectra of all Ir(dmppy)$_3$ derivatives are shown Fig. 4. From the absorption data of all Ir(dmppy)$_3$ complexes, ligand-centred π- π* transitions were observed between 250 nm and 310 nm and However, ^{1}MLCT and ^{3}MLCT peaks of all the complexes appeared at lower energies except Ir(4,4′dmppy)$_3$ complex which showed at higher energies. This shows that the methyl groups on the dmppy rings have a significant effect on the electronic transitions energies. The emission spectra of all complexes show the phosphorescent emission between 509 nm and 534 nm in solution and thin films at room temperature (Ichimura et al. 1987; King et al. 1985). The absorption and emission data of all the Ir(dmppy)$_3$ derivatives in CH_2Cl_2 solution are shown in Table. 1.

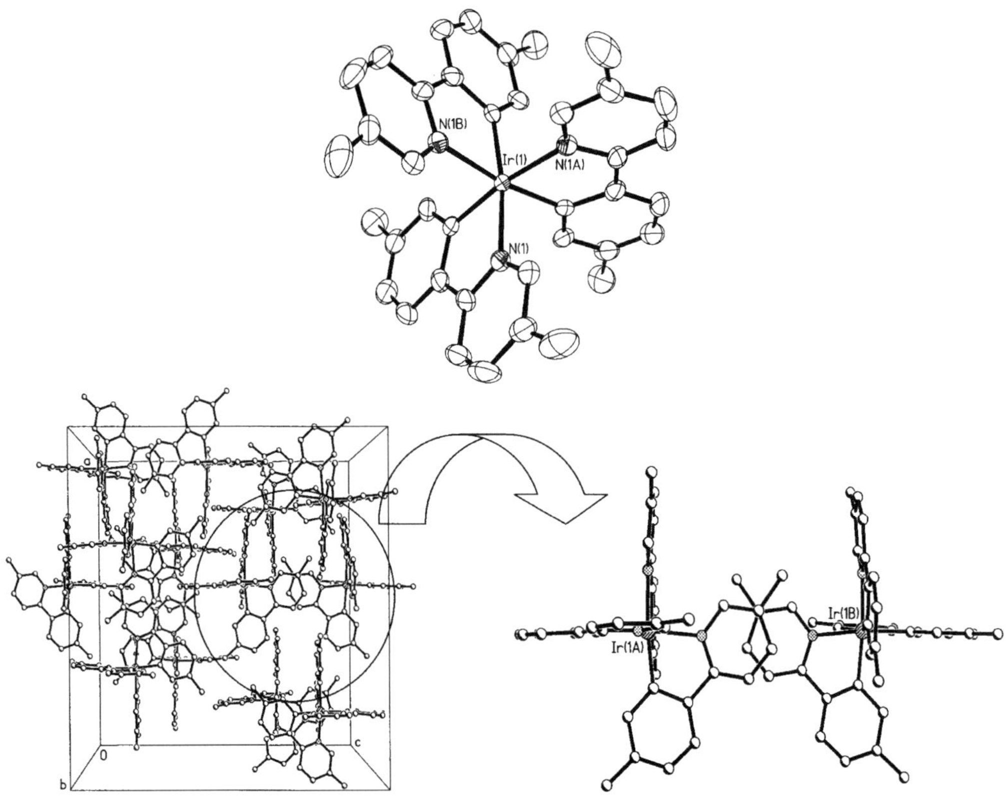

Fig 3. Top: Molecular structure of Ir(4,5′dmppy)$_3$ with atom labeling schemes and 50% thermal ellipsoids; all hydrogen atoms have been omitted for clarity; bottom: crystal packing diagram between two adjacent molecules of Ir(4,5′dmppy)$_3$ showing the intermolecular interaction in the solid state.

The blue-shifted absorption and emission spectra of Ir(4,4′dmppy)$_3$ complex compared with other complexes and Ir(ppy)$_3$ can be attributed to the reduction of conjugation between phenyl and pyridine rings as well as the lack of intermolecular interactions in the solid state (Sapochak et al. 2001). This is due to steric factors with a contribution from the strong electron donating effect of methyl groups. It has been reported that the donating effect from the methyl groups at either the 4′- or 6′- positions does not have any significant influence on the HOMO energy levels, while methyl groups at the 3′- or 5′- positions effectively donate electrons to dmppy moiety.

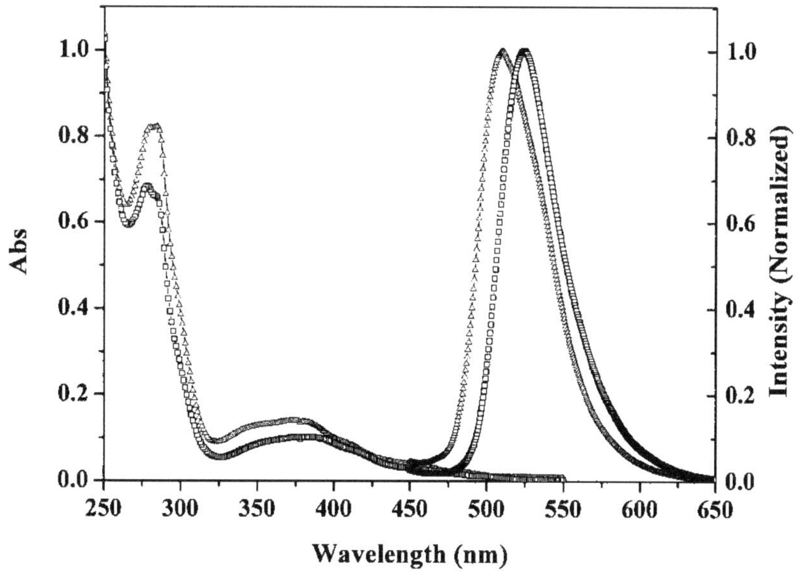

Fig. 4. Absorption and emission spectra of Ir(4,4′dmppy)$_3$ (triangles) and Ir(5,4′dmppy)$_3$ (squares) in CH_2Cl_2 at room temperature.

Ir Complex	UV-Vis Absorption nm(ε)	Emission (λ_{max}) Sol.	Emission (λ_{max}) Film	Φ_{PL}	Redox $E_{1/2}$ (V)
Ir(ppy)$_3$	283(4.9),380(4.1),405(4.0),455(3.5),490(3.1)	514	516	0.40	0.71
Ir(3,4′dmppy)$_3$	284(4.9),384(4.1),408(4.0),455(3.6),493(3.2)	522	529	0.41	0.51
Ir(4,4′dmppy)$_3$	278(4.9),374(4.2),407(4.0),452(3.6),484(3.3)	509	512	0.52	0.54
Ir(4,5′dmppy)$_3$	284(4.9),391(4.1),414(4.0),456(3.6),493(3.2)	524	534	0.32	0.49
Ir(5,4′dmppy)$_3$	284(4.9),383(4.1),412(4.0),459(3.5),490(3.2)	524	536	0.34	0.52
Ir(5,5′dmppy)$_3$	284(4.9),384(4.0),414(3.9),458(3.4),497(3.0)	524	532	0.29	0.47

Table 1. Photophysical and electrochemical data of Ir(dmppy)$_3$ complexes.

2.2 Electroluminescence properties

The phosphorescent organic light emitting diodes (PHOLEDs) based on Ir(dmppy)$_3$ complexes were fabricated by the vacuum deposition technique with the following configuration: ITO/copper phthalocyanine (CuPc, 10 nm) as hole injection layer/4,4′-bis[(1-naphthyl)(phenyl)-amino]-1,1′-biphenyl (NPD, 40 nm) as hole transport layer/CBP:Ir(dmppy)$_3$ (8%) (20 nm) as emissive layer/2,9-dimethyl-4,7-diphenyl-1,10-phenanthroline (BCP, 10 nm) as a hole blocking layer/tris-(8-hydroxyquinoline)aluminum (Alq_3, 40 nm) as an electron transport layer/LiF (1 nm) as electron injection layer/Al (100 nm) as cathode. All iridium complexes, Ir(4,4′dmppy)$_3$, Ir(4,5′dmppy)$_3$ and Ir(5,4′dmppy)$_3$, doped in 4,4′-bis(9-carbazolyl)-1,1′-biphenyl (CBP) host were used as emissive layer for the devices and the devices performances were compared with that of a similar structured Ir(ppy)$_3$ based device. The device structure and molecular structures of the materials used in the devices are shown in Fig. 5.

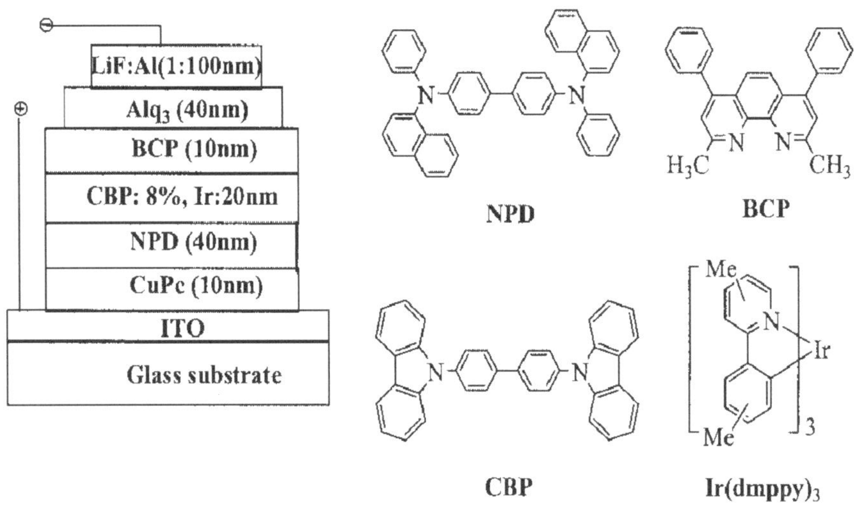

Fig. 5. Device structure and molecular structures of the materials used in the device.

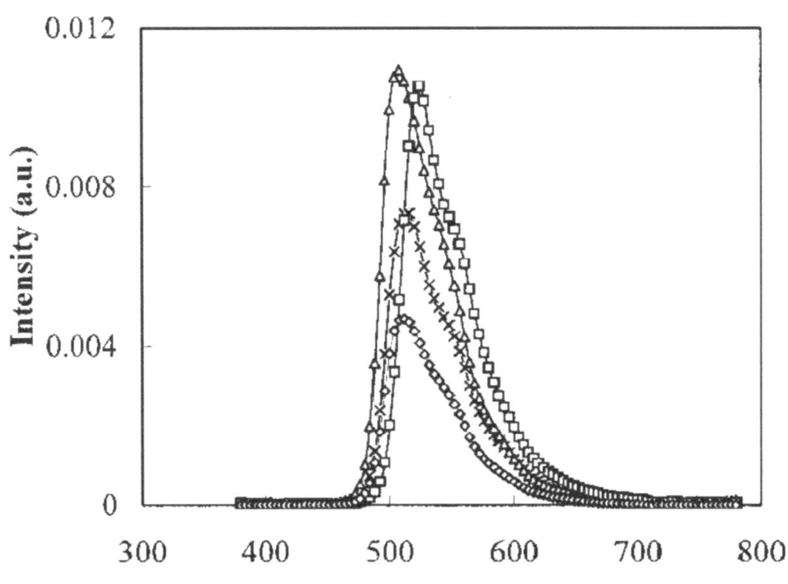

Fig. 6. Electroluminescence spectra of [Ir(ppy)3], (X), Ir(4,4′dmppy)$_3$ (triangles), Ir(4,5′dmppy)$_3$ (squares) and Ir(5,4′dmppy)$_3$ (diamonds) based devices.

The electroluminescence spectra of all devices showed green emission between 508 nm and 520 nm as shown in Fig. 6, which are quite similar to their PL emission spectra in solution. No host emission was observed, which indicates that the energy transfer from host to dopant is efficient in the EML of the device. The turn-on voltages (defined as 1 mA/cm^2) and device performances of all three Ir(dmppy)$_3$ and Ir(ppy)$_3$ complexes based devices are given in Table 2. The observed higher operating voltages for Ir(4,4'dmppy)$_3$- and Ir(5,4'dmppy)$_3$- based devices relative to Ir(ppy)$_3$ based device could be caused by the effective hole trapping. The observed decreasing device efficiencies of all the devices with increasing current densities were attributed to triplet-triplet annihilation. The device based on Ir(4,4'dmppy)$_3$ complex exhibits higher device performance compared with those of other devices.

Device	Luminescence (Cd/m^2) At 20 mA/cm^2	Operating Voltage (V)	External quantum efficiency (%)		
			at 1 mA/cm^2	at 10 mA/cm^2	At 50 mA/cm^2
Ir(ppy)$_3$	3156	4.9	6.0	5.4	4.7
Ir(4,4'dmppy)$_3$	4543	5.6	11	8.2	6.8
Ir(4,5'dmppy)$_3$	4470	5.1	10.9	7.6	5.8
Ir(5,4'dmppy)$_3$	3551	4.6	8.6	6.5	5.0

Table 2. Electroluminescent data of Ir(ppy)$_3$ and Ir(dmppy)$_3$ complexes based devices.

3. Photophysical and electroluminescence characteristics of bulky trimethylsilyl substituted Iridium(III) complex

The bulky trimethylsilyl substituted iridium(III) complex, *fac*-tris[4-methyl-2-(4'trimethylsilylphenyl)pyridine]iridium(III) [Ir(msippy)$_3$], has been synthesized by our research group and studied its photo-physical and electroluminescence properties in device (Jung et al. 2006). This bulky trimethylsilyl substitution is expected to have inhibited triplet annihilation due to the intermolecular interactions. The molecular structure, UV-Vis and Photoluminescence (PL) spectra of Ir(msippy)$_3$ complex are shown in Fig. 7. In the UV-Vis spectrum of Ir(msippy)$_3$ in dichloromethane solution at room temperature, the strong absorption band around 286 nm, which is assigned to π - π^* transitions on the cyclometalated ligand, and broad absorption bands from 380 nm to 480 nm assigned to spin allowed and spin forbidden metal-to-ligand charge transfer MLCT) transitions were clearly observed. PL spectrum of Ir(msippy)$_3$ in solution shows the emission at 510 nm, indicating the radiative decay from the triplet manifold to the ground state. PL quantum yield of Φ = 0.43 was observed for the Ir(msippy)$_3$, which is slightly higher than that of widely used green emitting phosphorescent Ir(ppy)$_3$ complex (Φ = 0.40) (Jung et al. 2006; King et al. 1985). This bulky trimethylsilyl substituted Ir(msippy)$_3$ complex is expected to give better device performances compared with that of Ir(ppy)$_3$ complex. The electrochemical studies were carried out by cyclic voltametry. The HOMO, LUMO and band gap of the Ir(msippy)$_3$ are 5.15 eV, 2.67 eV, and 2.47 eV, respectively, from UV absorption data.

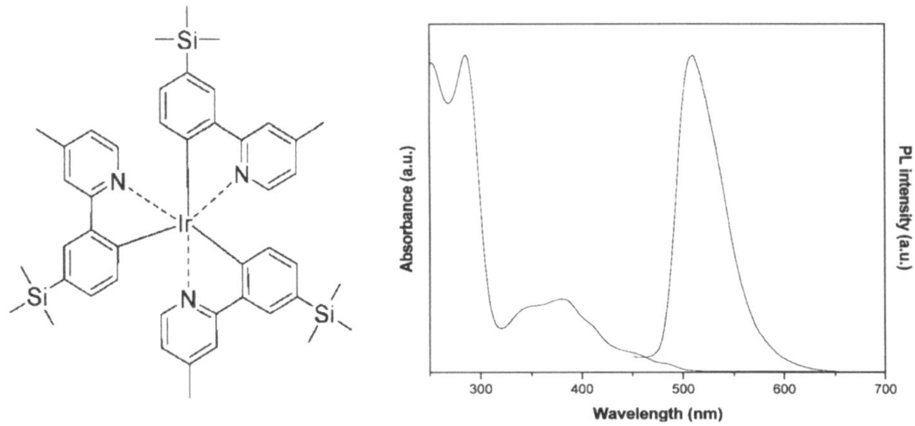

Fig. 7. The molecular structure (left), UV-Visible absorption and Photoluminescence spectra of Ir(msippy)3 complex (right) in dichloromethane solution.

The phosphorescent organic light emitting diodes (PHOLEDs) were fabricated using the Ir(msippy)3 complex as phosphorescent dopant by thermal evaporation process and characterized. The optimized device structure was indium-tin-oxide (ITO) (70 nm)/1,1-bis[di-4-tolylamino]phenyl]cyclohexane (TAPC) (50 nm) as a hole transporting layer/ 4,4′,4″-tris(carbazole-9-yl)tri-phenylamine (TcTa) (10 nm) as exciton blocking layer/Ir(msippy)3 or Ir(ppy)3 (5%) doped in 4,4′-N,N′-dicarbazole)biphenyl (CBP) host as EML (30 nm) / 1,3,5-tris(m-pyrid-3-yl-phenyl)benzene (TmPyPB) (60 nm) as a hole blocking layer as well as electron transport layer/ LiF (1 nm) as an electron injection layer/ Al (120 nm) as cathode. The device structure and electroluminescence (EL) spectra of the fabricated PHOLEDs are shown in Fig. 8. PHOLEDs based on Ir(msippy)3 shows the EL emission peak at 521 nm with shoulder peak around 549 nm, which confirms the yellowish-green emission originating from the triplet excited states of the Ir(misppy)3 dopant in the EML of the device. No emission from host and/or adjacent layers was observed, indicating the charge carriers and excitons are confined well within the EML. It has also been reported that the energy and/or charge transfer form CBP host to Ir(msippy)3 dopant is complete. The maximum external quantum efficiency (EQE) of 25.6% and current efficiency of 84.4 Cd/A were observed for Ir(msippy)3 based device with CIE color coordinates of (0.31, 0.64). PHOLEDs based on Ir(ppy)3 shows the green EL emission peak at 512 nm with shoulder peak around 539 nm originating from the triplet excited states of the Ir(ppy)3 dopant in the EML of the device, which is consistent with earlier report (Zhang et al. 2005; Cheng et al. 2003; Kim et al. 2007). The small additional emissions from CBP host and TcTa layers were observed for the devices based on Ir(ppy)3 complex, which indicates that the energy and/or charge transfer from host to dopant is incomplete. The PHOLEDs based on Ir(ppy)3 showed a maximum EQE of 18.7 % and current efficiency of 60.3 Cd/A. The device performances are shown in Table 3.

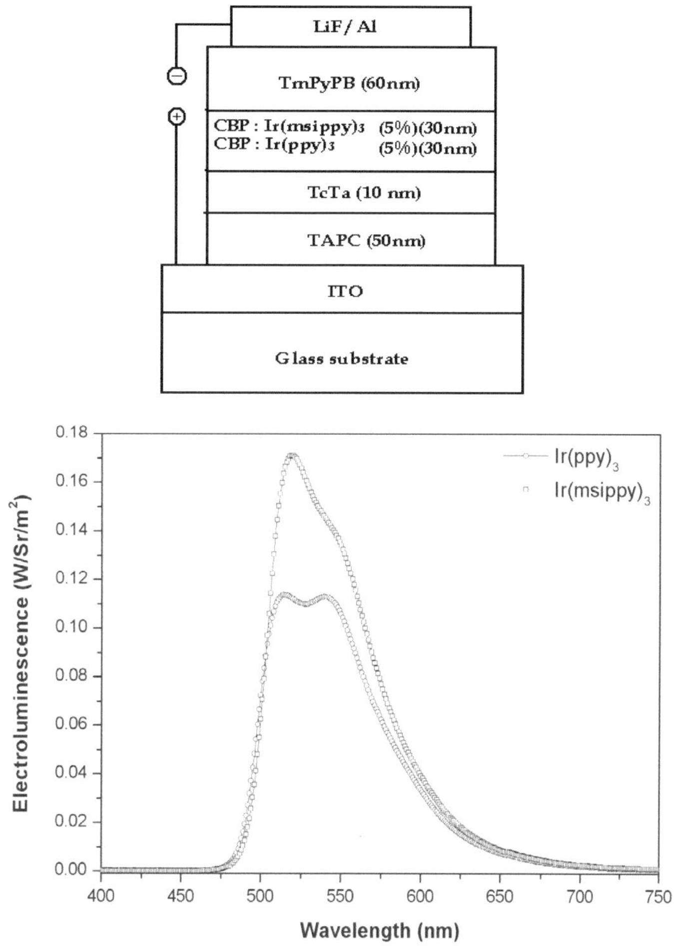

Fig. 8. The device structure (top), Electroluminescence spectra (bottom) of the fabricated PHOLEDs based on $Ir(msippy)_3$ and $Ir(ppy)_3$ complexes.

PHOLEDs based on	Luminescence (Cd/m^2)	EQE (%)	Current efficiency (Cd/A)
$Ir(msippy)_3$	77,910 (18V)	25.6	84.4
$Ir(ppy)_3$	44,700 (18V)	18.7	60.3

Table 3. The device efficiencies of PHOLEDs based on $Ir(msippy)_3$ and $Ir(ppy)_3$ dopants.

4. A narrow band green emitting bulky trimethylsilyl substituted Iridium(III) complex and PHOLED characteristics

The homoleptic iridium(III) complex, *fac*-tris[2-(3′-trimethylsilylphenyl)-5-trimethylsilylpyridinato]iridium [$Ir(disppy)_3$], has been synthesized by Suzuki coupling reaction. The effect of the substitution of bulky silyl groups on the photophysical and electroluminescence properties of $Ir(disppy)_3$ based device has also been investigated (Jung et al. 2009). The trimethyl functional groups provide to the molecules such higher vapour pressure, higher thermal stability, good solubility and steric bulk via higher volume. These properties of silyl moieties effectively hinder the aggregation and excimer formation of ppy based iridium(III) complex (Liu et al. 2005). Ir(disppy)$_3$ complex is highly soluble in common organic solvents and slightly soluble in hexane due to the introduction of bulky trimethylsilyl groups. This complex is very stable up to 290 °C without degradation under N_2 atmosphere. Differential scanning calorimetry (DSC) showed the glass transition temperature of 184 °C. The observed Tg value of Ir(disppy)$_3$ complex is considerably higher than that of Ir(ppy)$_3$ complex. This suggests that the introduction of silyl moieties in the ppy ring leads to higher thermal stability.

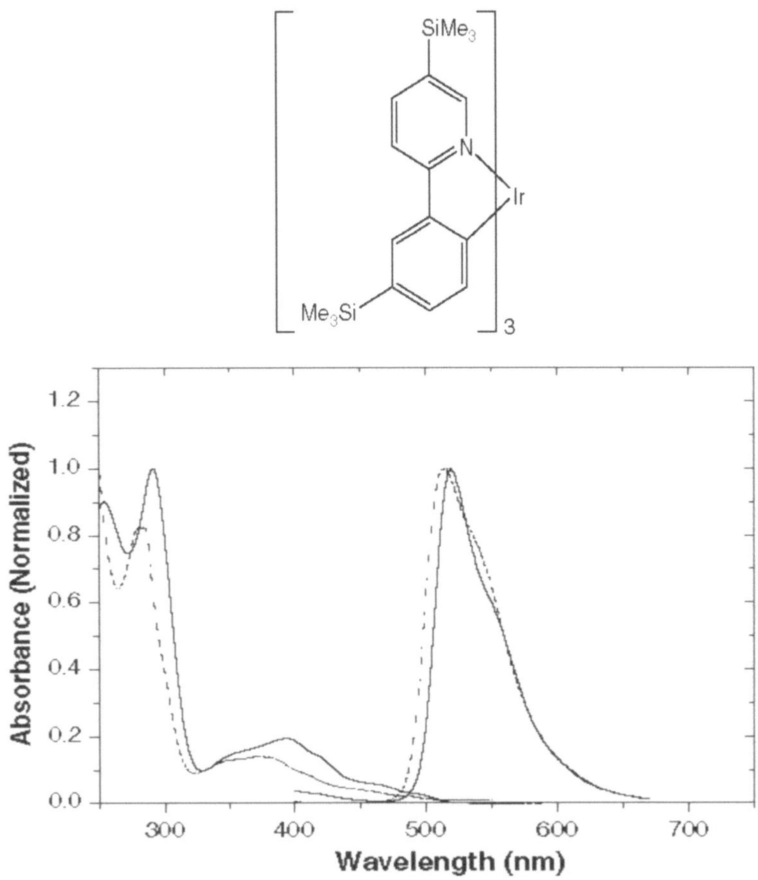

Fig. 9. The molecular structure (top), UV-Visible and photoluminescence spectra of Ir(disppy)$_3$ complex (bottom) in solution (solid line) and in film (dashed line).

The UV-Visible spectra of Ir(disppy)$_3$ and Ir(ppy)$_3$ complexes in solution show the absorption band around 288 nm corresponding to ligand centered (LC) π - π^* transitions and the absorption in the region of 350 to 480 nm corresponding to spin-allowed (singlet) and spin-forbidden (triplet) metal-to-ligand charge transfer (MLCT) transistions. The larger extinction coefficient of the singlet and triplet MLCT states of Ir(disppy)$_3$ complex relative to that of $Ir(ppy)_3$ complex indicates the substituent silyl groups play a key role in the enhancement of the spin-orbit coupling. The PL specta of $Ir(disppy)_3$ complex show the phosphorescence emission at 519 nm in solution and 513 nm in the film. Interestingly, the emission spectrum of $Ir(disppy)_3$ complex in the solid state is blue shifted in comparison to the diluted solution (10^{-5}-10^{-6} M) and the emission spectrum in solution shows the narrow band with a small full width at half maximum of 50 nm for solution spectrum and of 60 nm for film spectrum. The blue shifted emission of solution spectrum may indicate that bulky trimethylsilyl groups hamper intermolecular interactions even in the solid state. The molecular structure, UV-Vis and PL spectra of $Ir(disppy)_3$ complex are shown in Fig. 9. The HOMO level of $Ir(disppy)_3$ complex was estimated from cyclic voltametry to be -5.30 eV, which is compared with the value (-5.2 eV) obtained from ultraviolet photoelectron spectroscopy (UPS) and this is slightly higher than that of $Ir(ppy)_3$ complex . The LUMO level and optical band gap of $Ir(disppy)_3$ were estimated from its absorption data to be -2.71 eV and 2.59 eV, respectively. From the PL efficiency measurements, it is observed that the bulky silyl group on the ppy ring seems to play a key role in preventing self-quenching.

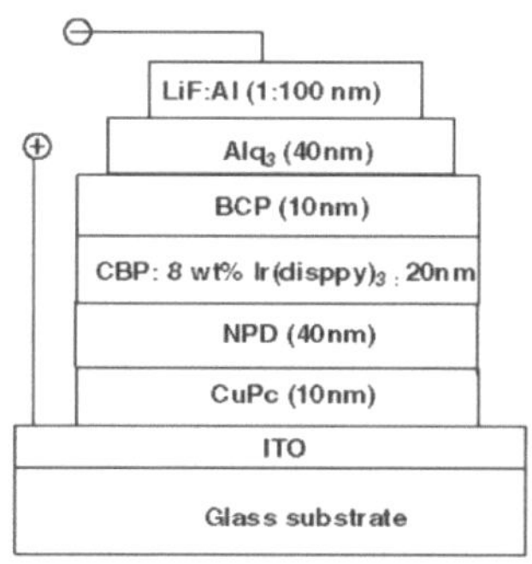

Fig. 10. The device structure of $Ir(disppy)_3$ or $Ir(ppy)_3$ based PHOLEDs.

The phosphorescent organic light emitting diodes (PHOLEDs) based on $Ir(disppy)_3$ complex were fabricated by the vacuum deposition process. The devices were made using ITO as anode/copper phthalocyanine (CuPc, 10 nm) as hole injection layer/4,4'-bis[(1-naphthyl)(phenyl)-amino]-1,1'-biphenyl (NPD, 40 nm) as hole transporting layer/CBP host : $Ir(disppy)_3$ or $Ir(ppy)_3$ (8%) (20 nm) as phosphorescent dopant/2,9-dimethyl-4,7-diphenyl-1,10-phenanthroline (BCP, 10 nm) as hole blocking layer/tris-(8-hydroxyquinoline)aluminum (Alq_3, 40 nm) as electron transport layer/LiF (1 nm) as electron injection layer/Al (100 nm) as cathode. The device structure is shown in Fig. 10. The fabricated PHOLEDs show the yellowish green EL emission at 524 nm for $Ir(disppy)_3$ based device and at 516 nm for $Ir(ppy)_3$ based device, which are well matched with the solution photoluminescence (PL) spectra. $Ir(disppy)_3$ based PHOLED device exhibits the lower operating voltage (7.4 V), higher brightness and power efficiency compared with that of $Ir(ppy)_3$ based device as shown in Table 4. The higher device efficiencies observed for the $Ir(disppy)_3$ based device were compared with that of $Ir(ppy)_3$ device.

PHOLEDs based on	EL (λ_{max}) (nm)	Operating voltage (V)	Luminous efficiency (Cd/A)	Power efficiency (lm/W)
Ir(disppy)$_3$	524	7.4	39.2	17.3
Ir(ppy)$_3$	516	8.7	32.5	11.7

Table 4. The device performances (at 10 mA/cm^2) of PHOLEDs based on Ir(disppy)$_3$ and Ir(ppy)$_3$ (reference) (8%) dopant in CBP host as EML.

5. Bulky cycloalkene substituted Iridium(III) complexes and PHOLED characteristics

The electroluminescence (EL) efficiency and the emission energy of iridium(III) complex based devices are greatly influenced by the organic ligand chromophores (Tang & VanSlyke 1987; Tang & VanSlyke 1989; Baldo et al. 1989). In the way to improve and tune the emission colors, we synthesized and reported the iridium(III) complexes using 2-cycloalkenylpyridine derivatives as cyclometalated ligands for OLEDs (Kang et al. 2008). Cyclic alkene is expected to give better stability than alkene in the complexes (Takiguchi et al. 2002). The molecular structures of the iridium complexes, tris-[2-(1-cyclohexenyl)pyridine]iridium [Ir(chpy)$_3$] and tris-[2-(3-methyl-1-cyclohexenyl)pyridine]iridium [Ir(mchpy)$_3$], are shown in Fig. 11.

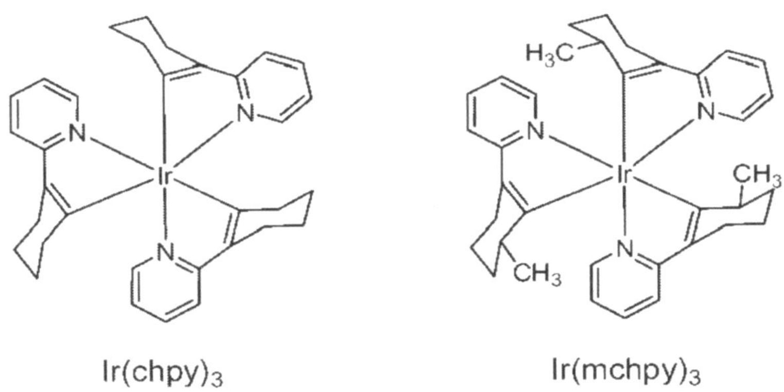

Fig. 11. The molecular structures of 2-cycloalkenylpyridine substituted iridium(III) complexes, Ir(chpy)$_3$ or Ir(mchpy)$_3$.

The introduction of rigid and bulky cycloalkene unit in these iridium complexes is expected to provide high device efficiencies as well as the suppressed triplet-triplet (T-T) annihilation in the OLED devices. These iridium complexes, Ir(chpy)$_3$ or Ir(mchpy)$_3$, with cycloalkenylpyridines have higher HOMO and lower LUMO energy levels than iridium(III) complex, Ir(ppy)$_3$. We synthesized 2-cycloalkenylpyridine substituted iridium complexes, Ir(chpy)$_3$ or Ir(mchpy)$_3$, in 44-74% yields and reported (Kang et al. 2008).

Ir(III) complex	UV-Vis Absorption (λmax) (nm)	PL		Φ_{PL}	(τ_{ph}) (μs)	HOMO/LUMO (eV)
		Sol	Film			
$Ir(chpy)_3$	336, 394, 447, 517	536	550	0.68	2.0	5.0/2.5
$Ir(mchpy)_3$	336, 394, 447, 517	535	543	0.61	1.3	5.1/2.6

Table 5. Photophysical and electrochemical data of $Ir(chpy)_3$ and $Ir(mchpy)_3$ complexes.

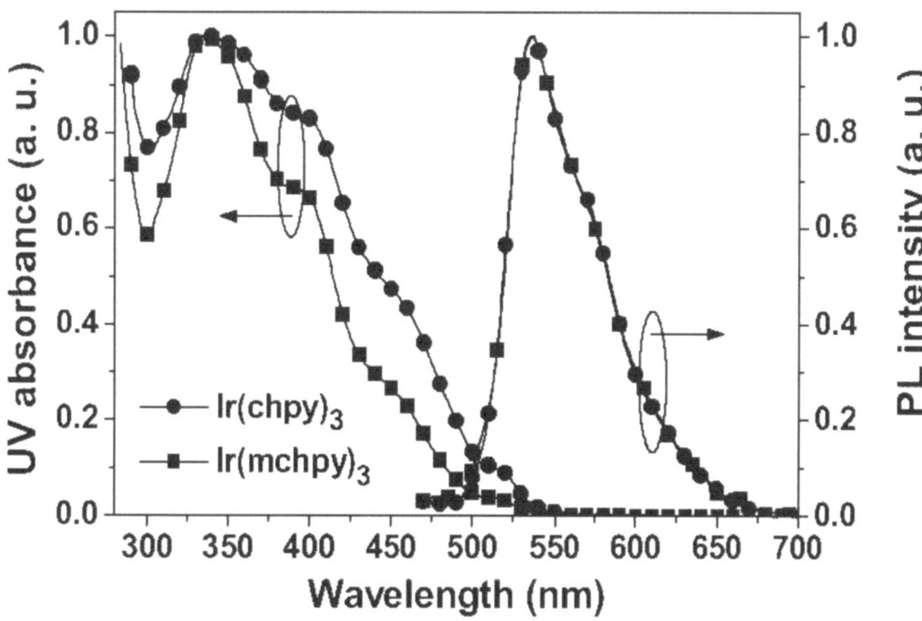

Fig. 12. The UV-Visible absorption and Photoluminescence spectra of iridium(III) complexes, $Ir(chpy)_3$ or $Ir(mchpy)_3$ in toluene.

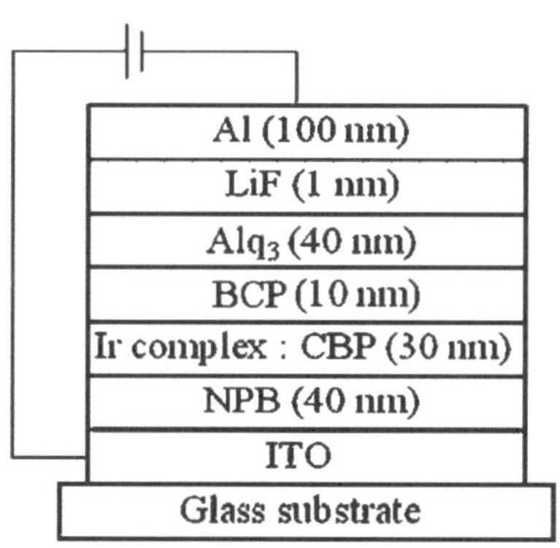

Fig. 13. The device structure of PHOLEDs based on Ir complex, $Ir(chpy)_3$ or $Ir(mchpy)_3$.

The iridium(III) complexes, $Ir(chpy)_3$ or $Ir(mchpy)_3$, show similar UV-Visible absorption and photoluminescence (PL) characteristics as can be seen in Fig. 12. The photophysical properties of these complexes are summarized in Table 5. In UV-Vis absorption spectra, the absorption maxima were observed for both the complexes in solution at 336 nm and 394 nm, which are assigned to ligand based transitions and at 447 nm and 517 nm, assigned to spin allowed and spin forbidden metal-to-ligand charge transfer (MLCT) transitions. The photoluminescence (PL) spectra exhibited the emission at 536 nm (in solution) and 550 nm (in film) for Ir(chpy)$_3$ and at 535 nm (in solution) and 543 nm (in film) for Ir(mchpy)$_3$. The photoluminescence lifetime (τ_{ph}) were measured in toluene solution to be 2.0 μs for Ir(chpy)$_3$ and 1.3 μs for Ir(mchpy)$_3$, which are consistent with emission from a triplet excited state (Lamansky et al. 2001). The electrochemical properties were estimated by cyclic voltametry using Ag/AgCl with reference of 4,4′-bis[N-(1-naphthyl)-N-phenylamino]-biphenyl (NPB) (HOMO) and estimated using absorption edge (LUMO). The determined HOMO and LUMO energy levels are -5.0 eV and -2.5 eV for Ir(chpy)$_3$ and -5.1 eV and -2.6 eV for Ir(mchpy)$_3$, respectively. These HOMO and LUMO energy levels are higher than those of Ir(ppy)$_3$ (HOMO: -5.4 eV and LUMO: -2.9 eV).

The PHOLEDs were fabricated using Ir(chpy)$_3$ and Ir(mchpy)$_3$ complexes. The device structure has the following configuration: ITO (anode)/NPB (40 nm) as hole transporting layer/phosphorescent dopant Ir complex (6%), Ir(chpy)$_3$ or Ir(mchpy)$_3$ or Ir(ppy)$_3$ (for reference), doped in CBP host (30 nm) as EML/2,9-dimethyl-4,7-diphenylphenanthroline (BCP) (10 nm) as a hole blocking layer/Alq$_3$ (40 nm) as an electron transport layer/LiF (1 nm)/Al (100 nm) as cathode and the device structure can be seen in Fig. 13. The EL spectra of Ir complexes in devices are the same as the PL spectra of those iridium complexes, indicating that the most of the excitons recombine at the dopant Ir complex in the device. The Ir(chpy)$_3$ and Ir(mchpy)$_3$ complexes based devices exhibit yellow green emission with CIE color coordinates of (0.40, 0.59) for both Ir complexes. Ir(chpy)$_3$ based PHOLEDs showed a maximum external quantum efficiency (EQE) of 18.7%, a current efficiency of 69 cd/A, and a power efficiency of 62 lm/W, which is much higher than the Ir(ppy)$_3$ based device, while Ir(mchpy)$_3$ based device exhibited a little lower device performances than Ir(chpy)$_3$ based device but still it exhibited a much better performances than the Ir(ppy)$_3$ based device. The device performances are summarized in Table 6. The high efficiency of the Ir(chpy)$_3$ and Ir(mchpy)$_3$ based devices has been explained by more balanced injection and transport of electrons and holes in(to) the emitting layer. Because of the HOMO and

PHOLEDs With EML	Turn-on Voltage (V)	EQE (%)	Current efficiency (cd/A)	Power efficiency (lm/W)	CIE, 8 V, (x, y)
CBP:Ir(chpy)$_3$	3.4	18.7	69.0	62.0	(0.40, 0.59)
CBP:Ir(mchpy)$_3$	3.7	17.1	62.5	53.1	(0.41, 0.58)
CBP:Ir(ppy)$_3$	3.7	14.6	50.2	47.8	(0.32, 0.61)

Table 6. PHOLED characteristics of Ir(chpy)$_3$, Ir(mchpy)$_3$ and Ir(ppy)$_3$ (reference) (6%) doped in CBP host as EML.

LUMO levels of Ir(chpy)$_3$ and Ir(mchpy)$_3$ are higher than those of CBP host (HOMO/LUMO: 6.0 eV/2.9 eV), the dopants are behaving as hole traps and electron scattering centers so that both electron and hole mobility in the EML will be retarded by the doping. In contrast, Ir(ppy)$_3$ has almost the same LUMO level (2.9eV) as CBP so that Ir(ppy)$_3$ will have little effect on electron mobility of EML. The better hole trapping and balanced hole and electron transporting ability in $Ir(chpy)_3$ in comparison with $Ir(ppy)_3$ resulted in better recombination of electrons and holes in EML, resulted in higher devices performances.

The substituents such as methyl, bulky trimethylsilyl, and cycloalkene groups substituted iridium(III) complexes have been investigated on their photophysical and electrochemical properties. The PHOLEDs based on these iridium(III) complexes have been presented. Among those, the methyl groups substituted $Ir(dmppy)_3$ based devices exhibited the green electroluminescence emission in the range of 508 nm to 520 nm, the bulky trimethyl substituted Ir(III) complexes based devices showed the yellowish green emission between 521 nm and 524 nm and the cycloalkene substituted iridium(III) complexes based devices showed the yellow green emission between 543 nm and 550 nm as summarized in Table 7.

PHOLEDs Based on	EL emission (nm)	PHOLEDs Based on	EL emission (nm)
$Ir(dmppy)_3$ complexes	508 -520	$Ir(disppy)_3$	524
$Ir(msippy)_3$	521	$Ir(chpy)_3$ & $Ir(mchpy)_3$	550-543

Table 7. Electroluminescence data of PHOLED based on various substituents (methyl, bulky trimethylsilyl, and cycloalkene groups) substituted iridium(III) complexes.

6. Conclusion

We have presented the effect of various substituents on the photo-physical, electrochemical and electroluminescence properties of green emitting iridium(III) complexes and phosphorescent organic light emitting diodes.

(a) The methyl groups were substituted on the ppy ligand of $Ir(ppy)_3$ and prepared a series of *fac*-[$Ir(dmppy)_3$] complex derivatives. All $Ir(dmppy)_3$ derivatives are very stable up to 300°C without degradation in air. The crystal structures of $Ir(4,4'dmppy)_3$ and $Ir(4,5'dmppy)_3$ complexes exhibit only the *fac*- configuration with a distorted octahedral geometry around the Ir atom and indicated the decreased conjugation of 4,4'-ppy ligands. These derivatives show the emission between 509 nm and 534 nm in solution as well as in thin films at room temperature. The electroluminescence spectra of all derivatives in devices showed green emission between 508 nm and 520 nm. The device based on $Ir(4,4'dmppy)_3$ complex exhibited higher device external quantum efficiency of 10.9% at 4470 cd/m^2 compared with those of other devices.

(b) The bulky trimethylsilyl substituted iridium(III) complex showed the PL emission at 510 nm in solution with higher PL quantum yield (Φ = 0.43). PHOLEDs exhibited the yellowish-green EL emission at 521 nm. The maximum external quantum efficiency (EQE) of 25.6% and current efficiency of 84.4 Cd/A were observed for $Ir(msippy)_3$ based device with CIE color coordinates of (0.31, 0.64). It has been reported that the charge carriers and excitons are confined within the EML of device and the energy and/or charge transfer form host to $Ir(msippy)_3$ dopant is efficient.

(c) The homoleptic iridium(III) complex, *fac*-tris[2-(3′-trimethylsilylphenyl)-5-trimethylsilylpyridinato]iridium [Ir(disppy)$_3$], is very stable up to 290 °C without degradation under N_2 atmosphere. Differential scanning calorimetry (DSC) showed the glass transition temperature of 184 °C. The introduction of silyl moieties in the ppy ring leads to higher thermal stability. The PL specta of Ir(disppy)$_3$ complex showed the emission between 519 nm and 513 nm and showed the narrow band with FWHM of 50 nm. PHOLEDs based on Ir(disppy)$_3$ complex showed the yellowish green EL emission at 524 nm and exhibited the lower operating voltage (7.4 V), higher efficiencies of 39.2 cd/A and 17.3 lm/W.

(d) Iridium(III) complexes using 2-cycloalkenylpyridine derivatives as cyclometalated ligands, Ir(chpy)$_3$ or Ir(mchpy)$_3$, exhibited the PL emission between 536 nm (solution) and 550 nm (film) for Ir(chpy)$_3$ and 535 nm (Solution) and 543 nm (film) for Ir(mchpy)$_3$. The Ir(chpy)$_3$ based PHOLED showed a maximum external quantum efficiency (EQE) of 18.7%, a current efficiency of 69 cd/A, and a power efficiency of 62 lm/W than Ir(mchpy)$_3$ device.

7. References

Adachi, C.; Baldo, M. A.; O'Brien, D. F.; Thompson, M. E. & Forrest, S. R. (2001). Nearly 100% internal phosphorescence efficiency in an organic light-emitting device. *Journal of Applied Physics*, Vol. 90, pp. 5048-5052.

Baldo, M. A.; O'Brien, D. F.; You, F; Shoustikov, A.; Sibley, S.; Thompson, M. E. & Forrest, S. R. (1998). Highly efficient phosphorescent emission from organic electroluminescent devices. *Nature*, Vol. 395, pp. 151-154.

Baldo, M.A.; O'Brien, D. F.; Thompson, M. E. & Forrest, S. R. (1999). Excitonic singlet-triplet ratio in a semiconducting organic thin film. *Physical Review B: Condensed Matter and Material Physics*, Vol. 60, pp. 14422-14428.

Cheng, G.; Li, F.; Duan, Y.; Feng, J.; Liu, S.; Qiu, S.; Lin, D.; Ma, Y. & Lee, S. T. (2003). White organic light-emitting devices using a phosphorescent sensitizer. *Applied Physics Letters*, Vol. 82, pp. 4224-4226.

Duan, J.-P.; Sun, P.-P. & Cheng, C.-H. (2003). New Iridium Complexes as Highly Efficient Orange-Red Emitters in Organic Light-Emitting Diodes. *Advanced Materials*, Vol. 15, pp. 224-228.

Forrest, S. R. (2004). The path to ubiquitous and low-cost organic electronic appliances on plastic. *Nature*, Vol. 428, pp. 911-918.

Grushin, V. V.; Herron, N.; LeCloux, D. D.; Marshall, W. J.; Petrov, V. A. & Wang, Y. (2001). New, efficient electroluminescent materials based on organometallic Ir complexes. Chemical Communications, Vol. 16, pp. 1494-1495.

Ichimura, K.; Kobayashi, T.; King, K. A. & Watts, R. J. (1987). Excited-state absorption spectroscopy of ortho-metalated iridium(III) complexes. *Journal of Physical Chemistry*, Vol. 91, pp. 6104-6106.

Ikai, M.; Tokito, S.; Sakamoto, Y.; Suzuki, T. & Taga, Y. (2001). Highly efficient phosphorescence from organic light-emitting devices with an exciton-block layer. *Applied Physics Letters*, Vol. 79, pp. 156-158.

Jung, S.-O.; Kang, Y.; Kim, H.-S.; Kim, Y.-H.; Lee, C.-L.; Kim, J.-J.; Lee, S.-K. & Kwon, S.-K. (2004). Effect of substitution of methyl groups on the luminescence performance of Ir$^{(III)}$ complexes: Preparation, structures, Electrochemistry, Photophysical properties and their applications in organic light emitting diodes (OLEDs). *European Journal of Inorganic Chemistry*, Vol. 16, pp. 3415-3423.

Jung, S.-O.; Kim, Y.-H.; Kim, H.-S. & Kwon, S.-K. (2006). Effective electrophosphorescence emitting devices by using new type Ir(III) complex with bulky substistuent spaces. *Molecular crystals and liquid crystals*, Vol. 444, pp. 95-101.

Jung, S.-O.; Zhao, Q.; Park, J.-W.; Kim, S. O.; Kim, Y.-H.; Oh, H.-Y.; Kim, J.; Kwon, S.-K. & Kang, Y. (2009). A green emitting iridium(III) complex with narrow emission band and its application to phosphorescence organic light emitting diodes (OLEDs). *Organic Electronics*, Vol. 10, pp. 1066-1073.

Kang, D. M.; Kang, J.-W.; Park, J.-W.; Jung, S. O.; Lee, S.-H.; Park, H.-D.; Kim, Y.-H.; Shin, S. C.; Kim, J.-J. & Kwon, S.-K. (2008). Iridium complexes with cyclometalated 2-cycloalkenyl-pyridine ligands as highly efficient emitters for organic light emitting diodes. *Advanced Materials*, Vol. 20, pp. 2003-2007.

Kim, S. H.; Jang, J. & Lee, J. Y. (2007). High efficiency phosphorescent organic light-emitting diodes using carbazole-type triplet exciton blocking layer. *Applied Physics Letters*, Vol. 90, pp. 223505-223507.

King, K. A.; Spellane, P. J.; Watts, R.J. (1985). Excited-state properties of a triply ortho-metalated iridium(III) complexe. *Journal of American Chemical Society*, Vol. 107, pp. 1431-1432.

Lamansky, S.; Djurovich, P.; Murphy, D.; Abdel-Razzaq, F.; Lee, H.-E.; Adachi, C.; Burrows, B. E. & Forrest, S. R. (2001). Highly Phosphorescent Bis-Cyclometalated Iridium Complexes: Synthesis, Photophysical Characterization, and Use in Organic Light Emitting Diodes. *Journal of American Chemical Society*, Vol. 123, pp. 4304-4312.

Lee, S. J.; Park, K.-M.; Yang. K. & Kang. Y. (2009). Blue Phosphorescent Ir(III) Complex with High Color Purity: *fac*-Tris(2',6'-difluoro-2,3'-bipyridinato-N,C$^{4'}$)iridium(III). *Inorganic Chemistry*, Vol. 48, pp. 1030-1037.

Liu, X.-M., Xu, J.; Lu, X. & He, C. (2005). Novel Glassy Tetra(*N*-alkyl-3-bromocarbazole-6-yl)silanes as Building Blocks for Efficient and Nonaggregating Blue-Light-Emitting Tetrahedral Materials. *Organic Letters*, Vol. 7, pp. 2829-2832.

Lo, S.-C.; Male, N. A. H.; Markham, J. P. J.; Magennis, S. W.; Burn, P. L.; Salata, O. V. & Samuel, I. D. W. (2002). Green Phosphorescent Dendrimer for Light-Emitting Diodes. *Advanced Materials*, Vol. 14, pp. 975-979.

Neve, F; Crispini, A.; Serroni, S.; Loiseau, F. & Campagna, S. (2001). Novel dinuclear luminescent compounds based on iridium(III) cyclometalated chromophores and containing bridging ligands with ester-linked chelating sites. *Inorganic Chemistry*, Vol. 40, pp. 1093-1101.

Noh, Y.-Y.; Lee, C.-L; Kim, J.-J. & Yase, K. (2003). Energy transfer and device performance in phosphorescent dye doped polymer light emitting diodes. *Journal of Chemical Physics*, Vol. 118, pp. 2853-2864.

O'Brien, D. F.; Baldo, M. A.; Thompson, M. E. & Forrest, S. R. (1999). Improved energy transfer in electrophosphorescent devices. *Applied Physics Letters*, Vol. 74, pp. 442-444.

Sapochak, L. S.; Padmaperuma, A.; Washton, N.; Endrino, F.; Schmett, G. T.; Marshall, J.; Fogarty, D.; Burrows, P. E. & Forrest, S. R. (2001), Effects of Systematic Methyl Substitution of Metal (III) Tris(*n*-Methyl-8-Quinolinolato) Chelates on Material Properties for Optimum Electroluminescence Device Performance. *Journal of American Chemical Society*, Vol. 123, pp. 6300-6307.

Shen, Z.; Burrows, P. B.; Bluovic V.; Forrest, S. R. & Thompson, M. E. (1997). Three-color, Tunable, Organic Light emitting diodes, *Science*, Vol. 276, pp. 2009-2011.

Takiguchi, T.; Okada, S.; Tsuboyama, A.; Noguchi, K.; Moriyama, T.; Kamatani, J. & Furugori, M. (2002). US Patent 20020094453A1.

Tang, C. W. & VanSlyke, S. A. (1987). Organic Electroluminescent diodes, *Applied Physics Letter*, Vol. 51, pp. 913-915.

Tang, C. W.; VanSlyke, S. A. & Chen, C. H. (1989). Electroluminescence OF Doped Organic Thin-Films. *Journal of Applied Physics*, Vol. 65. pp. 3610–3616.

Wang, Y.; Herron, N.; Grushin, V. V.; LeCloux, D. & Petrov, V. (2001), Highly efficient electroluminescent materials based on fluorinated organometallic iridium compounds. *Applied Physics Letters*, Vol. 79, pp. 449-451.

Xie, H. Z.; Liu, M. W.; Wang, O. Y.; Zhang, X. H.; Lee, C. S.; Hung, L. S.; Lee, S. T.; Teng, P. F.; Kwong, H. L.; Zheng, H. & Che, C. M. (2001). Reduction of Self-Quenching Effect in Organic Electrophosphorescence Emitting Devices via the Use of Sterically Hindered Spacers in Phosphorescence Molecules. *Advanced Materials*, Vol. 13, pp. 1245-1248.

Zhang, Y.; Cheng, G.; Zhao, Y.; Hou, J. & Liu, S. (2005). White organic light emitting devices based on 4,4′-bis(2,2′-diphenylvinyl)-1,1′-biphenyl and phosphorescence sensitized 5,6,11,12-tetraphenylnaphthacene. *Applied Physics Letters*, Vol. 86, pp. 011112-011114.

Material Issues in AMOLED

Jong Hyuk Lee, Chang Ho Lee and Sung Chul Kim
Samsung Mobile Display, San #24 Nongseo-Dong, Giheung-Gu, Yongin-City, Gyunggi-Do, Korea 446-711

1. Introduction

Since the first mass production of AMOLED (active matrix organic light emitting diode) for mobile display in 2007, many companies have dived into the market for mobile phones and the other potable displays based on extraordinary image qualities of AMOLED. In a mass production point of view, small-sized AMOLED almost attained a stage of technological maturity. However, it still needs some more improvements in terms of materials for lower power consumption, longer life time of AMOLED. Besides outstanding market expansion of AMOLED in mobile applications, AMOLED also can bring us new displays that are only shown in some SF movies, such as paper-thin, foldable, bendable and transparent displays.

In terms of power consumption, AMOLED is intrinsically superior to LCD, where the backlight should be always "on". If we consider that the on-ratio is usually less than 30% in most TV broadcastings, a big advantage exists for AMOLED because AMOLED turn on the light for each pixel individually. Moreover, AMOLED still have plenty of rooms to further reduce power consumption. Although low power consumption is the reason why AMOLED is a better choice for portable devices, recent trend of green business require consuming lower power for brighter display. In order to meet those stringent requirements, new materials with high efficiency and optimization of AMOLED device structure is necessary.

This article reviews current material issues of AMOLED for general application and for unique application such as transparent and bendable displays.

2. Organic material issues of amoled

2.1 Materials for hole transporting

Since the first report of multi-layered OLEDs, many studies have focused on improving device efficiency and enhancing the durability of OLEDs. Development of new materials for improving device performance such as device driving voltage, efficiency, and life-time is one of the major research subjects in the OLED research area. And there have been lots of progress in performance characteristics. Despite much improved device performance, an insufficient life-time remains one of the primary issues limiting the wide-spread commercial use of OLED. Life-time property is a major obstacle in the competing with liquid crystal display (LCD) as flat panel displays and, life-time related image sticking is an emerging issue of OLED operation.

For the fabrication of highly stable OLEDs, specific optical and electronic properties, such as fluorescence, energy levels, charge mobility etc, and high morphologic stability are required [1-4]. The electrochemical stability of materials used in OLED is very important to improve the device properties. Also, the thermal stability of hole-transporting material is one of the significant factors of the device durability. Under thermal stress, most organic glass transition materials tend to turn into the thermodynamically stable crystalline state, which leads to device failure [5, 6]. It is known that an amorphous thin film with a high glass transition temperature (T_g) is more stable to heat damage [7-11]. In general, high thermal stability, especially high T_g above 100 °C, good hole transporting ability, and excellent film formability are essentially needed for the hole-transporting materials. Various triarylamine derivatives have been utilized as hole-transporting materials (HTMs) because of their good film forming capabilities as well as good hole-transporting abilities [3, 12-13].
The radical cation is one of important reactive intermediate in organic molecules and it can be obtained by loss of single electron from neutral molecules. The chemical structures of common radical cation species are shown in Fig. 1.

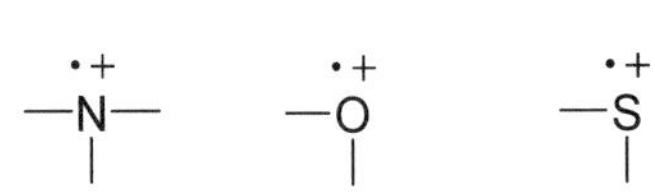

Fig. 1. The common radical cation species

Both hole and charge are not necessary to be localized together on one atom and they can be delocalized over the whole molecule. In fact, the delocalization of the unpaired electron in conjugated system can lead to stable radical cations such as the Wurster salt. This compound is isolable and the chemical structure including its resonance forms are shown in Scheme 1. Aryl amine moieties are thought to be a main core structure in HTMs because amine atom is relatively easy to lose one electron and the resulting radical cation can be stabilized by resonance effect of adjacent aryl substituent. It is worth to note that the Wurster salt mentioned above is stabilized by two factors. One is a resonance effect by aryl substituents and the other is stabilized by counter ion, perchlorate.

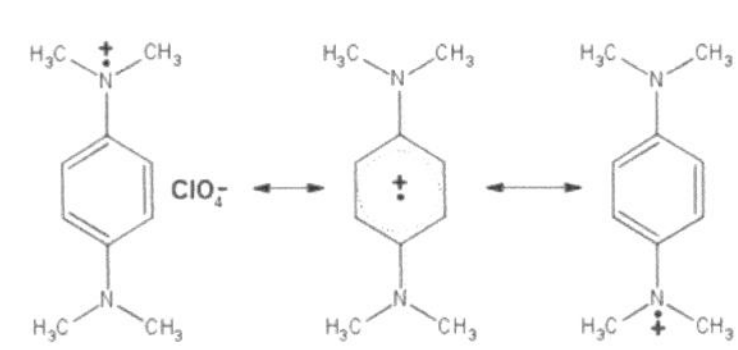

Scheme 1. The possible resonance forms of Wurster salt

However, there is no such stabilization by counter anion in OLED devices. The stability of radical cation mainly depends on its adjacent substituent. Therefore the HTMs stabilized by their substituents are one of important factors to improve the OLED performance.
There are several factors contributing to the stability of radicals. Those are hyperconjugation, resonance, hybridization, captodative effect, and steric effect [24]. Among them resonance and steric effect are important in aryl aminyl radical cations. These aminyl radical cations can be reactive and there are many possible reactions such as fragmentation, dimerization,

disproportionation, and, oxidation. However, the first three reactions are not favorable in OLED device because these give rise to change of the original HTMs via formation or cleavage of σ bond. It is thought to be one of plausible reasons for OLED degradation. However, the oxidation is desirable in OLED device because this single electron transfer process between adjacent molecules results in a hole-transporting process, a fundamental reaction of HTMs. Fragmentation and disproportionation reaction is relatively less important in solid state because the interactions between each radical cations are small but the interactions between a molecule and solvent are strong in solution. In contrast, the dimerization and the oxidation reaction are more important in solid state owing to their strong interaction between radical cation each other. The following Scheme 2 summarizes the important reactions in OLED device.

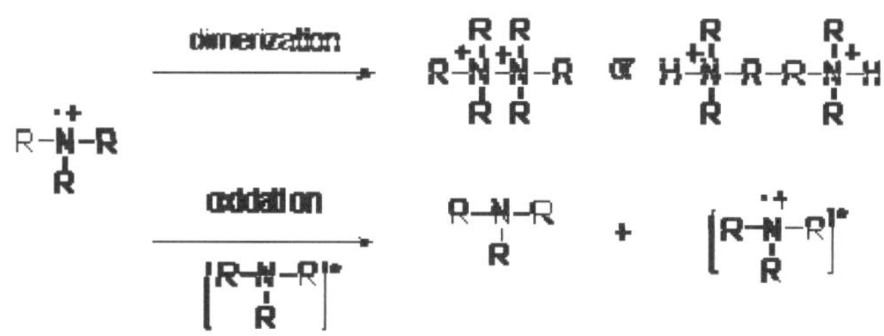

Scheme 2. Dimerization vs oxidation reaction

Therefore HTMs have to be modified to increase the stability of aminyl radical cation which can result in minimizing the cleavage of σ bond in molecules. Furthermore resonance effect and steric factor have to be considered to minimize the dimerization reaction in solid state. Recently, considerable efforts have been devoted to the development of new amorphous triarylamines possessing high morphologic stability [14-20]. We have already reported that the device employing thermally stable hole-transporting materials showed high efficiencies [21, 22]. However, it is thought that these hole-tansporting materials cannot meet high efficiency and long lifetime simultaneously. Therefore, we will discuss how to modify the structure of HTMs in order to increase their radical cationic stabilities. In addition, device performance with these modified molecules will be discussed.

2.1.1 Physical properties of hole transporting materials

Tested molecules having hole-transporting properties are shown in Fig. 2.

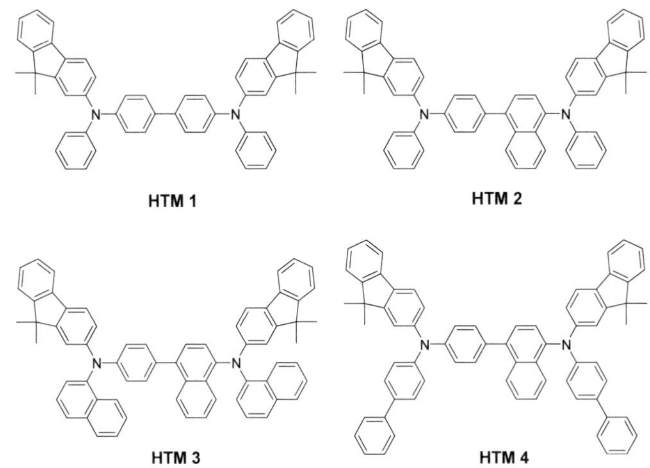

Fig. 2. Hole-transporting materials

Absorption spectra were measured with a HITACHI U-3000 UV spectrophotometer. ^{1}H NMR and ^{13}C NMR spectra were recorded with a JEOL JNM-ECP 400 FT NMR spectrometer. Differential scanning calorimetry (DSC) was performed on a TA Instruments, DSC-2910 unit using a heating rate of 10 °C/min and a cooling rate of 40 °C/min. Samples were scanned from 30 to 300 °C, cooled to 0 °C, and then scanned again from 30 to 300 °C. The glass transition temperatures (T_g) were determined from the second heating scan. Thermogravimetric analysis (TGA) was undertaken on a TA instrument, TGA-2050. The thermal stability of phenylnaphthyldiamine derivatives was determined under a nitrogen atmosphere, by measuring weight loss while heating at a rate of 20 °C/min. The results are summarized in Table 1 along with literature data of common hole-transporting materials for comparison. The ionization potentials (IPs) of materials used in device fabrication were determined by ultraviolet photoelectron spectroscopy (UPS) (Riken Keiki, AC-2) using the samples prepared by PMMA polymer binder on glass and the energy levels of lowest unoccupied orbital (LUMO) were approximately defined as differences between IPs and long wavelength cutoffs of the absorption spectra of 0.2 mM CH_2Cl_2 solution.
As mentioned before, a radical cation of **HTM 2** is more stable than that of **HTM 1** by two factors. As shown in structure A and B, the naphthyl amine radical cation B is preferred because it has two more resonance forms than cation A. It is well known that molecules having more resonance form are more stable. In addition, cation B can be stabilized further by steric effect. A bulky naphthyl moiety which is bigger than phenyl moiety can retard dimerization of radical cations.

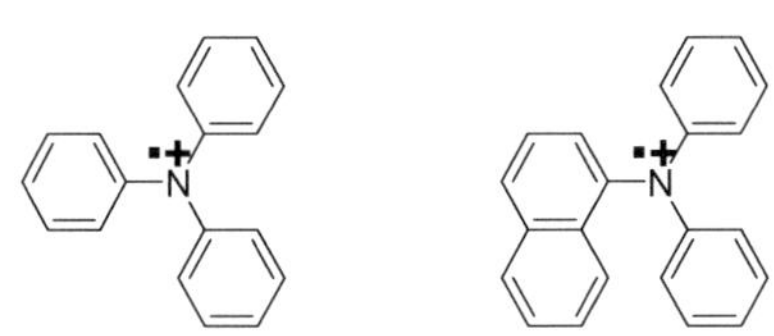

A B

The thermal stability data of these three phenylnaphthyldiamine derivatives, **HTM 2-4**, were investigated by differential scanning calorimetry and thermogravimetric analysis; the results are summarized in Table 1 with the well-known hole-transporting material **HTM 1** for comparison. As shown in Table 1, all three phenylnaphthyldiamine-cored HTMs (**HTM 2-4**) have higher value of T_g relative to their biphenyldiamine analog **HTM 1**, proofing the high morphologic stability of the amorphous phase in a deposited film, which is a prerequisite for the application in organic electroluminescent devices.

According to Shirota [1], a non-planar molecular structure preventing easy packing of molecules and an increased number of conformers in the molecule are preconditions in the design and synthesis of amorphous molecular glasses. Glass formation is enhanced by incorporation of bulky substituents. The incorporation of bulky substituents also hinders translational, rotational and vibrational motions of molecules, leading to an increase in the T_g. We attribute to the increase in the morphological stability of the biphenyl-based material to the presence of naphthalene substituents at the central phenylnaphthyl core, which may hinder the crystallization process. It is important that OLEDs be constructed from materials having a relatively high value of T_g to avoid problems associated with crystallization, which leads to device degradation.

The HOMO and LUMO levels of these phenylnaphthyldiamine derivatives are also listed in Table 1. The HOMO was determined using a photoelectron spectrometer, while LUMO was calculated based on the HOMO energy level and the lowest energy absorption edge of the UV absorption spectrum. The HOMO and LUMO levels of these compounds were measured at ca. 5.40 - 5.45 eV and 2.43 - 2.53eV, respectively.

Compounds	$T_d{}^a$ (°C)	$T_g{}^b$ (°C)	$T_m{}^b$ (°C)	$T_c{}^b$ (°C)	$\lambda_{max}{}^c$ (nm)	HOMO[d] (eV)	LUMO[e](eV)
HTM 1	380	121	264	NA	360	5.40	2.38
HTM 2	395	159	296	NA	342	5.40	2.43
HTM 3	430	167	225	NA	355	5.45	2.53
HTM 4	423	174	255	NA	353	5.45	2.48

[a] Obtained from TGA measurement. [b] Obtained from DSC measurement. [c] Measured in CH_2Cl_2 solution. [d] Determined by ultraviolet photoelectron spectroscopy (UPS). [e] Calculated based on the HOMO level and the lowest energy absorption edge of the UV spectrum.

Table 1. Physical properties of the phenylnaphthyldiamine derivatives and biphenyldiamine derivative **HTM 1**

2.1.2 Device fabrication and characteristics

Prior to device fabrication, ITO with a resistance of 15 Ω/□ on glass was patterned as an active area of 4mm^2 (2mm x 2 mm) square. The substrates were cleaned by sonication in deionized water, boiled in isopropylalcohol for 20 min, and dried with nitrogen. Finally, the substrates were treated with UV/ozone for 20 min. Organic layers were deposited sequentially by thermal evaporation from resistively heated alumina crucibles onto the substrate at a rate of 0.5 – 1.0 Å/sec in the organic chamber. The base pressure at room temperature was 3 x 10^{-6} Torr. The deposition rate was controlled using a ULVAC crystal monitor that was located near the substrate. After organic film deposition, the substrate was transferred to another chamber maintaining the base pressure of 3 x 10^{-6} Torr. Before the deposition of metal cathode, LiF was deposited onto the organic layers with the thickness of

10 Å. A high-purity aluminum cathode was deposited at a rate of 4–8 Å/sec with the thickness of 3000 Å as the top layer. After the metal chamber was vented with N_2 gas, the device was immediately transferred to an N_2-filled glove-box upon fabrication. A thin bead of epoxy adhesive was applied from a syringe around the edge of a clean cover glass. To complete the package, a clean cover glass was placed on the top of the device. The epoxy resin was cured under intense UV radiation for 5 min. The current–voltage characteristics of the encapsulated devices were measured on a programmable electrometer having current and voltage sources, Source Measure Unit, model 238, (Keithley). The luminance and EL spectra were measured with a PR650 system (Photo Research). The current–voltage, EL spectra and luminance measurements were carried out in air at room temperature. Only light emitting from the front face of the OLED was collected and used in subsequent efficiency calculations.

To evaluate hole-transporting ability of newly synthesized phenylnaphthyldiamine derivative **HTM 2**, we fabricated the hole-dominant device using **HTM 2** as a hole-transporting material with structures as follows: ITO/2-TNATA/**HTM 2**/EML/Al (device II). On ITO substrate, 4´,4´´-tris(*N*-(naphth-2-yl)-*N*-phenyl-amino)tri- phenylamine (2-TNATA) was previously deposited as a hole-injecting material. IDE 215 doped with 3 % of IDE 118 (host and dopant materials by Idemitsu Co., LTD) was used as blue emitting layer.

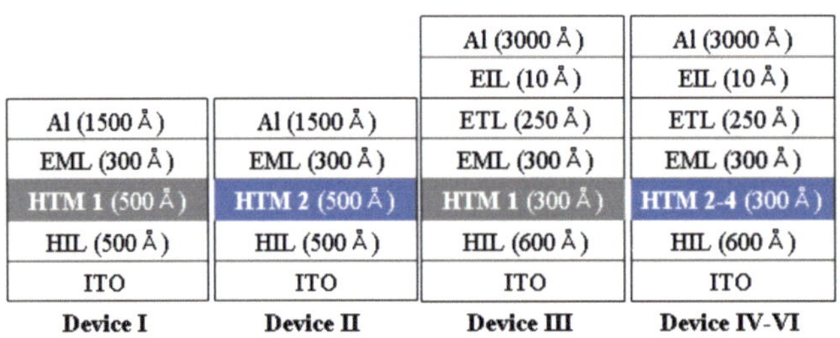

Fig. 4. Structures of EL devices used in this study

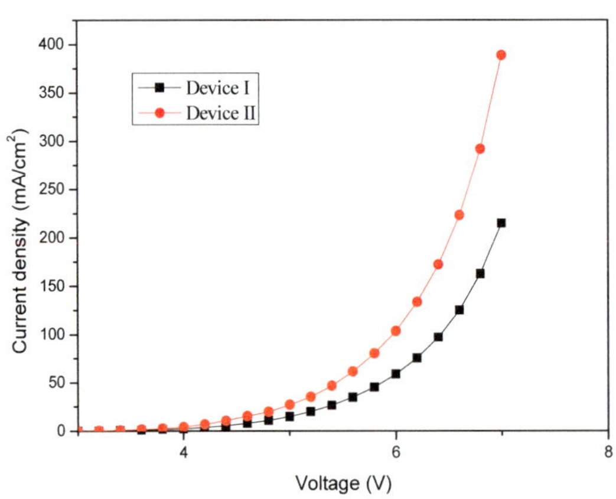

Fig. 5. Current density-applied voltage characteristics of device I and II

A reference device I composed of **HTM 1** as a hole-transporting material with the same thickness was also constructed for comparison (Fig. 4). The current-voltage (I-V) characteristics of the devices are shown in Fig. 5. The current density of the device II prepared with **HTL 2** is almost twice higher than that of the reference device I (103.4 mA/cm^2 vs 58.5 mA/cm^2 at 6 V). These outcomes showed that the hole-transporting ability of a phenylnaphthyldamine-cored **HTM 2** was highly improved than that of biphenyldamine-cored **HTL 1** due to its more resonance form in the radical cation as well as the steric effect of a naphthyl moiety resulting in efficient carrier transportation.

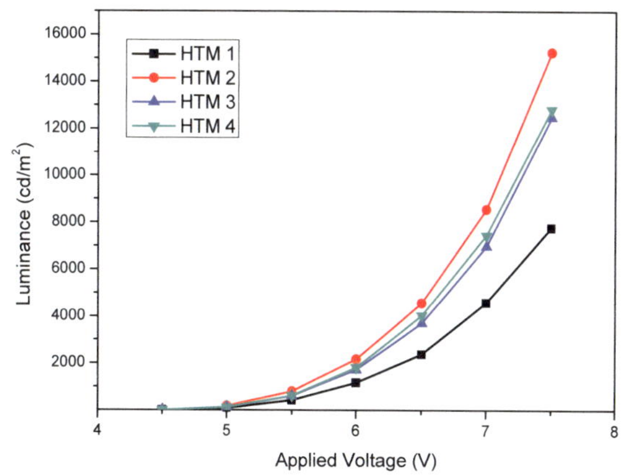

Fig. 6. Luminance-applied voltage characteristics of devices III-VI.

We also expected the stability of the phenylnaphthyl core is better than that of the biphenyl core since it has more resonance structure and higher radical stability. Three EL devices: ITO/2-TNATA/**HTM 2**/EML/Alq_3/LiF/Al (device IV), ITO/2-TNATA/**HTM 3**/EML/Alq_3/LiF/Al (device V), and ITO/2-TNATA/**HTM 4**/EML/Alq_3/LiF/Al (device VI), were fabricated in order to estimate their suitabilities as a hole transporting material in comparison with the reference device; ITO/2-TNATA/**HTM 1**/EML/Alq_3/LiF/Al (device III). The structures of EL devices are shown in Fig. 4. Fig. 6 shows the luminance and the applied voltage characteristics in the four devices. The luminance of the device IV reached 4,561 cd/m^2 at 6.5V.

Surprisingly, the devices IV-VI with **HTM 2-4** as a hole-transporting material showed a significant enhancement of the luminous efficiency compared to reference device III. The luminous efficiency of the device IV is about 40% higher than that of the device III. The luminous efficiencies of other two devices were also higher than that of the device III. The luminous efficiencies of the device IV-VI with **HTM 2-4** were 8.52, 7.98 and 7.50 cd/A, respectively. Table 2 shows the EL performances of all devices at 6.5V and Fig. 7 shows the current efficiency-applied voltage characteristics.

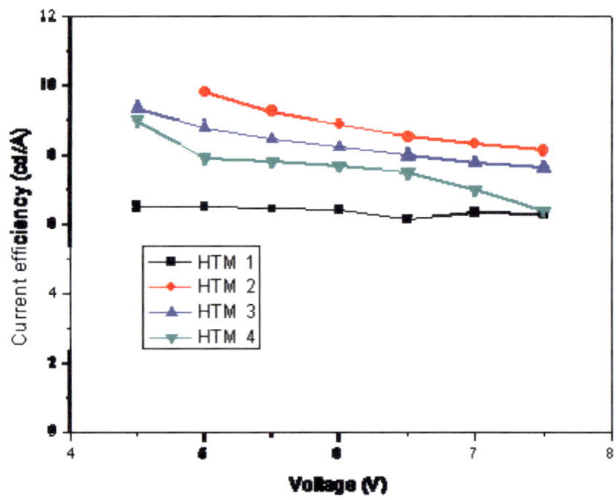

Fig. 7. Current efficiency-applied voltage characteristics of devices III-VI.

Device	Current density (mA/cm^2)	Luminance (cd/m^2)	Current efficiency (cd/A)
Device III	39.01	2383.2	6.11
Device IV	53.54	4561.6	8.52
Device V	46.48	3708.9	7.98
Device VI	53.87	4041.0	7.50

Table 2. EL performance of four devices III-VI at 6.5V

As shown in Table 2, the devices IV-VI using **HTM 2-4** as a hole-transporting material showed remarkable current density and current efficiency performance compared to the reference device III. Fig. 6 shows the devices III needs higher electric power than the devices IV-VI at the same luminance. In other words, it is thought that phenylnaphthydiamine derivatives **HTM 2-4** have superior hole-transporting abilities than biphenyldiamine derivative **HTM 1**. As we mentioned before, these high performances of the devices IV-VI, using phenylnaphthyldiamine derivatives might be attributed to the more efficient hole transportation and higher T_g of those compounds compared to that of biphenyldiamine derivative.

Fig. 8 shows the life-time characteristics of the device IV and the reference device III. The life-time of the device IV is about two times longer than that of the standard device III within the measured current density, indicating more effective recombination at the emitting layer of device IV. These results indicated that phenylnaphthyldiamine derivatives have higher hole-transporting abilities and stabilities toward electric current than that of biphenyldiamine derivative.

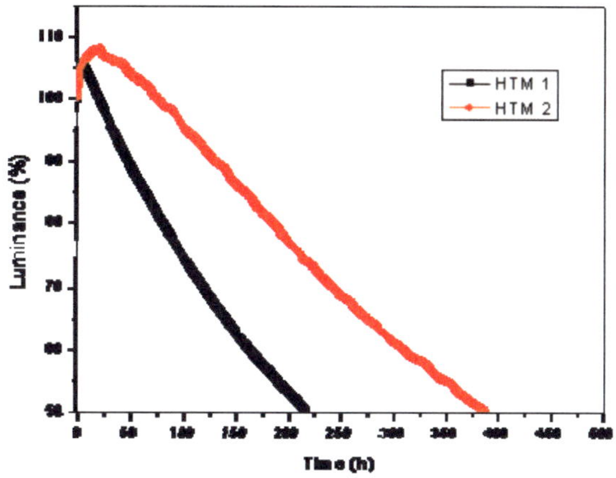

Fig. 8. The life-time characteristics of the two devices at 100mA/cm^2; device III (■), device IV (•)

Consequently, hole transporting materials having naphthalene moiety are more stable in radical cation state because of more resonance form and sterically more favored than that of phenyl moiety because of retardation of dimerization reaction. These two factors can contribute to the enhancement of the hole transporting ability resulting in better OLED performance. The device of fluorescent blue OLED using phenylnaphthyldiamines as the hole-transporting layer have much better overall EL performance and longer life-time than the reference device with biphenyldiamine layer.

2.2 Materials for electron transporting

Above mentioned, although OLEDs have shown high brightness and vivid color, long-term stability and image sticking remains a critical issue for practical applications. Device degradation is also largely attributed to the delamination between different layers and the crystallization of organic materials due to electrochemical reaction on interfaces. In general, degradation in OLEDs essentially appears in the form of a decrease in device luminance with time. The decrease in luminance can proceed through three independent and visually distinct modes. These modes are referred as (i) dark-spot degradation, (ii) catastrophic failure, and (iii) intrinsic degradation [23]. These well known degradation mechanism is related with intrinsic material properties as well as device structure.

Recently, it was elucidated that dipole moment of ETM (electron transport material) could be an important factor of initial luminance decrease in OLED. It is well known that luminance decrease at initial phase is a main reason for image sticking. In this chater, we will focus on electron transporting material-dependent life-time pattern and found the relationship between dipole moment of electron transpoting material and initial life-time tendency. it is explained how dipole moment of ETM affected initial luminance decrease in OLED [24-26].

To elucidate a cause of luminance decrease, it has been designed and synthesized a series of ETMs with different value of dipole moment and evaluated the initial life-time of the OLED device using them as ETM.

2.2.1 Dipole moment of electron transporting materials

Dipole moment values for each ETM were calculated by using GAUSSIAN 03 program package. We generated the optimized geometric structure by means of time-dependent density functional theory (TD-DFT) [27],[28] and each dipole moment value was included in the calculation summary. Dipole moments of ETMs were enlisted in Fig. 9 (ADN (Anthracene dinaphthalene) derivatives), Fig. 10 (Phenanthroline derivatives) and there were dipole moment differences among the ETMs according to the arrangement of heteroatom in the molecules. The ETMs were designed in order to minimize the effect of molecular size (Induced dipole-induced dipole interaction or London force) and dipole-dipole interaction is predominant intermolecular force among them.

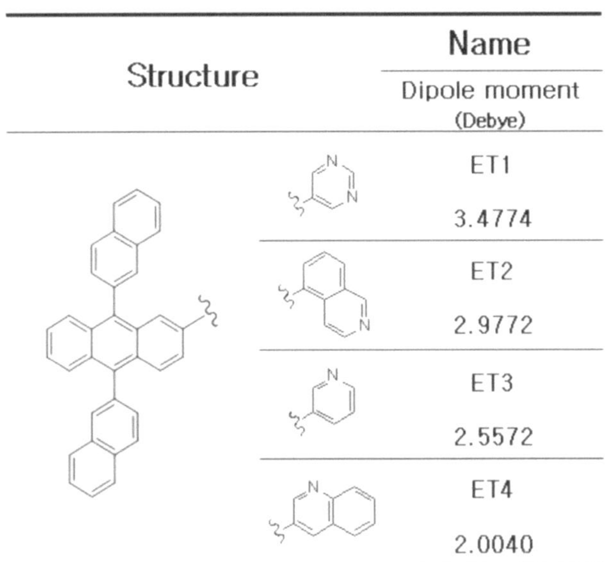

Fig. 9. Molecular structure and calculated dipole moment of ADN series ETMs

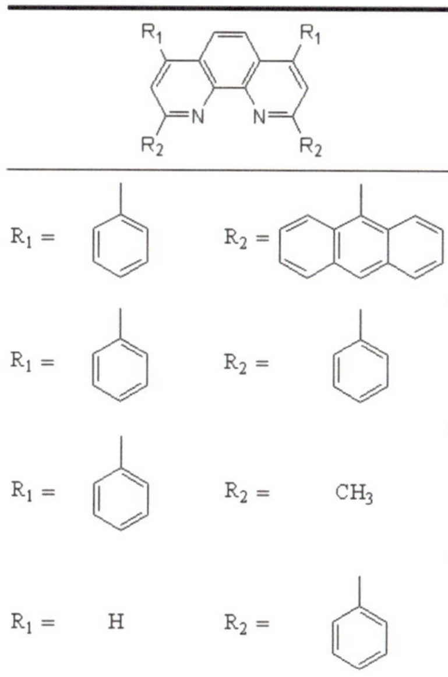

Name	Dipole moment (Debye)
ET5	3.6353
ET6	3.1120
ET7	2.9257
ET8	2.1860

Fig. 10. Molecular structure and calculated dipole moment of Phenanthroline series ETMs

2.2.2 Device fabrication and characteristics

All OLEDs were fabricated on indium tin oxide precoated onto glass substrate. Organic layers were vacuum deposited via thermal evaporation in the high-vacuum chamber. Fig. 11 shows the structure of blue OLED device and its energy diagram. The thickness and materials of each layer are same for fabricated devices except ETMs to eliminate another possible luminance attenuating factor.

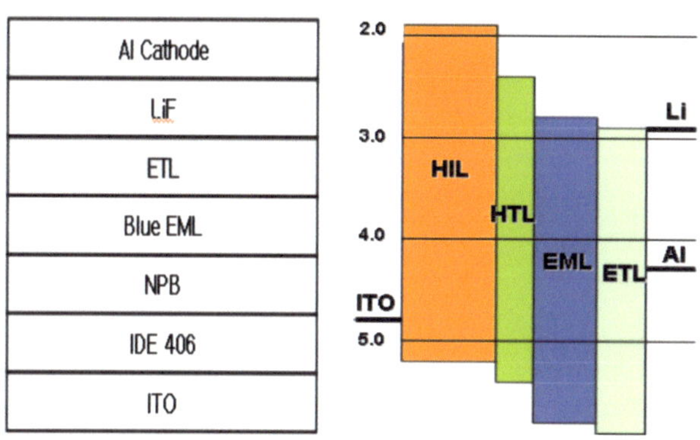

Fig. 11. Blue OLED structure and Energy diagram

The IVL (current density, applied voltage, luminance) characteristics of OLEDs were measured using a Photo Research Inc. PR-650 spectrometer. The operational life-time characteristics were determined from a series of measurements of changes in luminance and drive voltage as function of time under DC driving conditions.

Devices performance properties depended on ETMs and their ability to transport electron [29]. Fig. 12 and Fig. 13 showed I-V and L-Efficiency characteristics of ADN series ETMs, respectively. Regardless of dipole moment differences, device performances depended on the electron transporting ability of the hetero-aromatic rings attached to ADN backbone, and similar property tendency was observed when we applied those hetero-aromatic rings to another framework before. **ET4** (quinoline moiety) gave the best electron transporting performance, but **ET2** (iso-quinoline) showed poor electron transporting ability. In spite of different electronic structures of hetero-aromatic rings, Pyrimidine(**ET1**) and pyridine(**ET3**) moieties exhibited similar properties.

Similar to the results of the ADN series ETMs, it was difficult to explain the result in dipole moment aspect, and device performance characteristics mainly due to the property of phenanthroline stacking. **ET7** and **ET8** are simple structure with minimum appendages and almost flat 3-D structure, so they can be effectively stacked through π-π interaction in deposited layer. On the other hand, relatively bulky side aromatic rings obstruct stacking of phenanthroline moiety. It is well known that π-π stacking among ETMs in deposited layer can enhance electron transporting ability, and that is corresponded with the results of the phenanthroline ETMs [30].

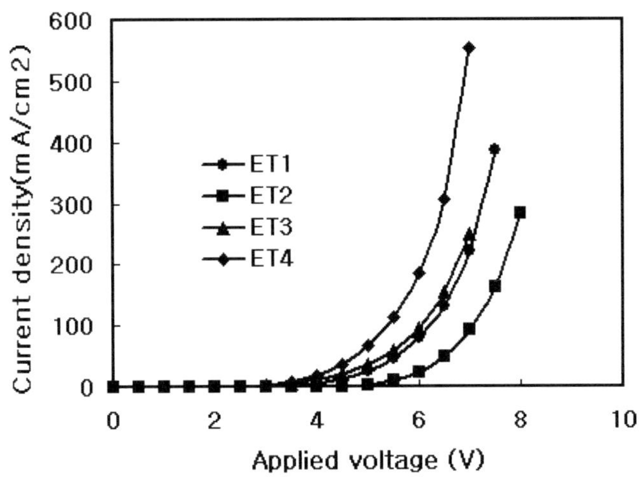

Fig. 12. Voltage-Current curve of ADN series ETMs

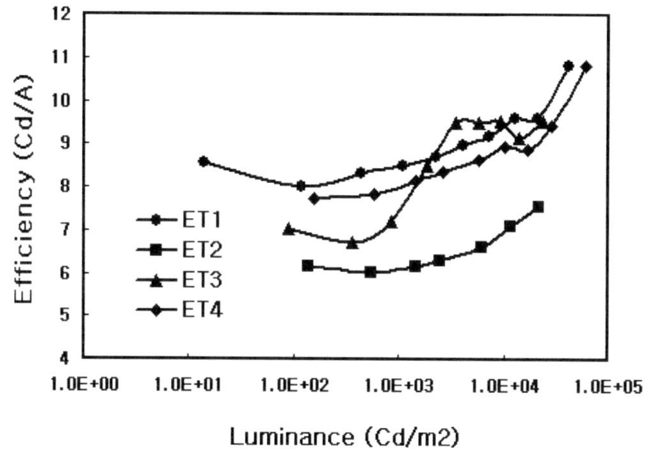

Fig. 13. Luminance-Current efficiency curve of ADN series ETMs

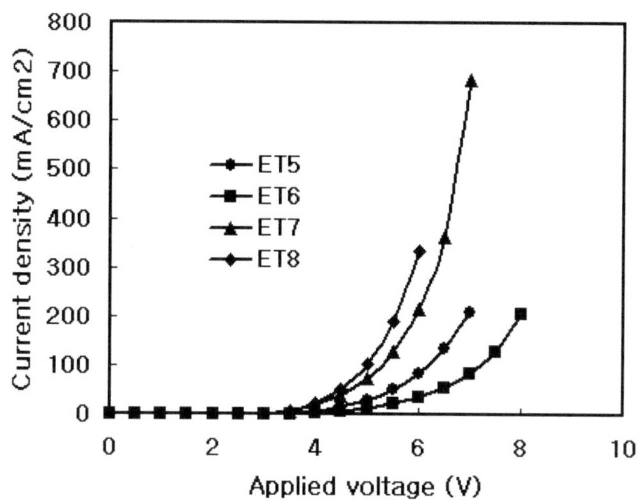

Fig. 14. Voltage-Current curve of Phenanthroline series ETMs

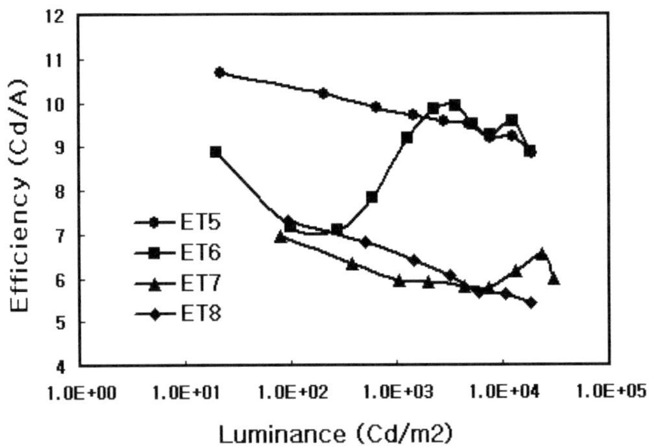

Fig. 15. Luminance-Current efficiency curve of Phenanthroline series ETMs

2.2.3 Analysis of Initial Luminance Decrease

Contrary to the results of performance property, initial phase tendencies of life-time were well corresponded with the dipole moment values. As showed in Fig. 16, the rate of an initial luminance was decreased in the order of **ET1**, **ET2**, **ET3**, and **ET4**, which is contrary to the direction of the dipole moment increase in Fig. 9. **ET1** that can form a robust deposited layer through strong dipole-dipole interactions showed moderate luminance decrease tendency. It can stack regularly in order to form intermolecular network through localized charge distribution in the molecules. It is supposed that an electrically polarized material located under electric field is torqued by an applied electric force and tends to rotate (Fig 17).

When a high electric field is applied, if a material has great dipole moment, phase transformation is difficult to occur in the layer owing to strong intermolecular interactions among deposited molecules (Fig 18a). On the other side, because a low dipole moment material (**ET4**) could not have strong intermolecular force, it cannot stack in compact manner. So it forms less tight layer than materials with strong dipole moment. As depicted in Fig. 18b, deposited molecules rearrange along the electric field or bond strain could be generated in the molecule, and if there is a weak chemical bond, that could be broken when high electric field is applied. In the event, a device composed of small dipole moment ETM showed a steep slope in time-luminance graph at initial phase.

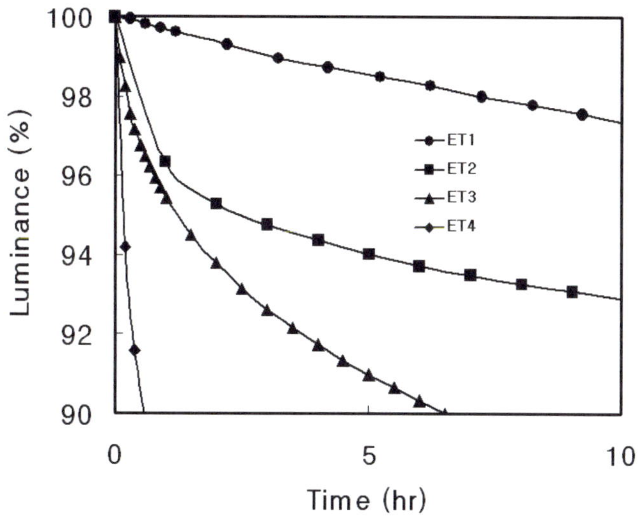

Fig. 16. Initial luminance decrease tendency of devices with ADN series ETMs at 100mA/cm2.

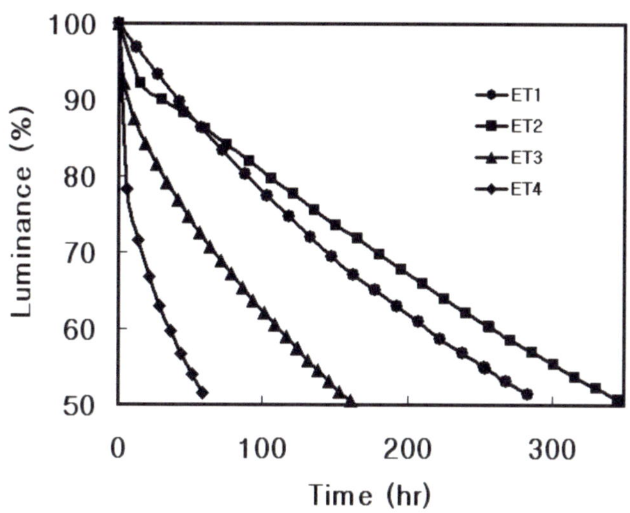

Fig. 16. Life time (half-life) tendency of devices with ADN series ETMs at 100mA/cm2

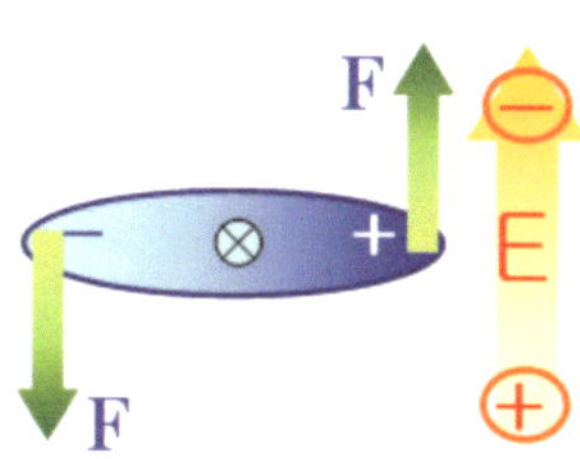

Fig. 17. Behavior of a polarized molecule when electric field is applied

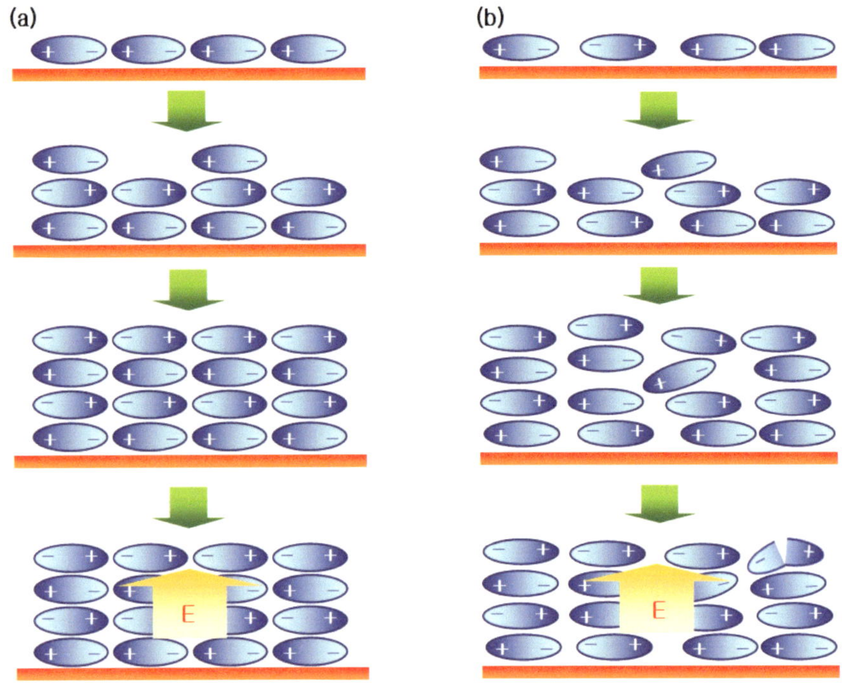

Fig. 18. Schematic descriptions of layer stacking and behavior pattern of polarized molecules under electric field. (a) Material with great dipole moment (b) Material with small dipole moment

Pattern of half-life was similar to that of initial phase, but there must be another factor that could affect life-time. There are some assumed mechanisms for luminance decrease or device degradation [31],[32]. For example, degra- dation of the interface between deposited layers, shift of exciton recombination zone from emitting layer, intrinsically weak chemical bond of used materials and there must be a lot of possible degradation mechanism we could not conceive.

For the test result (Fig. 16), **ET2** showed dramatic luminance decrease in the initial stage, but after initial rearrangement of molecules in the layer, luminance decreased in a moderate manner. On the other hand, luminance decrease rate of **ET1** device was almost constant over all in operation.

Same luminance decrease tendency was observed in phenanthroline series ETMs. Fig. 19 showed the initial luminance of the devices decreased according to the dipole moment order of ETMs, but half-life in Fig. 20 showed a little different degradation order like the case of the ADN series ETMs. **ET6** and **ET8** gave moderate overall life-time graph, and, **ET5** and **ET7** showed Different stacking pattern caused by dipole moment differences could affect density of deposited layer. If the deposited molecules can interact more tightly each other, density of the deposited layer is greater than less tightly interacting one. Actually, density of deposited layer could be influenced by molecular shape and dipole moment. And to avoid ambiguity from structural differences, we measured densities of **ET1** and **ET3** using XRD

[33] because these two materials are spatially same structure except C-H and N. Densities of **ET1** and **ET3** are 1.25 g/cm^3 and 1.21 g/cm^3 respectively, and this result reflects dipole moment differences. Density differences of the deposited layers also provided a clue for elucidating effect of dipole moment on the pattern of luminance decrease at initial phase.

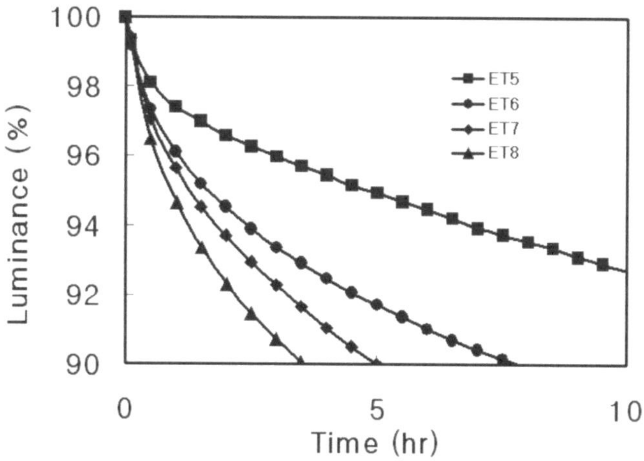

Fig. 19. Initial luminance decrease tendency of devices with Phenanthroline series ETMs at 100mA/cm^2 relatively rapid degradation appearance.

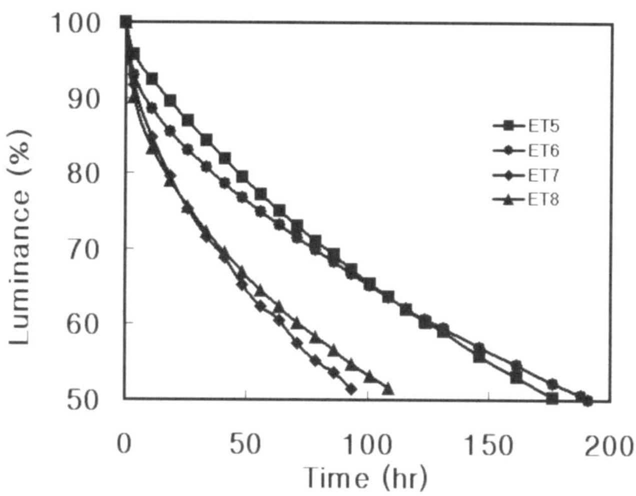

Fig. 20. Life time (half-life) tendency of devices with Phenanthroline series ETMs at 100mA/cm^2

It is concluded that the life-time properties at initial phase were controlled by dipole moment differences of ETMs and great dipole moment materials can enhance initial luminance properties by means of strong dipole-dipole interactions among the molecules. In near future, it is expected to find out a general mechanism of dipole moment effect on life-time.

3. Material issues for new applications of amoled

For future display technologies, there is significant interest in providing display devices with mechanical flexibility and transparency. And also, there are continuous requirements for ultra-thin display and extendable or foldable display without seam line between the two panels. Definitely, it is hardly allowed to manufacture transparent, paper-thin, foldable and flexible displays with the other display technology except for AMOLED.

3.1 Electrode materials for transparent OLED

Since Tang's report on the two-layered organic light emitting diode (OLED), top-emitting organic light emitting diodes (TEOLEDs) with their major advantages of the self-emission of light, wide viewing angle and low power consumption have acquired considerable interest [34-38]. Recently, a number of full-color TEOLEDs have already been introduced into the display market. Especially, OLEDs with transparent cathode are expected to be the next generation displays because of their possible applications for transparent organic light emitting diodes (TOLEDs) which can emit light through both anode and cathode.

Transparent display can be defined as the display across which objects can be seen. Transparency is very attractive feature in the area of display device because it provides both new functional characteristic and plentiful imagination in design. PNS (Private Navigation System) with Augmented Reality technology can come true with the function of transparency in display. Transparency can enlarge its degree of freedom in the conceptual design of PID (Public Information Display) system. It is able to be an important factor which opens up new area of display applications. In addition, Transparency can be secured only with OLED until now on account of its structural simplicity. Because polarizer is not obligatory element in the OLED devices but in the LCD devices, OLED can be a unique display device with its transparency feature.

In most TEOLED, semi-transparent metal such as Mg:Ag is used as a cathode with good electron injection properties. However, resistance and transparence of Mg:Ag cathode is not enough to be applied to the TOLEDs. Recently, many efforts have been made to use indium tin oxide (ITO) and zinc oxide (ZnO) doped with impurities as transparent cathode by sputter deposition method [39-42]. However, the sputtered metal oxide films have some drawbacks on the device stability. It has already reported that the high sputtering power and their high work function led to failures of TOLEDs [43]. While double layer cathode structures of metal/metal have also been investigated, the transmittance of the layer reached at most around 70% which could not be used as the transparent cathode [44,45]. Another researchers tried to developed the ITO films with low work function by introducing a monolayer of strong electron-donor molecules such as tetrakis-(dimethylamino)ethylene, TDAE [46]. However, it is turned out for the efficiency of electron injection to be low.

Recently, Ryu and Baik have already created a TOLED using an indium tin oxide (ITO)/Ag(metal)/ITO (IMI) cathode with low resistance. However, an IMI cathode showed low transmittance due to the low substrateheating through e-beam deposition,2,6,7 and an ITO/Ag/tungsten oxide (WO3) (IAW) for higher transmittance was used in this instance. The Al 20 nm showed a transmittance of 43% and a sheet resistance of 13 ohm. The IMI, WO3 /Ag/WO3, SiO2/Ag/SiO2, and IAW exhibited a transmittance of 27%, 90%, 68%, and 40% at 550 nm respectively. Although the multilayer consisted of a single layer with poor transmita tance, they showed little decrease in transmittance because there is a light pathway due to multiple reflections. However, it could not be used as a cathode for TOLED due to poor electron injection properties. Therefore, in order to gain the good device characteristics and stability, it is necessary that damage-free cathode material with low work function, low resistance and high transparency is developed.

In spite of many problems as menstioned above, it could be overcomed obstacles by developing highly conducting oxide material without damage to organic layers and by newly designed thermal deposition method. Opimization of device structure can be also made a progress for the transparent cathode structure having electron injection property that were superior to a conventional OLED.

Unlike heat-deposited atoms, sputtered atoms with a high energy could damage the underlying organic layers during the deposition process. To avoid any damage, our cathode films were deposited by the ion beam assisted thermal deposition (IBAD) process using reactive technique. First, we could obtain single-component oxide cathode, indium oxide (InOx) with high work-function by using the thermal process. InOx based films are widely used as transparent cathodes for OLEDs because of their excellent properties of both a high transmittance in the visible region and a low electrical resistivity. In the case of single-component oxide cathode, InOx, although we obtained enough transmittance (Fig. 21(a)) and low electrical resistivity of 20 Ω/, there remains difficulty in using the thin film as a cathode electrode. The difference between LUMO of typical electron transporting layer (ETL) materials, 2.8~3.2eV, and work function of the InOx film, 4.8~5.0eV, is too big to overcome the energy barrier, even if electron injection layer (EIL) such as LiF is used to assist electron injection from the cathode. Therefore, in order to improve electron injection efficiency, new cathode structure with excellent charge injection is required. The electron injection could be improved dramatically without decreasing the transmittance of the InOx cathode structure by using a thin Mg and C60 layers. However, even if our cathode structure enhanced electron injection, it was still important to improve the work function of transparent cathode itself because our device performances were not enough to reach the performance of conventional OLED using cathode of Al/LiF.

3.1.1 Physical properties of new transparent electrode materials

The most suitable cathodes are metals having low work functions ranging from 2.63 to 4.30 eV [47]. In most case, these include Al, Mg, Ca, Ag, Mg:Ag. Especially, Ca has the lowest work function value of 2.63 eV. For improving the work function of transparent cathode, therefore, an InOx electrode doped with Ca was fabricated by the IBAD technique at different deposition rates. The process employed reactive evaporation to grow the InCaOx (ICO) films. The ICO films were deposited by the ion beam assisted thermal evaporation technique using In (wire, 99.99 %) and Ca (3 mm random piece, 99.9%) reactive evaporation at the substrate temperature of 85 °C. The evaporation of In and Ca is performed steadily at the Ar and O2 gas flow rate of 10 sccm. The time of evaporation recorded, in order to calculate the evaporation rate. The deposition pressure is below 5×10^{-7} torr and typical deposition rates are 0.3 ~ 1.0 □ /sec. The film thickness was measured by an alpha-step system. The optical transmission spectra of the films and the sheet resistances were measured using a UV-VIS spectrophotometer (Perkin Elmer, Lambda 950) and a four-point probe, respectively. The spectral region used in this work was 400-700 nm. The work function was measured using a UV spectrometer (RIKEN KEIKI, AC-2) in the atmospheric environment. The structure and atomic composition of the thin films were studied by SEM, XRD and X-ray photoelectron spectroscopy (XPS).

The typical transmission spectra as a function of the Ca deposition rate are shown in Fig. 21(a). A high transmittance above 92% in the visible range was obtained in the ICO films deposited on the substrates above approximately 85□. Significantly, it was found that the transmittance in the visible range could be improved in the ICO films deposited on the low-temperature substrates by introducing O_2 gas. The enhancement of the transparency of the ICO film is attributed to the partial oxidation of Ca during the deposition. Furthermore, it is possible that the high energy level of CaO (1.7 eV) which is built up through the reactive IBAD process enhance the injection efficiency of electron.

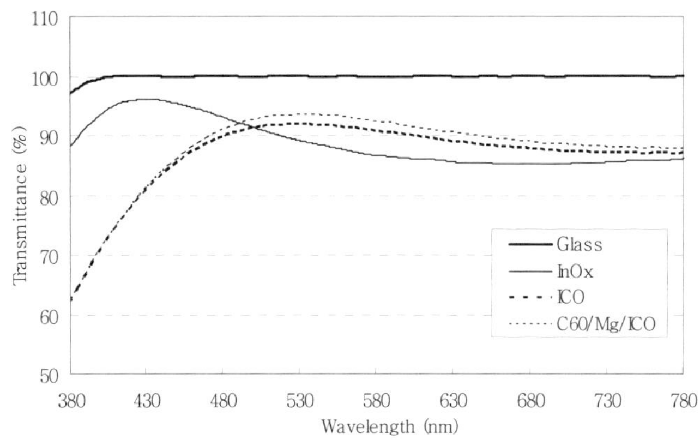

(a)

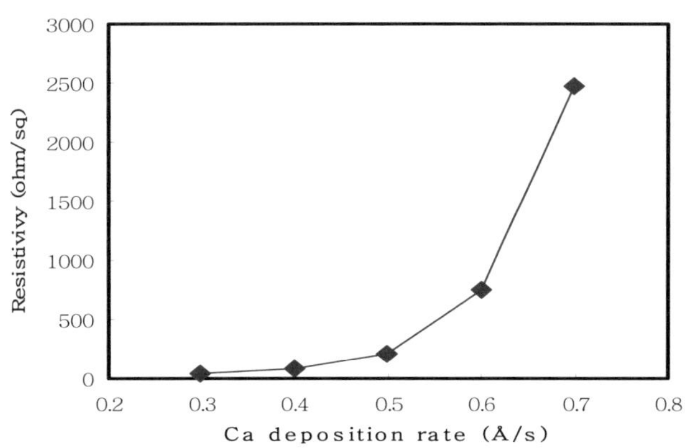

(b)

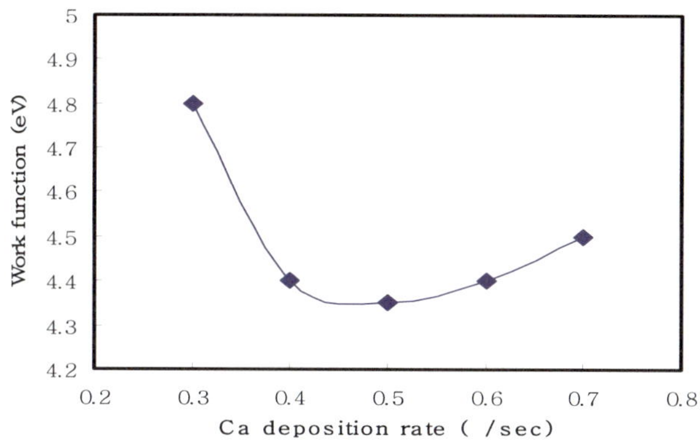

(c)

Fig. 21. (a) Transmittance spectra of ITO and ICO films. (b) work function and (c) resistivity of the ICO film as a function of Ca deposition rate.

The work function and resistivity of the ICO films are shown in Fig. 21(b). In this figure, the measured work function and resistivity of the transparent conducting ITO films prepared by r.f. magnetron sputtering were plotted. The work function of obtained ITO was about 4.9 eV. This value was approximately equal to that measuring other ITO. ICO film had the lowest work function and resistivity between the rate of 0.4 and 0.5 □/sec. The work function of the ICO film was very stable in the ambient atmosphere environment over an extended period of exposure. It can be concluded that the work function of the ICO films, which are possible to use, is in the range from 4.35 to 4.80 eV. Therefore, these films could be selected for use as the cathode. The lowest electrical resistivity value of the film obtained at a Ca deposition rate of 0.1 □/sec while keeping the other deposition parameters. It can be seen that the Ca deposition rate has a strong influence on the conductivity of the films, with the conductivity ranging between 45 and 2469 Ω/ for the films deposited at 0.3 and 0.7 □/sec, respectively. Therefore, the electron injection efficiency is primarily due to Ca metal having a low work function (2.9 to 3 eV), while the Ca metal increases the lateral electrical conductance of the ICO film. Consequently, ICO film which was deposited at 0.4 □/sec was choose owing to the appropriate value of work function and resistivity as a cathode for TOLEDs.

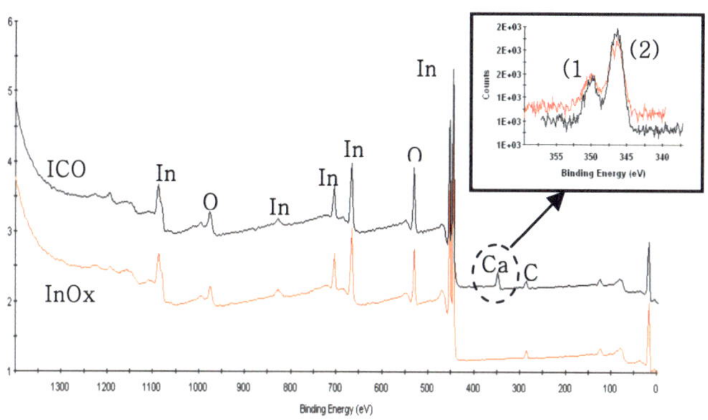

(a)

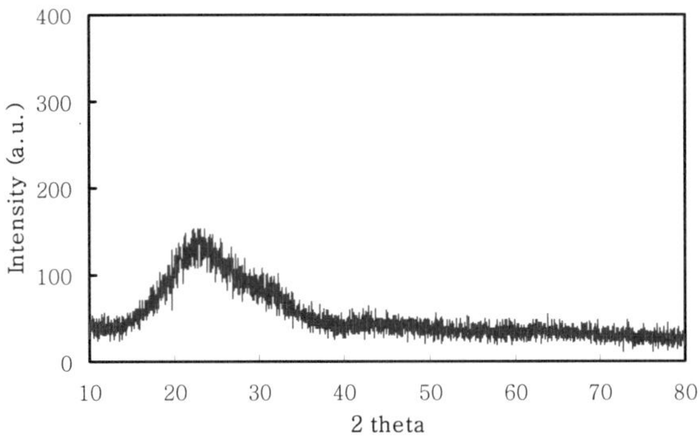

(b)

Fig. 22. (a) XPS wide scan and (b) XRD diffraction spectra of the ICO film.

The XPS wide scan spectra in Fig. 22(a) is from a typical InOx and ICO film on a Si wafer. In addition to the In and O features, a Ca $2_{P3/2}$ peak at 346 eV is clearly observed. The inset figure shows the narrow scans of the spectra around (1) Ca $2_{P1/2}$ and (2) Ca $2_{P3/2}$. The XPS data showed that only the ICO films have definitive Ca peaks for $2_{P3/2}$ whose estimated concentration was approximately 3 %. The surface chemistry and morphology are known to have dramatic impact on charge transport across the organic-TCO interface. The XRD pattern also showed no diffraction peaks, so that the ICO film should be amorphous with no visible pores or larger interface defects, as well as having a smoother surface (Fig. 22(b)). The surface morphology also has a strong influence on the optical and electrical properties of the films, since the textured surfaces lead to the enhancement of the light collected. Fig. 23 shows the SEM images of (a) the commercial ITO and (b) ICO film, in which the ICO film is observed to exhibit a more smooth and homogeneous surface, compared with the ITO film. In the case of commercial ITO, crystallization leads to the formation of a polycrystalline structure. The ICO films are amorphous with low surface roughness (RMS, Ra=1.1 nm), since they are fabricated using a slower deposition rate or lower temperature.

We investigated for IBAD process to induce damage to organic layers. In general, leakage current is observed during reverse sweep, if the organic bond in ETL is broken or damaged by energetic particles during the deposition. While typical sputtered ITO film gave a serious damage to organic layer, as is shown in Fig. 24, there is no leakage current in the transparent conducting film, keeping the current below 10^{-5} mA in the reverse bias region in J-V curve.

(a) (b)

Fig. 23. SEM (Scanning Electron Microscope) images of (a) Commercial ITO and (b) ICO film fabricated by the ion beam assisted thermal evaporation.

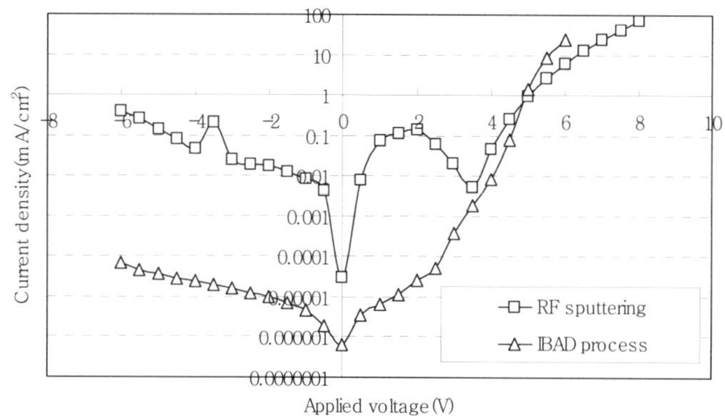

Fig. 24. Current density-voltage curves of ITO-deposited OLED with RF sputtering and ICO-deposited with IBAD thermal process at 85°C.

3.1.2 Device fabrication and characteristics

To study the cathode dependence to the device performance, various device structures were manufactured. The device configuration of the TEOLEDs was Ag (100 nm)/ITO (10 nm)/*N,N'*-diphenyl-*N,N'*-bis-[4-(phenyl-m-tolyl-amino)-phenyl]-biphenyl-4,4'-diamine (DNTPD, 60 nm)/*N, N'*-di(1-naphthyl)-*N,N'*-diphenylbenzidine (NPB, 30 nm)/distyryl anthracene(DSA):tetra(*t*-butyl)perylene (30 nm, 100:5)/bis(10-hydroxybenzo[h]quinolinato)beryllium (Bebq2, 20 nm)/cathode. A basic electrode structure of transparent cathode (100 nm)/Mg (7 nm)/LiF (1 nm)/C60 (2 nm) was used in the TEOLEDs after optimizing the thickness of Mg and C60 layers (Fig. 25).

The devices were fabricated on glass substrates precoated with a high-work-function anode such as indium–tin-oxide. The ITO substrates was cleaned in deionized water, and given an UV-ozone treatment prior to use. In the case of an anode with Ag reflector, EL was observed through a transparent ICO cathode. The active area of our devices was 4 mm^2. Prior to use, all organic materials were purified by vacuum train sublimation. Deposition of the organic materials was carried out in a high vacuum system (Sunic) by thermal evaporation from resistively heated alumina crucible. The base pressure in the chamber ranged between $4x10^{-7}$ and $1x10^{-6}$ mbar. Typical deposition rates were 1 Å/s. The evaporation chamber was attached directly to a nitrogen glove-box system, which allowed devices to be fabricated, characterized, and encapsulated under inert conditions. *I–V* and luminance–voltage characteristics were measured with a Kiethley 237 programmable electrometer and Photo research PR650 spectroradiometer.

As shown in Fig. 26(a), the current density-voltage properties of the devices having the IBAD-processed InOx cathodes were correlated with the electrode structure. The Mg and C60 thin layer was incorporated to ensure efficient electron injection. It is believed that the work function of Mg, 3.6 eV, can decrease the energy barrier between indium oxide and ETL material. Feng et al. have already found that metal/LiF /C60 interface exhibits ohmic type junction characteristics whereas metal/C60 interface exhibits rectifying Schottky-type junction characteristics [48]. They described the role of C60 in the cathode structure as like the followings. First, $F^-(-Li^+)$ anions introduce an *n*-type doping zone near the interface. The *n*-type doping was found by x-ray photoelectron spectroscopy analysis of LiF-doped C60 films. Second, a recent calculation suggests that LiF-C60 interaction leads a reduced energy gap and thus lead to a metallic interface.

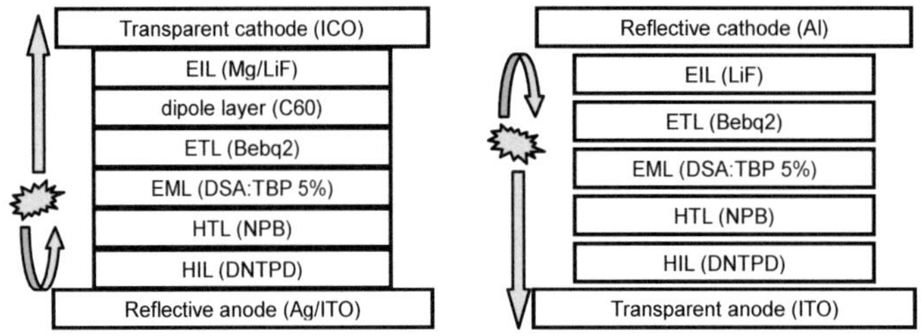

(a) (b)

Fig. 25. Schematic device structures of (a) TEOLED with transparent cathode and (b) conventional OLED.

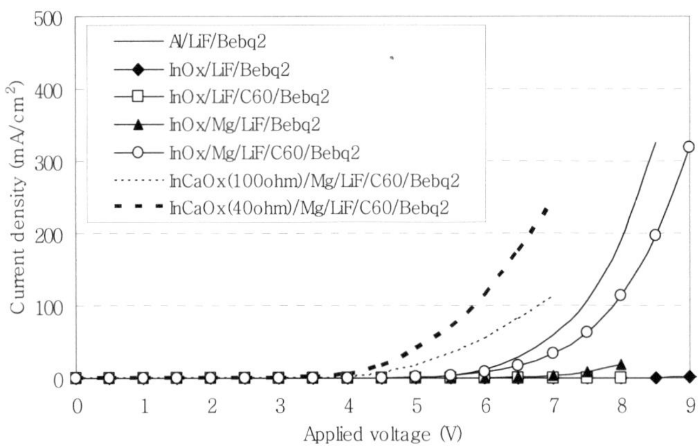

(a)

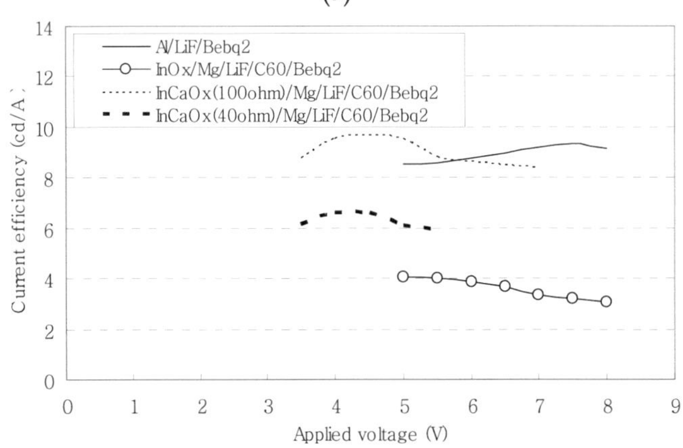

(b)

Fig. 26. The performances of OLEDs as the various cathode structures; (a) current density-voltage and (b) current efficiency-voltage.

In order to compare the characteristics of ICO with those of InOx, we also investigated the performance of devices having ICO. The resistance of ICO, 30 Ω/□, is slightly higher than that of InOx, 20 Ω/□. However, operational voltage of device having ICO is lower than that of InOx at the same current density. This indicated that the insertion of Ca metal between ICO and Mg is effective reducing the drive voltages and improving device efficiency. Moreover, as shown in the Fig. 26(a), ICO which is reduced the resistance of 30 Ω/□ presented lower operational voltage than ICO with the resistance of 100 Ω/□. These results clearly demonstrate that J-V performance is better as the resistance is lower in the case of same electron injection efficiency. Consequently, transparent cathode should have low resistance and low work function for good J-V performance of OLED device.

In spite of good J-V performance, however, we can not always expect for the OLED device with high current density to present high current efficiency. Even though a device which have so much electron carrier or hole carrier shows high current density, the recombination zone of the device could not be formed in emitting layer. Fig 26(b) shows the current efficiency of the devices having the different transparent cathode. The highest current efficiency value of 10 cd/A was obtained from the device having the 100 ohm ICO. The hole injection efficiency is very similar to the every devices because of the same structure of ITO/HIL/HTL/EML. From this assumption, it can be thought that the current efficiency of OLED having ICO is greatly affected by electron injection efficiency between ETL and cathode structure. Therefore, it can be also supposed that the charge balance of both device having Al/LiF and 100 ohm ICO is superior to other cathode structure.

For the better J-V characteristics and the decrease of IR drop, it is needed to acquire lower resistance of cathode. So we fabricated the gradient Ca-doped ICO which was not doped with Ca to the whole range of InOx, but doped with Ca to the only shallow interface (about 50 Å) between InOx and Mg layer. We could improve the resistance of the newly designed cathode structure with this gradient Ca-doped ICO (gradient type ICO).

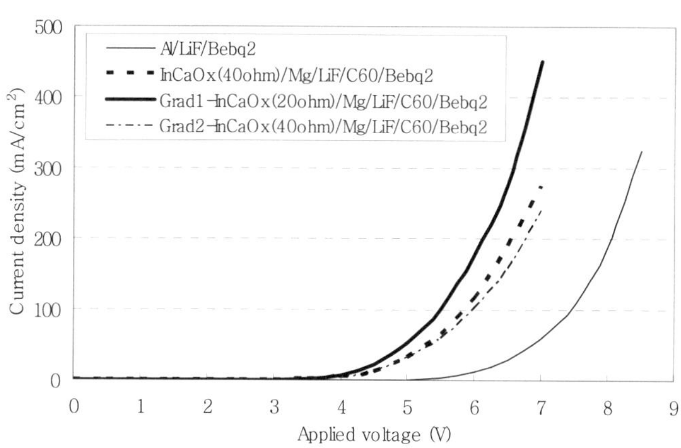

(a)

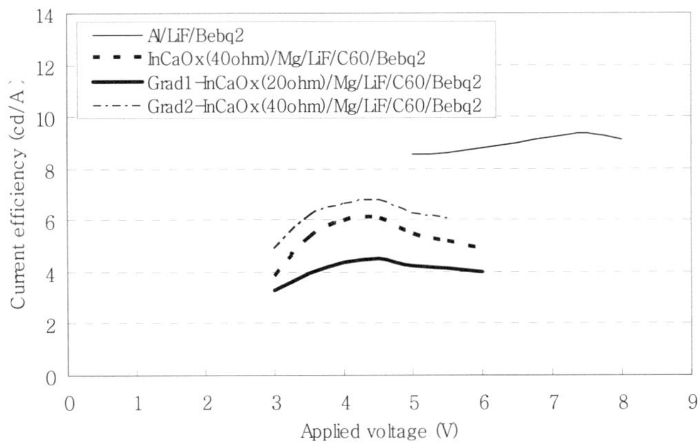

(b)

Fig. 27. The performances of OLEDs with ICO and gradient doped ICO; (a) current density-voltage and (b) current efficiency-voltage.

Fig 27 shows the J-V performance of the device having gradient type ICO. As like our initial intention to the correlation between the resistance of ICO and J-V characteristics of device, the lower resistance of ICO lead to the lower applied voltage at the same current density. Simultaneously, in the case of current efficiency, the similar J-V characteristics lead to the similar current efficiency. This result is very concurrent with the former result of Fig 26. From the same value of resistance and J-V curve, it can be thought that the electron injection efficiency is very similar for the both cathode of ICO and gradient type ICO with similar resistance. Furthermore, the similar J-V current and resistance derived the similar current efficiency from each devices. Then it is also supposed that J-V and current efficiency is determined by electron injection in the case of the similar hole injection properties and electron injection is defined by the energy barrier of ICO-Mg interface. Hereafter, in order to acquire the better properties of OLED device with ICO, it is needed to improve the mobility and injection property of HIL and HTL for controlling the recombination zone and enhancing the J-V characteristics and current efficiency.

In conclusion, new transparent and conductive ICO films were fabricated by ion beam assisted thermal evaporation. From the results of this study, it is possible to conclude that the use of a dopant leads to significant changes in the optical and electrical properties of the ICO films. In fact, it was found that the electrical conductivity (30 Ω/□) and work function (4.35 eV) of the films thickness of 1000 Å can be controlled by adjusting the Ca deposition rate, thereby allowing us to attain films with desirable properties. Optical transmission higher than 92% was measured in the visible region. In addition, new top-emitting device structure with good device efficiency was proposed by using new cathode structure and transparent cathode materials. These films will be good candidates for transparent cathodes in TOLEDs.

3.2 Encapsulation materials for foldable and flexible OLED

Another unique displays using AMOLED are paper-thin, foldable and flexible displays. A number of technologies are developing towards flexible and thin displays that can be flapping, like paper (Fig. 28). In these cases, encapsulation technology is most important among various AMOLED technologies. For general AMOLED, edge sealing with the frit which is used for small-sized AMOLED has serious problems such as mechanical strength under external stress. In order to surmount those fatal flaws, new techniques such as filling the gap between two glasses are currently under development. Organic materials are susceptible to water, oxygen and other environmental elements present in ambient conditions. Furthermore, electrode metals deposited on the light-emitting layer are prone to oxidation resulting from exposure to water and oxygen, etc. While commercialized powder-type desiccants have been used for the encapsulation of bottom-emitting OLEDs, these desiccants could not be applied to top-emitting AMOLEDs because they blocked the emitted light due to their opaqueness. So, the transparent film desiccants were used by mixing solutions dispersed with calcium oxide powders and ultraviolet-curable resins. As the solid content in the solutions increased from 15 wt.% to 30 wt.%, the average particle size increased from 107 nm to 240 nm, whereas the transmittance of the films decreased from 98 % to 80 % in the visible range. The devices encapsulated with the transparent film desiccants which contained 20 wt.% CaO exhibited no dark spots and 97 % of the initial luminance, even after being stored for over 500 hrs at 70 ^{o}C and 90 % R.H. Also, the operational lifetime of these devices was 1850 hr, 10 times longer than that of device without desiccant. These results confirmed that the transparent film desiccants which absorbed the moisture that penetrated into the devices could be applied to the encapsulation of top-emitting AMOLEDs. However, Paper-thin, foldable, flexible even flapping AMOLEDs can't be easily made by normal encapsulation method which seals with hard encapsulation glass and desiccants. Instead of that, thin film encapsulation technology should be employed that protects the organic device but leaves it thin and flexible. Thin film encapsulation is very powerful solution for obtaining unique characteristics of AMOLED [49]. Instead of using upper encapsulation glass, TFE employs layer-by-layer deposition of thick films with compensating diffusion barrier properties. The biggest merit of TFE is that it enables single glass display, which makes extremely slim and flexible panels possible. The challenges for TFE include material optimization, minimization of stacking layers, and applicability for large size mother glasses. Figure 29 shows SEM image of thin film encapsulated OLED which consist of inorganic and organic alternative layers.

Fig. 28. 6.5" flexible AMOLED display by Samsung.ufigureLOGICAL ISSUES FOR SMALL-SIZED MOBILE OLEDD

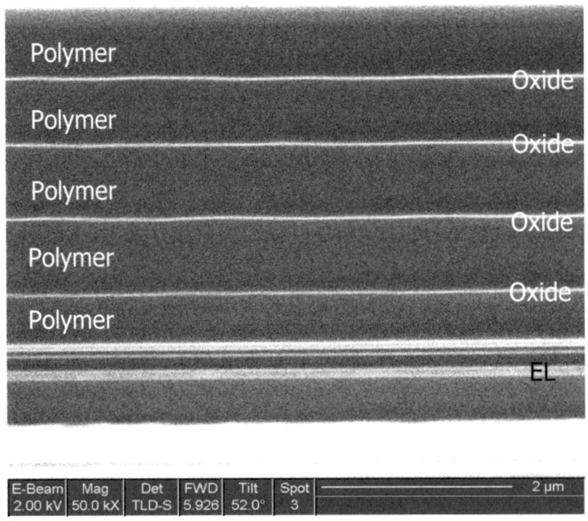

Fig. 29. SEM image of thin film encapsulated AMOLED.

4. Summary

In this article, we presented the organic material issues for common AMOLED. It was also reviewed that unique materials of AMOLED could create the new applications such as paper-thin, foldable, bendable and transparent displays. Although there remain some problems unsolved, we are convinced that AMOLED should be the leader of information display in the near future.

5. References

Y. Shirota, J. Mater. Chem. 10, 1 (2000).

T. Fuhrmann, J. Salbeck, Advances in Photochemistry (eds., D. C. Neckers, D. H. Volman, G. von Bünau), vol. 27, Wiley, New York (2002).

U. Mitschke, P. Bauerle, J. Mater. Chem. 10, 1471 (2000).

J. Kido, Bull. Electrochem. 10, 1 (1994).

S. Tokito, Y. Taga, Appl. Phys. Lett. 66, 673 (1995)

P. F. Smith, P. Gerroir, S. Xie, A. AM. Hor, Z. Popovic, M. L. Hair, Langmuir 14, 5946 (1998)

K. Naito, A. Miura, J. Phys. Chem. 97 (1993) 6240.

S. Tokito, H. Tanaka, K. Noda, A. Okada, Y. Taga, Appl. Phys. Lett. 70, 1929 (1997).

B.E. Konne, D.E. Loy, M.E. Thompson, Chem. Mater. 10, 2235 (1998).

F. Steuber, J. Staudigel, M. St¨ossel, J. Simmerer, A. Winnacker, H. Spreitzer, F. Weiss¨ortel, J. Salbe´ck, Adv. Mater. 12, 130 (2000).

P. Strohriegl, J.V. Grazulevicius, Adv. Mater. 14 (2002) 1439.

L. S. Hung and C. H. Chen, Mater. Sci. Eng. R39, 143 (2002).

P. Strohriegl and J. V. Grazulevicius, Adv. Mater. 14, 1439 (2002).

B. E. Konne, D. E. Loy and M. E. Thompson, Chem. Mater. 10, 2235 (1998).

M. Thelakkat and H. Schmidt, Adv. Mater. 10, 219 (1998).

K. Katsuma and Y. Shirota, Adv. Mater. 10, 223 (1998).

I. Y. Wu, J. T. Lin, Y. T. Tao, E. Balasubramaniam, Y. Z. Su and C. W. Ko, Chem. Mater. 13, 2626 (2001).
C. W. Ko and T. Tao, Synth. Met. 126, 37 (2002).
K. T. Wong, Z. J. Wang, Y. Y. Chien and C. L. Wang, Org. Lett. 3, 2285 (2001).
D. E. Loy, B. E. Konne and M. E. Thompson, Adv. Funct. Mater. 12, 245 (2002).
Y. K. Kim and S. H. Hwang, Chem. Lett. 35, 120 (2006).
Y. K. Kim and S. H. Hwang, Synth. Met. 156, 1028 (2006).
H.Aziz, Z.D. Popovic, Chem. Mater., 16, 4522 (2004).
P. M. Borsenberger, J. J. Fitzgerald J. Phys. Chem., 97(18), 4815 (1993).
R. H. Young, T. M. Kung, J. A. Sinicropi, N. G. Rule, J. J. Fitzgerald, J. E. Eilers, C. H. Chen, N. W. Boaz J. Phys. Chem., 100, 17923 (1996).
(a) K. L. Tong, S. K. So, H. F. Ng, L. M. Leung, M. Y. Yeung, C. F. Lo Syn. Met., 147, 199 (2004), (b) H. Bässler Phys. Status Solidi B, 175, 15 (1993) (c) A. Dieckmann, H. Bässler, P. M. Borsenberger J. Chem. Phys., 99, 8136 (1993) (d) R. H. Young Philos. Mag. B, 72, 435 (1995).
DFT calculation method was B3LYP, basis set was 6-31G (d) and charge calculation option was NBO (natural bonding orbital).
E. S. Replogle, G. W. Trucks, S. W. Staley J. Phys. Chem., 95(18), 6908 (1991).
(a) G. He et al ., Proc. SPIE, Vol. 5519, 42 (2004), (b) B. W. E-Andrade et al., Appl. Phys. Lett., 83(19), 3858 (2003), (c) R. H. Friend, R. W. Gymer, A. B. Holmes, J. H. Burroughes, R. N. Marks, C. Taliani, D. D. C Bradley, D. A. Cos Santos, J. L. Bredas, M Logdlund and W. R. Salneck, Nature, 397, 121 (1999).
A. P. Kulkarni, C. J. Tonzola, A. Babel, S. A. Jenekhe, Chem. Mater., 16, 4558 (2004).
J. R. Gong, L. J. Wan, S. B. Lei, C. L. Bai J. Phys.
Chem. B, 109, 1675 (2005).
C. C. Tong, K. C. Hwang J. Phys. Chem. C, 111, 3490 (2007).
Density of 1000Å organic deposited layer was measured by reflectivity simulation using XRD in AE center, Samsung Advanced Institute of Technology.
C. W. Tang, S. A. VanSlyke, Appl. Phys. Lett. 51, 913 (1987).
C. W. Tang, S. A. VanSlyke, C. H. Chen, Appl. Phys. Lett. 65, 3610 (1989).
L. S. Hung, C. W. Tang, M. G. Mason, Appl. Phys. Lett. 70, 152 (1997).
G. E. Jabbour, Y. Kawabe, S. E. Shaheen, J. F. Wang, M. M. Morrel, B. Kippelen, N. Peyghambarian, Appl. Phys. Lett. 71, 176 (1997).
Y. W. Ko, C. H. Chung, J. H. Lee, Y. H. Kim, C. Y. Son, B. C. Kim, C. S. Hwang, Y. H. Song, J. Lim, Y. J. Ahn, G. W. Kang, N. Lee, C. Lee, Thin Solid Films, 46, 246 (2003).
T. C. Gorjanc, D. Leong, C. Py, D. Roth, Thin Solid Film, 413, 181 (2002).
M. Ohyama, H. Kozuka, T. Yoko, Thin Solid Films, 306, 78 (1997).
R. Martins, P. Barquinha, I. Ferreira, L. Pereira, G. Goncalves, E. Fortunato, J. Appl. Phys. 101, 044505 (2007).
T. Yamada, A. Miyake, S. Kishimoto, H. Makino, N. Naoki, T. Yamamoto, Appl. Phys. Lett.
G. Gu, V. Bulovic, P. E. Burrows, S. R. Forrest, M. E. Thomson, Appl. Phys. Lett. 68, 2606 (1996).
R. B. Pode, C. J. Lee, D. G. Moon, J. I. Han, Appl. Phys. Lett. 84, 4614 (2004).
R. B. Pode, C. J. Lee, D. G. Moon, J. I. Han, Appl. Phys. Lett. 89, 123501 (2006).
W. Osikowicz, X. Crispin, C. Tengstedt, L. Lindell, T. Kugler, W.R. Salaneck, Appl. Phys. Lctt. 85, 1616 (2004).

K. S. Yook, S. O. Jeon, C. W. Joo, J. Y. Lee, Appl. Phys. Lett. 93, 013301 (2008).
X. D. Feng, C. J. Huang, V. Lui, R. S. Khangura, Z. H. Lu, Appl. Phys. Lett. 86, 1 (2005).
A. B. Chwang, M. A. Rothman, S. Y. Mao, R. H. Hewitt, M. S. Weaver, J. A. Silvernail, K Rajan, M. Hack, J. J. Brown, X. Chu, L. Moro, T. Krajewski, and N. Rutherford, Symposium Digest Tech Papers, 34, 868-871 (2003).

Nanocomposites for Organic Light Emiting Diodes

Nguyen Nang Dinh
University of Engineering and Technology, Vietnam National University Hanoi
Vietnam

1. Introduction

Recently, both the theoretical and experimental researches on conducting polymers and polymer-based devices have strongly been increasing (Salafsky, 1999, Huynh, 2002, Petrella et al., 2004, Burlakov et al., 2005), due to their potential application in optoelectronics, organic light emiting diode (OLED) displays, solar flexible cells, etc. Similar to inorganic semiconductors, from the point of energy bandgap, conducting polymers also have a bandgap - the gap between the highest occupied molecular orbital (HOMO) and the lowest unoccupied molecular orbital (LUMO). When sufficient energy is applied to a conducting polymer (or a semiconductor), it becomes conducting excitation of electrons from the HOMO level (valence band) into the LUMO level (conduction band). This excitation process leaves holes in the valence band, and thus creates "electron-hole-pairs" (EHPs). When these EHPs are in intimate contact (i.e., the electrons and holes have not dissociated) they are termed "excitons". In presence of an external electric field, the electron and the hole will migrate (in opposite directions) in the conduction and valence bands, respectively (Figure 1).

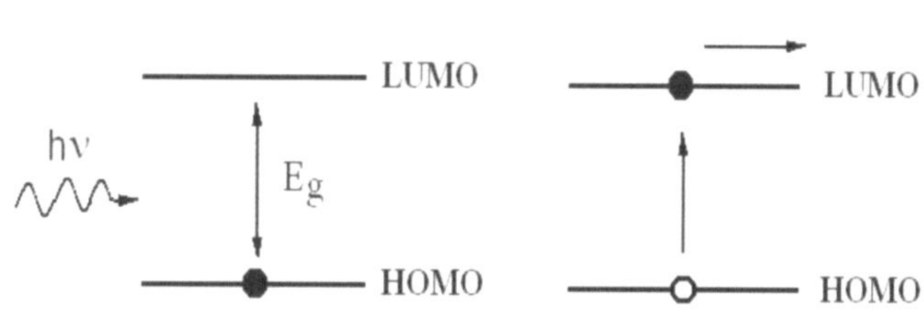

Fig. 1. Formation of "electron-hole pair" induced by an excitation from an external energy source (Klabunde, 2001)

On the other hand, inorganic semiconductors when reduce to the nanometer regime possess characteristics between the classic bulk and molecular descreptions, exhibiting properties of quantum confinement. These materials are reflected to as nanoparticles (or nanocrystals), or

"quantum dot". Thus, adding metallic, semiconducting, and dielectric nanocrystals into polymer matrices enables enhance the efficiency and service duration of the devices. The inorganic additives usually have nanoparticle form. Inorganic nanoparticles can substantially influence the mechanical, electrical, and optical (including nonlinear optical as well as photoluminescent, electroluminescent, and photoconductive) properties of the polymer in which they are embedded. The influence of nanocrystalline oxides on the properties of conducting polymers has been investigated by many scientists in the world. A very rich publication has been issued regarding the nanostructured composites and nano hybrid layers or heterojunctions which can be applied for different practical purposes. Among these applications one can divide two scopes, those concern to interaction between electrons and photons such as OLED (electricity generates light) and solar cells (light generates electricity).
In this chapter there are presented two types of the nanocomposite materials: the first one is the nanostructured composite with a structure of nanoparticles embedded in polymers, abbreviated to NIP, the second one is the nanocomposite with a structure of polymers deposited on nanoporous thin films, called as PON.

2. NIP nanocomposite

2.1 The role of Ti oxide nanoparticles in NIP

It is known that a basic requirement for a photovoltaic material is to generate free charge carriers produced by photoexcitation (Petrella et al., 2004, Burlakov et al., 2005). Subsequently, these carriers are transported through the device to the electrodes without recombining with oppositely charged carriers. Due to the low dielectric constant of organic materials, the dominant photogenerated species in most conjugated polymer is a neutral bound electron-hole pair (exciton). These neutral excitons can be dissociated from Coulomb attraction by offering an energetically favorable pathway for the electron from polymer (donor) to transfer to electron-accepting specie (acceptor). Charge separation in the polymer is often enhanced by inclusion of a high electron affinity substance such as C_{60} (Salafsky, 1999) organic dyes (Huynh et al., 2002, Ma et al., 2005), or nanocrystals (Burlakov et al., 2005). Nanocrystals are considered more attractive in photovoltaic applications due to their large surface-to-bulk ratio, giving an extension of interfacial area for electron transfer, and higher stability. The charge separation process must be fast compared to radiative or non-radiative decays of the singlet exciton, leading to the quench of the photoluminescence (PL) intensities. In addition, electron transport in the polymer/nanoparticle hybrid is usually limited by poorly formed conduction path. Thus, one-dimensional semiconductor nanorods are preferable over nanoparticles for offering direct pathways for electric conduction. It has been demonstrated that the solar cell based on the CdSe nanorods/poly(3-hexylthiophene)(P3HT) hybrid material exhibits a better power conversion efficiency than its CdSe nanoparticle counterpart. The environmental friendly and low-cost TiO_2 nanocrystal is another promising material in hybrid polymer/nanocrystal solar cell applications (Haugeneder, 1999, Dittmer et al., 2000).
The influence of nanooxides on the photoelectric properties of nanocomposites is explained with regard to the fact that TiO_2 particles usually form a type-II heterojunction with a polymer matrix, which essentially results in the separation of nonequilibrium electrons and holes. Embedding SiO_2 particles results in stabilization of the nanocomposite properties and

an increase in the lifetime of polymer-based electroluminescent devices. It is usually assumed that embedding semiconducting or dielectric nanocrystals creates additional potential wells and/or barriers for carriers and does not influence the energy spectrum of the polymer itself, except for a possible implicit influence through a change of the polymer conjugated length. However, it is also known that, in a conducting polymer with very low carrier mobility, the energy of carriers is determined to a considerable degree by the polarization of the material, which influences the position of the HOMO and LUMO levels as well as the exciton energy. The influence can be considerable, and can result in energy shifts of the order of 1 eV for free (unbound) electrons and holes in a polymer. In a uniform polymer medium this component of energy is determined by the molecular structure of the polymer and the fabrication technology. In nonuniform media, such as polymer–nanocrystal mixtures, the picture may change. In that case the polarization energy component may additionally depend on the relative position of carriers and inorganic inclusions.
Results in time-resolved PL measurements were reported (Dittmer et al., 2000). It is seen that time evolution of PL intensity of poly[2-methoxy-5-(2'-ethyl-hexyloxy)-1,4-phenylene vinylene] (MEH–PPV) on quartz shows mono-exponential decrease due to natural decay of excitons with a characteristic time constant 300 ps. PL intensity of MEH–PPV on TiO_2 decreases at initial time much quicker than that for MEH–PPV on quartz due to exciton quenching at the interface with TiO_2 substrate (Figure 2).

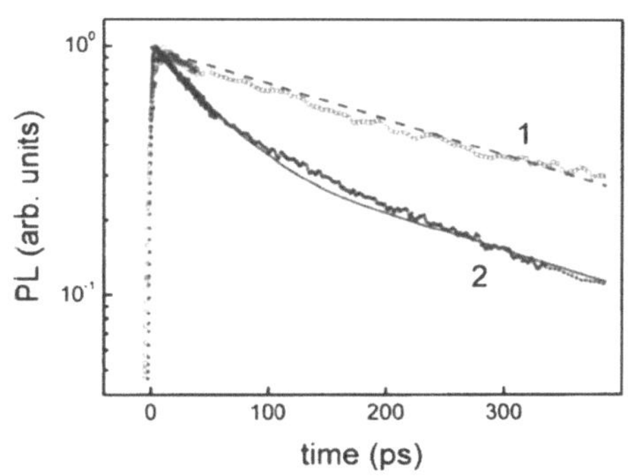

Fig. 2. PL intensity as a function of time in logarithmic scale. The symbols are experimental data for MEH-PPV film deposited on quartz (1) and TiO_2 (2) substrates, respectively. The dashed curve corresponds to monoexponential decay enabling determination of exciton life-time τ. The solid curve is theoretically calculated (Burlakov, et al., 2005)

TiO_2 nanocrystals - MEH-PPV composite thin films have also been studied as photoactive material (Petrella et al., 2004). It has been shown that MEH-PPV luminescence quenching is strongly dependent on the nature of nanostructral particles embedded in polymer matrix. Fluorescence quenching is much higher with rod titanium dioxide. In principle, rod particles can be expected to exhibit higher photoactivity with respect to spherical particles. In fact, when compared with the dot-like shape, rod-like geometry is advantageous for a more efficient packing of the inorganic units, owing to both a higher contact area and more intensive van der Waals forces. Actually, the higher quenching of the polymer fluorescence observed in presence of titania nanoparticles (Figure 3) proves that transfer of the photogenerated electrons to TiO_2 is more efficient for rods.

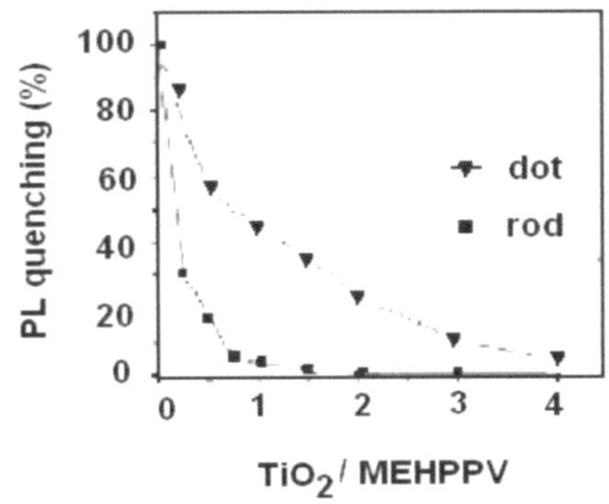

Fig. 3. MEH-PPV luminescence quenching vs. TiO_2/polymer volume ratio at λ = 480 nm (Petrella et al., 2004)

Chronoamperometric measurements have been performed on films of MEH-PPV, nanocrystalline TiO_2 and their blends. Thin films were deposited onto ITO from $CHCl_3$ solutions by spin-coating and immersed into an acetonitrile solution of tetrabutylammonium-perchlorate. As the authors showed, the light absorption and electron-hole pair photogeneration occur exclusively in MEH-PPV. The electron is then injected into the conduction band of the inorganic material, while the hole is transferred to the interface with electrolyte solution. Figure 4 indicates a higher photoactivity in blends when compared to the single components; the anodic photocurrents are higher with respect to the currents measured for MEH-PPV thin films, and are very reproducible. High film photostability was observed under longterm operative conditions.

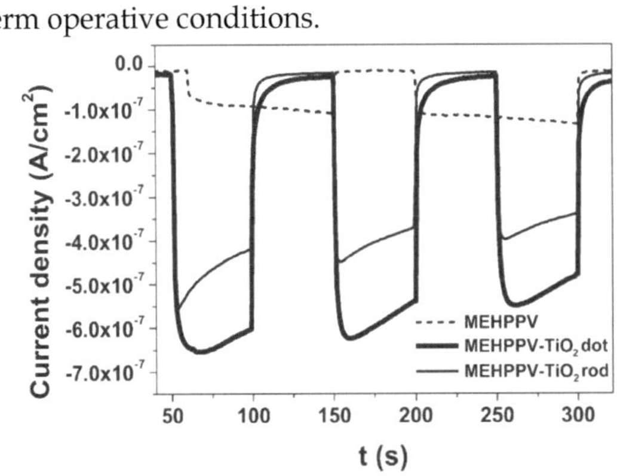

Fig. 4. Chronoamperometric measurements of MEH-PPV (...., blends of MEH-PPV TiO_2 dots (—) and MEH-PPV TiO_2 rods (thin solid line) in a photoelectrochemical cell. Ag/AgCl is chosen as reference electrode, while ITO and platinum as working and counter-electrode, respectively. A halogen lamp is used. The films were deposited onto ITO and immersed into acetonitrile solution of tetrabutyl-ammonium-perchlorate 0.1 M (Petrella et al., 2004)

From the obtained results it is known that the deposited composites film showed a higher photoactivity when compared to the single components due to the availability of numerous interfaces for enhanced charge transfer at the hetero-junction. Effective transport of excitons in conjugated polymers is extremely important for performances of organic light emitting diodes and of plastic excitonic solar cells. A crucial step in the photovoltaic process, for instance, is the conversion of photogenerated excitons into charge carriers at the polymer-inorganic interfaces. High quantum yield of charge carriers could be achieved if the excitons would travel far enough from their generation points to appropriate interfaces where they can dissociate, injecting electrons into the electrode. The holes remaining in the polymer diffuse to the opposite electrode, completing charge separation. Only a fraction of the photogenerated excitons reach relevant interfaces while many of them decay by emitting light or exciting vibrations of the polymer molecules.

Besides, a limited lifetime, the length scale of the exciton migration is restricted by the spatial dependence of the exciton energy - i.e., inhomogeneous broadening of exciton energy level. A conjugated polymer chain, for example, can be thought of as series of molecular segments linked with each other at topological faults. Each segment has certain LUMO and HOMO levels depending in part on its conjugation length. While migrating, excitons on average lose their energy by predominantly hopping to lower-energy sites.

Therefore the migration of excitons slows down when they reach the low-energy sites where they find fewer sites with lower energy in its neighborhood. Due to such dispersive migration, the exciton diffusion cannot be described using a constant diffusion coefficient, but a time-dependent one.

Photoluminescence efficiency was observed as a function of the content of nanocrystalline TiO_2 (nc-TiO_2) embedded in PPV, as demonstrated in figure 5 (Salafsky, 1999).

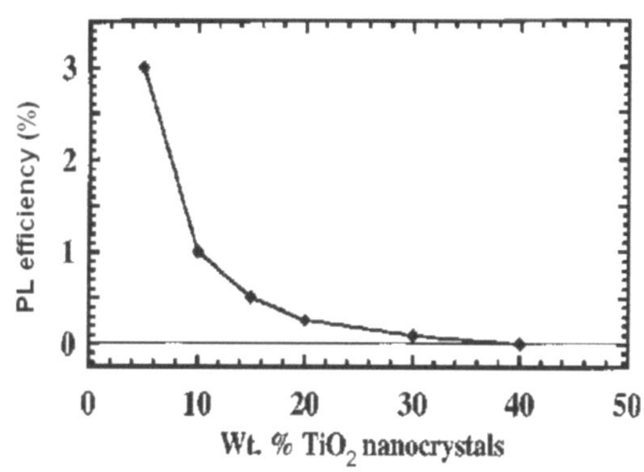

Fig. 5. Absolute photoluminescence (PL) efficiency of PPV:TiO_2 composites as a function of wt% TiO_2 nanocrystals (Salafsky, 1999)

The PL efficiency for PPV alone was measured to be 20%. This proves the PPV luminescence quenching. From point of review of photoactive materials, such a composite as PPV+nc-TiO_2 can be used for excitonic solar cells. The mechanism of the PPV luminescence quenching effect has been elucidated by energy diagram of polymer/oxide junctions (Figure 6).

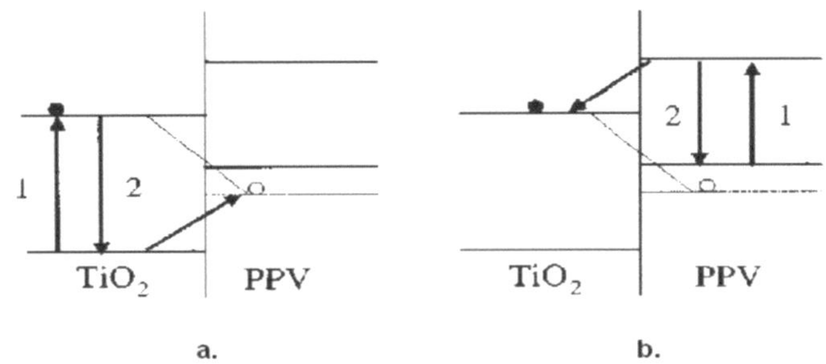

Fig. 6. Schematic diagram of the various excitation, charge transfer, and decay pathways available in a conjugated polymer nanocrystal composite (Salafsky, 1999)

The filled circles indicate electrons, and the open circles represent holes. Process 1 indicates photoexcitation; process 2 indicates decay of the electronic excited state; the dark slanting lines with arrows indicate a hole or electron transfer process (left and right sides, respectively); and the thin lines connecting the conduction band of TiO_2 with the hole level in PPV indicate an interfacial recombination process. The state levels are depicted as in this figure, with the holes placed at slightly lower energy than the polymer LUMO.

Absorbed photon-to-conducting-electron conversion efficiency (APCE) of solar devices based on the conjugated polymer-TiO_2 composite was obtained (Salafsky, 1999, Burlakov et al., 2005). It shows that the APCE is as a function of incident photon energy obtained. The quantum efficiency (QE) of light absorption, a fraction of photons absorbed within 50-nm-thick MEH-PPV with respect to the incident photons onto a device is also plotted which shows the photo-harvesting ability of the device (Figure 7).

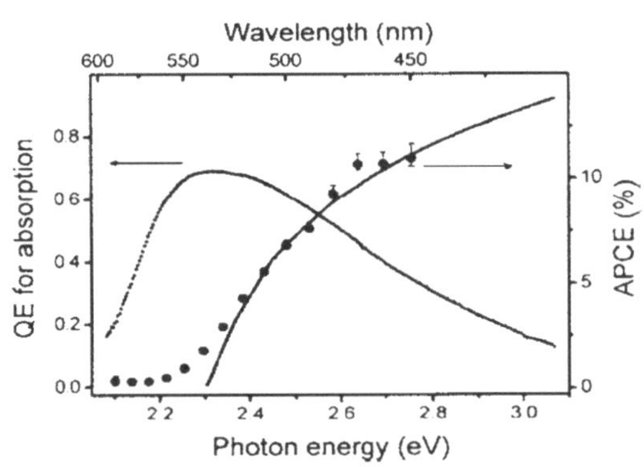

Fig. 7. Comparison of APCE curves obtained experimentally (solid circles) and theoretically (solid line) for 50-nm-thick MEH-PPV (Burlakov et al., 2005)

In a recent work (Lin et al., 2006), the authors have reported morphology and photoluminescent properties of MEH-PPV+nc-TiO_2 composites. The last is strongly dependent the excitation energy of photons. The samples were prepared with a large content of TiO_2, such as from 40 to 80 wt% of TiO_2 nanorods. The PL curves showed that the pristine MEH-PPV exhibits a broad absorption spectrum peaked at about 490 nm and TiO_2 nanorods have an absorption edge at about 350 nm. Due to the nature of indirect

semiconductor of TiO_2 nanorods, absorption and emission probabilities of indirect transition in pristine TiO_2 are much lower than for direct transitions. The inset shows the luminescence spectrum of TiO_2 nanorods excited at 280 nm. The broad emission band is mainly attributed to radiative recombination between electrons in the shallow trap states below the conduction band, the relative natural radiative lifetime resulted from oxygen vacancies and surface states, and holes in the valence band. Similar luminescence features of colloidal TiO_2 nanocrystals have been investigated previously (Ravirajan et al., 2005). For the excitation wavelengths in the range of 400-550 nm where only polymer is excited, the fluorescence intensities are further quenching, indicating that more efficient charge separation takes place with increasing TiO_2-nanorod content. In contrast, the intensities of fluorescence from polymer increase instead for the excitation wavelengths shorter than 350 nm. Due to the large absorption coefficient for TiO_2 nanorods at wavelengths less than 350 nm, the non-radiative Förster resonant energy transfer from TiO_2 nanorods to polymer may be responsible for the enhancement of fluorescence intensities. Enhancement in PL intensities in polymer suggests that absorption by TiO_2 nanorods leads to emission in the MEH-PPV by the non-radiative Förster resonant energy transfer (FRET) (Heliotis et al., 2006).
Cater et al have shown that the incorporation of nanoparticles inside an electroluminescent MEH-PPV thin film results in order of magnitude increases in current and luminance out-put (Figure 8). The nanoparticles appear to modify the device structures sufficiently to enable more efficient charge injection and transport as well as inhibiting the formation of current filaments and shorts through the polymer thin film. The composite nanoparticle/MEH-PPV films result in exceptionally bright and power efficient OLEDs (Cater et al., 1997). However, improvements are still needed in the device lifetime and homogeneity of the light output for these materials to be commercially viable.

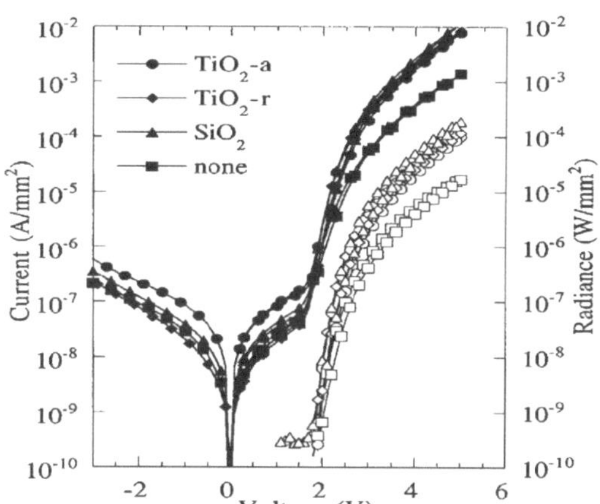

Fig. 8. Current-voltage and radiance-voltage curves for 1:1 TiO_2 (anatase)/MEH-PPV(circles), 1:1 TiO_2 (rutile)/MEH-PPV (diamonds), 1:1 SiO2/MEH-PPV (triangles), and for MEH-PPV film with no nanoparticles (squares). Close symbols are for current. Open symbols are for radiance. 1W/mm^2 = 7.3 ×10^7 cds/m^2 (Carter et al., 1997)

2.2 NIP composites for OLED

Polymer-based electroluminescent materials are very prospective for many applications, for instance, OLEDs are now commercialized in display fields. The efficient device operation requires optimization of three factors: (i) equalization of injection rates of positive (hole) and negative (electron) charge carriers (ii) recombination of the charge carriers to form singlet excitons and (iii) radiative decay of the excitons. Of the two carriers, holes have the lower mobility in general and may limit the current conduction process. By adding a hole transport layer (HTL) to the three-layer device one can expect equalization of injection rates of holes and electrons, to obtain consequently a higher electroluminescent efficiency of OLED. However, both the efficiency and the lifetime of OLEDs are still lower in comparison with those of inorganic LED. To improve these parameters one can expect using nanostructured polymeric/inorganic composites, instead of standard polymers for the emitting layer.

2.2.1 NIP films for hole transport layer

To prepare a NIP of polypropylene carbazone (PVK) and CdSe quantum dots (QD), a solution of PVK was made by dissolving PVK and in pure chloroform, then CdSe-QDs were added to this solution, stirred by ultrasonic bath. The solution then was spin-coated onto both glass and tin indium oxide (ITO) substrates with spin rates ranging from 1200 rpm to 2000 rpm for 1 to 2 min (Dinh et al., 2003). Under an excitation of short wavelength laser, the intensity of the PVK-NIP much increased, as seen in figure 9. Replacing CdSe-QDs by nc-TiO_2 the feature of the PL-enhancement is the same. Although the PVK-NIP can be used as HTL in OLED, polyethylenedioxythiophene (PEDOT) seemed to be much better candidat for the hole transoport, because it has a high transmission in the visible region, a good thermal stability and a high conductivity (Quyang et al., 2004; Tehrani et al., 2007). To enhance the interface contact between ITO and PEDOT, TiO_2 nanoparticles were embedded into PEDOT (Dinh et al., 2009)

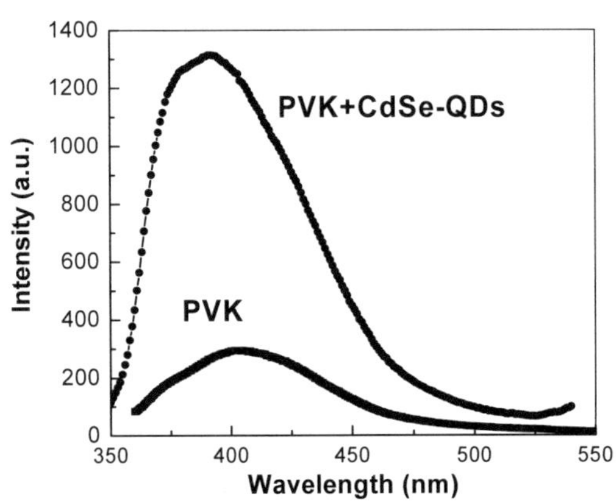

Fig. 9. Photoluminescence spetra of PVK and PVK+CdSe nanocomposite under a large photon energy excitation

Figure. 10 shows the atom force microscope (AFM) of a PEDOT composite with a percentage of 20 wt. % TiO_2 nanoparticles (about 5 nm in size). With such a high resolution of the AFM one can see a distribution of nanoparticles in the polymer due to the spin-coating process. For the pure polymeric PEDOT, the surface exhibits smoothness comparable to the one of the area surrounding the nanoparticles. The TiO_2 nanoparticles

contributed to the roughness of the composite surface and created numerous TiO_2/ PEDOT boundaries in the composite film.
Transmittance spectra respectively for a pure PEDOT and a nanocomposite films are plotted in Figure 11. From this figure one can see that nanoparticles of TiO_2 made the polymer film more absorbing in the violet range while making it more transparent in the near infrared range. At the range of the emission light of MEH-PPV, namely from 540 nm to 600 nm, the two samples have about a same transmittance of 82%. This transmittance is a bit lower, but still comparable to the transmittance of the ITO anode. Since PEDOT has a good conductivity, the electrical conductivity of this conducting polymer blend reaching up to 80 S/cm (Quyang et al., 2005), in the infrared wavelength range it reflects the IR light better resulting in a decrease in the transmittance. The presence of TiO_2 nanoparticles in PEDOT results in a cleavage of the polymer conjugation pathway, consequently leading to a decrease in film conductivity. This is why in the IR range the PEDOT composite has a higher transmittance than that of a pure PEDOT. However, this small decrease in conductivity does not affect much the performance of a OLED that uses the composite as a hole transport layer.

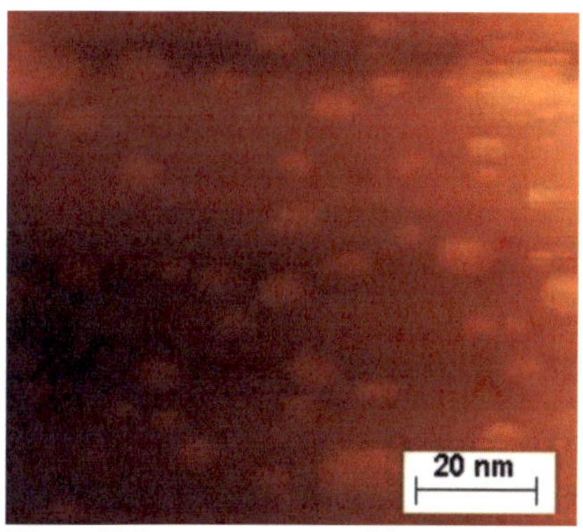

Fig. 10. AFM of a PEDOT+nc-TiO_2 composite film with embedding of 20 wt.% TiO_2 nanoparticles

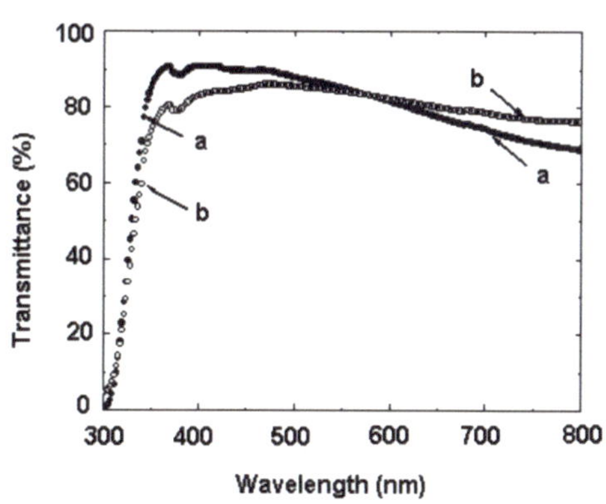

Fig. 11. Transmittance spectra of PEDOT (curve "a") and PEDOT composite films (curve "b")

2.2.2 NIP films for emitting layer

To deposit MEH-NIP composite layers, MEH-PPV solution was prepared by dissolving MEH-PPV powder in xylene with a ratio of 10 mg of MEH-PPV in 1 ml of xylene. Then,

TiO2 nanoparticles were embedded in these solutions according to a weight ratio TiO2/MEH-PPV of 0.15 (namely 15 wt. %), further referred to as MEHPPV+TiO_2. The last deposit was used as the emitter layer (EL). To obtain a homogenous dispersion of TiO_2 in polymer, the solutions were mixed for 8 hours by using magnetic stirring. These liquid composites were then used for spin-coating and casting. The conditions for spin-coating are as follows: a delay time of 120 s, a rest time of 30 s, a spin speed of 1500 rpm, an acceleration of 500 rpm and finally a drying time of 2 min. The films used for PL characterization were deposited by casting onto KBr tablets having a diameter of 10 mm, using 50 μl of the MEH-PPV solution. To dry the films, the samples were put in a flow of dried gaseous nitrogen for 12 hours (Dinh et al., 2009).

Surfaces of MEH-PPV+TiO_2 nanocomposite samples were examined by SEM. Figure 12 shows SEM images of a composite sample with embedding of 15 wt.% nanocrystalline titanium oxide particles (about 5 nm in size). The surface of this sample appears much smoother than the one of composites with a larger percentage of TiO_2 particles or with larger size TiO_2 particles. The influence of the heat treatment on the morphology of the films was weak, i.e. no noticeable differences in the surface were observed in samples annealed at 120°C, 150°C or 180°C in the same vacuum. But the most suitable heating temperature for other properties such as the current-voltage (I-V) characteristics and the PL spectra was found to be 150 °C. In the sample considered, the distribution of TiO_2 nanoparticles is mostly uniform, except for a few bright points indicating the presence of nanoparticle clusters.

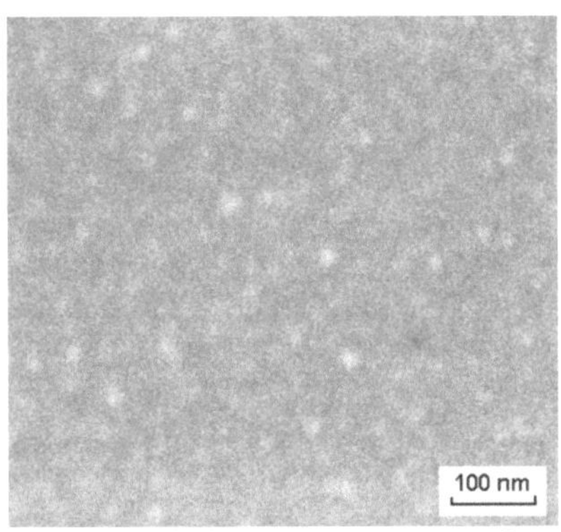

Fig. 12. SEM of a MEH+PPV-TiO_2 annealed in vacuum at 150 °C

The results of PL measurements the MEHPPV+TiO_2 nanocomposite excited at a short wavelength (325 nm) and at a standard one (470 nm) are presented. Figure 13 shows plots of the photoluminescence spectra measured on a pure MEH-PPV and a composite sample, using the FL3-2 spectrophotometer with an He-Ne laser as an excitation source (λ = 325 nm). With such a short wavelength excitation both the polymer and the composite emitted only one broad peak of wavelengths. From this figure, it is seen that the photoemission of the composite film exhibits much higher luminescence intensity than that of the pure MEH-PPV. A blue shift from 580.5 nm to 550.3 nm was observed for the PL peak. This result is consistent with currently obtained result on polymeric nanocomposites (Yang et al., 2005), where the blue shift was explained by the reduction of the chain length of polymer, when nanoparticles were embedded in this latter. Although PL enhancement has been rarely mentioned, one can suggest that the increase in the PL intensity for such a composite film can be explained by the large absorption coefficient for TiO_2 particles. Indeed, this phenomenon would be attributed to the non-radiative FRET from TiO_2 nanoparticles to polymer with excitation of wavelength less than 350 nm.

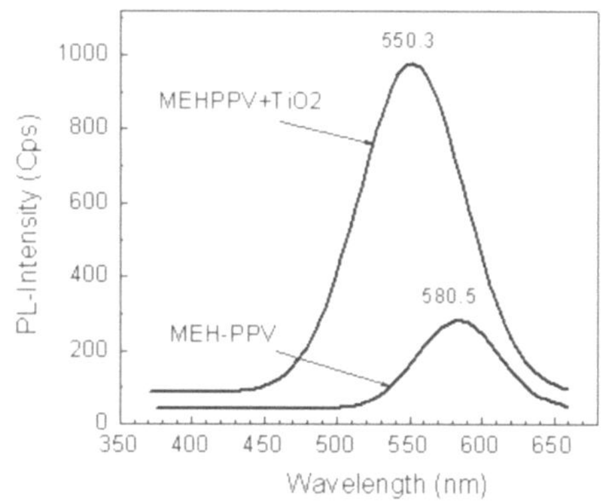

Fig. 13. PL spectra of MEH-PPV+nc-TiO_2. Excitation beam with λ = 325 nm

In figure 14 the PL spectra for the MEH-PPV and the composite films with excitation wavelength of 470 nm are plotted. In this case, the MEH-PPV luminescence quenching was observed. For both samples, the photoemission has two broad peaks respectively at 580.5 nm and 618.3 nm. The peak observed at 580.5 nm is larger than the one at 618.3 nm, similarly to the electroluminescence spectra plotted in the work of Carter et al (1997). As seen (Petrella et al., 2004) for a composite, in the presence of rod-like TiO2 nanocrystals, PPV quenching of fluorescence is significantly high. This phenomenon was explained by the transfer of the photogenerated electrons to the TiO_2. It is known (Yang et al., 2005) that the fluorescence quenching of MEH-PPV results in charge-separation at interfaces of TiO_2/MEH-PPV, consequently reducing the barrier height at those interfaces.

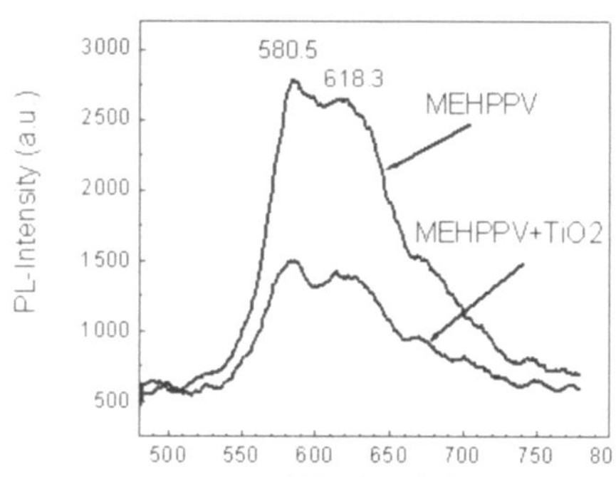

Fig.14. PL spectra of MEH-PPV+nc-TiO2. Excitation beam with λ = 470 nm

The effect of nanoparticles in composite films used for both the emitting layer (EL) and HTL in OLEDs was revealed by measuring the I-V characteristics of the devices made from different layers, such as a single pure EL diode (ITO/MEH-PPV/Al, abbreviated as SMED), a double pure polymer diode (ITO/PEDOT/MEH-PPV/Al or PPMD), a double polymeric composite layer diode, where a MEH-PPV+TiO2 composite was used as a EL and a PEDOT-composite film was used as a HTL (ITO/PEDOT+TiO_2/MEH-PPV+TiO_2/Al or PMCD), and a multilayer OLED, where a super thin LiF layer as ETL was added (ITO/PEDOT+TiO_2/MEH-PPV+TiO_2/LiF/Al or MMCD). A 10 nm-thick LiF layer used for the SCL was e-beam deposited onto the MEH-PPV+TiO_2; it was then covered by an Al coating prepared by evaporation. A detailed characterization of the SCL was however not carried out here, except for a comparison of the I-V characteristics (see figure 15). From this figure one can notice the following:

(i) The turn-on voltages for the diodes SMED, PPMD, PMCD to MMCD are found to be 3.4 V, 2.6V, 2.2 V and 1.7 V, respectively. For the full multilayer diode (MMCD), not only the turn-on voltage but also the reverse current is the smallest. This indicates the equalization of injection rates of holes and electrons due to both the HTL and the SCL which were added to the OLED.

(ii) A pure PEDOT used as HTL favors the hole injection from ITO into the organic layer deposited on the HTL, resulting in an enhancement of the I-V characteristics. Thus the turn-on voltage decreased from 3.4 V to 2.6 V (see the curve "b" for the PPMD diode).

(iii) Nanoparticles in both the EL and HTL films have contributed to significantly lowering the turn-on voltage of the device (see the curve "c" for the PMCD diode).

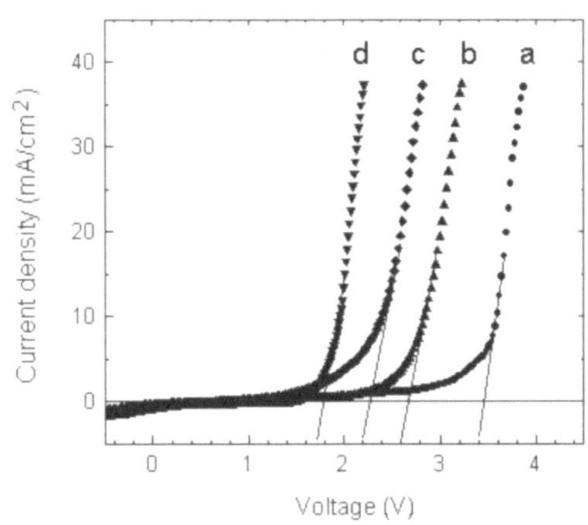

Fig. 15. I-V characteristics of OLED with different laminated structure. (a) - Single MEH-PPV, SMED; (b) - with HTL layer, PPMD; (c) - with HTL and EL composite layers, PMCD and (d) - with LiF, MMCD

The effect of HTL, ETL and/or SCL on the enhancement of the I-V characteristics was well demonstrated, associated with the equalization process of injection rates of holes and electrons. But the reason why the nanoparticles can improve the device performance is still open for discussion. For instance, in (Scott et al., 1996) the authors attributed this enhancement to the stimulated emission of optically-pumped MEH–PPV films when TiO_2 particles were embedded in. Whereas, in (Carter et al., 1997) the authors indicated that no evidence of line narrowing or changes in the line shape was observed at different voltages, implying that the mechanism for improved performance was distinctly different from that found in optically-pumped TiO_2/MEH–PPV films. These latter concluded that optical scattering phenomenon was not causing an enhancement in the device performance. Another possible explanation is that the nanoparticle surfaces increase the probability of electron-hole recombination; however, this would result in a change in the external quantum efficiency, rather than the current density as it was observed.

From the data of PL spectra for the MEH-PPV and the transmittance for PEDOT composites, we have observed both the improvement in PL intensity and the luminescence quenching of the composite (see figure 13 and 14). Similar phenomena obtained for nanohybrid layers were explained due to the TiO_2/polymer boundaries causing a difference in bandgap between the oxide nanoparticles and the conjugate polymer (Thuy et al., 2009). Based on these results, one can advance a hypothesis for the improved performance which supports the suggestion by Carter et al (1997). A change in the device morphology would be caused by the incorporation of nanoparticles into the solution. During the spinning process in the spin-coating technique, the nanoparticles can adhere by strong electrostatic forces to the HTL and between themselves, and capillary forces can then draw the MEH–PPV solution around the nanoparticles into cavities without opening up pinholes through the device. This will result in a rough surface over which the LiF (SCL) is evaporated and subsequently, a large surface area interface between the SCL and the electroluminescent composite material is formed. At a low voltage, charge-injection into MEH–PPV is expected to be cathode limited; the very steep rise in the I–V curves for the composite diodes however suggests that more efficient injection at the cathode through the SCL is occurring which would be caused by the rougher interface of the nanocomposites. At a higher voltage, transport in MEH–PPV appears to be space-charge limited.

The electroluminescence quantum efficiency can be caculated by using a well-known expression:

$$\eta_{\phi} = \gamma \times \eta_r \times \phi_f \tag{1}$$

where γ is a double charge injection factor which is dependent on the processes of carrier injection and is maximal ($\gamma = 1$) if a balanced charge injection into the emission layer of the device is achieved, i. e. the number of injected negative charges (electrons) equals the number of injected positive charges (holes); η_r quantifies the efficiency of the formation of a singlet exciton from a positive and a negative polaron, and ϕ_f is the photoluminescence quantum efficiency. From the PL spectra and the I-V characteristics obtained one can see that γ for the MMCD is the largest due to the addition of both the HTL and SCL into the device. Since nc-TiO2 particles embedded in MEH-PPV constitute a factor favouring electrons faster move in the EL, the intrinsic resistance of the OLED is lowered. This results in an improvement of the I-V characteristics of the device. Moreover, the more mobile electrons can create a larger probability of the electron-hole pairs formation in the emitting layer, resulting in an increase in ηr for the MMCD. Thus the electroluminescence quantum efficiency of the multilayer polymeric composite diodes can be evaluated from (1) and appears to be much larger than the one for the single polymeric layer device. As a result of the enhanced carriers injection and transport in the polymer composites, the electroluminescence quantum efficiency is roughly estimated to be improved by a factor exceeding about 10.

3. PON composites for inverse OLEDs

3.1. PVK/MoO$_3$ hybrid structrure

Polypropylene carbazone (PVK) deposited on a nanostructued MoO$_3$ (PVK/MoO$_3$), as the PON composite, can be seen as a hybrid structure between a polymer and an inorganic oxide. To prepare a hybrid structure of PVK/nc-MoO$_3$, Mo metallic substrate was annealed in oxygen, at temperature of 550 ^{o}C for ca. 2 hours to get a nanostructured MoO$_3$ layer, and then PVK was deposited by spin-coating, followed by vacuum annealing. Surface morphology and nano-crystalline structures of MoO$_3$ were checked, respectively by using Scanning Electron Microscopy (FE-SEM) and X-Rays Diffraction (XRD). I-V characteristics were measured using an Auto-Lab. Potentiostat PGS-30.

The thickness of the annealed Mo substrate layers was found to be dependent of the annealing conditions such as the temperature and time. The samples used for devices were prepared at 500 ^{o}C, for 2 hours. The structure of the films was checked by performing X-ray incident beam experiment. For thin annealed layers, three XRD peaks of the Mo substrate are obtained with a strong intensity (denoted by Mo-peaks in figure 15) indicating bulk Mo crystalline structure of the substrate.

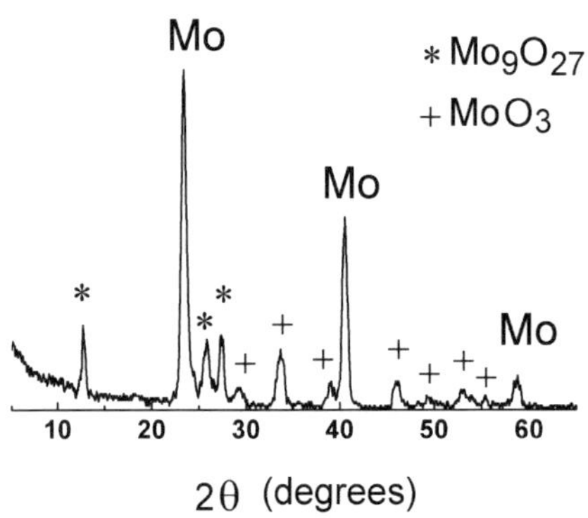

Fig. 15. XRD patterns of an annealed Mo-substrate showing, beside Mo structure, there are two structures of Mo oxides, namely MoO$_3$ and Mo$_9$O$_{27}$

Three other sharp peaks denoted by a star symbol in figure 15 characterize a crystalline structure of Mo$_9$O$_{27}$ that has been formed upon annealing. In the XRD diagram, there are seven diffraction peaks corresponding MoO$_3$. The fact that the peak width is rather large shows that the MoO$_3$ layer was formed by nanocrystalline grains. To obtain the grain size τ we used the Scherrer formula:

$$\tau = \frac{0.9\lambda}{\beta.\cos\theta} \tag{2}$$

where λ is X-ray wavelength, β is the full width at half maximum in radians and θ is the Bragg angle of the considered diffraction peak (Cullity, 1978). The values of β were found from 0.008 to 0.010, consequently the average size of the grains was determined as $\tau \approx$ 7-10 nm. This result is in a good agreement with the data obtained by FE-SEM for the average size of grains. The MoO$_3$ layer further would be spin-coated by PVK to get a heterojunction of PVK/nc-MoO$_3$.

Current-voltage characteristics of Ag/Mo/nc-MoO_3/PVK/Al and Ag/ITO/PVK/Al (Figure 16) show that the onset voltage of the hybrid junction is lowered in comparison with that of the standard junction. This may be explained by: (i) the workfunction of nc-MoO_3 is higher than that of ITO and (ii) the Mo substrate is metallic, thus Ag/Mo contact is more ohmic than Ag/ITO contact.

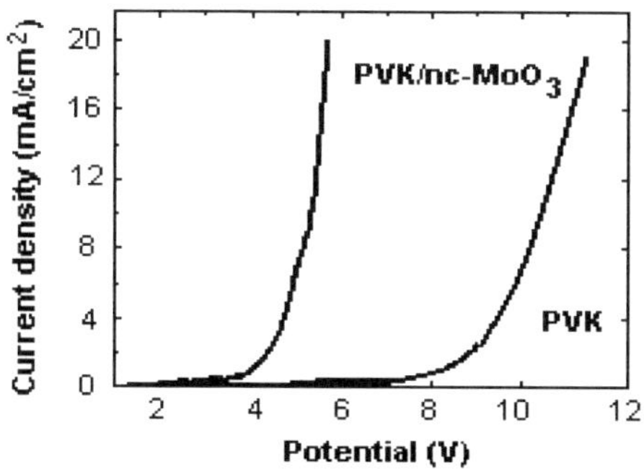

Fig. 16. I-V characteristic of PVK/MoO_3/Ag junction (left curve) and PVK/Ag junction (right curve)

3.2. MEH-PPV/TiO_2 hybrid structure

As seen in above mentioned PVK/nc-MoO_3/Mo hybrid layer, both the photoluminescence and I-V characteristics of the layer have been enhanced in comparison with those of the pure polymer based OLED. Lin et al (2007) showed that when a nanorod-like NIP composite of MEH-PPV+TiO_2 was excited by photons of a large energy, its photoluminescence was enhanced in comparison with that of MEH-PPV alone. As far as we know, the photoluminescencent properties of MEH-PPV/nc-TiO_2 hybrid PON films have been rarely studied. The aim of our work is to study the photoluminescent behavior of PON hybrid layers, when nanorod-like TiO_2 were grown on a flat titanium bar.

To grow nanocrystalline titanium oxide (nc-TiO_2) on metallic titanium, a 2-mm thick Ti wafer with a size of 5 mm in width and 10 mm in length were carefully polished using synthetic diamond powder of 0.5μm in size. The polished surface of Ti was ultrasonically cleaned in distilled water, followed by washing in ethylene and acetone. Then the dried Ti wafer was put in a furnace, whose temperature profile could be controlled automatically. We used three different annealing temperature profiles as follows: from room temperature, the furnace was heating up to 700°C for two hours and kept at this temperature respectively for one hour (the first profile), for one and a half hour (the second profile) and for two hours (the third profile), and these processes were followed by a cooling down to room temperature during three hours. To deposit hybrid layers, MEH-PPV solution was prepared by dissolving MEH-PPV powder (product of Aldrich, USA) in xylene with a proportion of 10 mg of MEH-PPV in 1 ml of xylene. The spincoating was carried-out in gaseous nitrogen with a set-up procedure described in the following. The delay time was 120s, the rest spin time 30s, the spin speed 1500 rpmin, the acceleration 500 rpmin and the relaxation time 5 min. After spincoating the samples were put into a vacuum oven for drying at 120°C at 1.33 Pa for 2 hours. For I-V testing, a silver-aluminum alloy coating was evaporated on the polymer to make diodes with the structure of AgAl/MEH-PPV/nc-TiO_2/Ti (Thuy et. al, 2009).

3.2.1 Morphology and crystalline structure of nanoporous TiO2 layer

Samples which were annealed respectively according to the first, second and third temperature profile are referred to by TC1, TC2 and TC3. The hybrid films having a structure of MEH-PPV/Ti-substrate, MEH-PPV/TC1, MEH-PPV/TC2 and MEH-PPV/TC3 are respectively abbreviated to MEHPPV, PON1, PON2 and PON3 for photoluminescence measurements. Similar symbols are adopted for the heterojunctions samples used in I-V tests, as follows:

MEHPPV: Ag-Al/MEH-PPV/Ti-substrate/Ag
PON1: Ag-Al/PON1/Ti-substrate/Ag
PON2: Ag-Al/PON2/Ti-substrate/Ag
PON3: Ag-Al/PON3/Ti-substrate/Ag

Figure 17 shows the FE-SEM images of three samples (TC1, TC2 and TC3). For all the samples TiO_2 was grown in form of nanorods whose size was strongly dependent on conditions of the thermal treatment. These pictures reflect a very high resolution of the FE-SEM: one can determine approximately both the size on the surface and the depth (or length) of TiO_2 rods grown in the titanium wafer. Thus, TiO_2 rods in TC2 (annealing time is 1.5 h) were estimated to have a width of about 70 nm on average and a length of about 200 nm. Moreover, a large number of the rods have orientation close to the vertical direction (see figure 17b). For the TC1 (Figure 17a) and TC3 (Figure 17c) samples, TiO2 rods were randomly orientated, TCl being thinner than TC3. The annealing time of TC3 was larger than that of TC2, and TC1, TC2 and TC3 had thicknesses respectively equal to ca. 100, 200 and 150 nm. We also annealed Ti wafers at the temperature of 500°C or 800°C. Even with a different annealing process, pictures of the nanorods on titanium substrate were similar to those for TC1 and TC3. This shows that for growing a nanorod-like TiO_2 on titanium surfaces, the temperature can be maintained at 700°C for 1.5 h.
Figure 18 shows XRD patterns of TC1 (top), TC2 (middle) and TC3 (bottom) samples. Although the annealed layers of the samples are thin (i.e. ~ 200 nm), in the XRD patterns all the key characteristic peaks of a rutile TiO_2 crystal are revealed. These peaks correspond to space distances of 0.322, 0.290, 0.217, 0.205, 0.168, 0.162 and 0.115 nm for all the samples. The fact that two intensive peaks of the titanium crystal (0.245 and 0.224 nm) occurred proves that X-ray went through the TiO_2 layer and interacted with the titanium crystalline lattice. Using formula (2) for the determination of crystalline grain size of the TiO_2, an average value calculated for all the TiO_2 peaks was found to be around 100 nm for the TC2 sample. This value is fairly different for TC1 and TC3 samples. However, these results are in a good agreement with the results by FE-SEM.

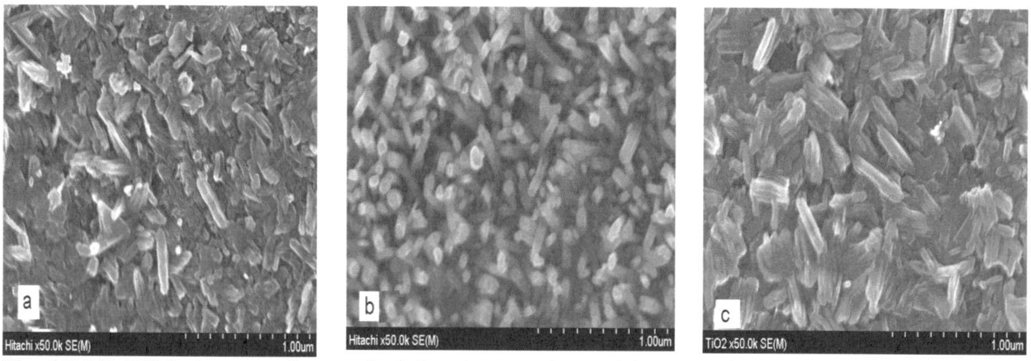

Fig. 17. FE-SEM pictures of annealed titanium surfaces: (a) 700 °C for 1 h (TC1), (b) 700°C for 1.5 h (TC2) and (c) 700°C for 2 h (TC3). The thickness of nc-TiO_2 layers is of 100 nm, 200 nm and 150 nm, respectively for TC1, TC2 and TC3 samples

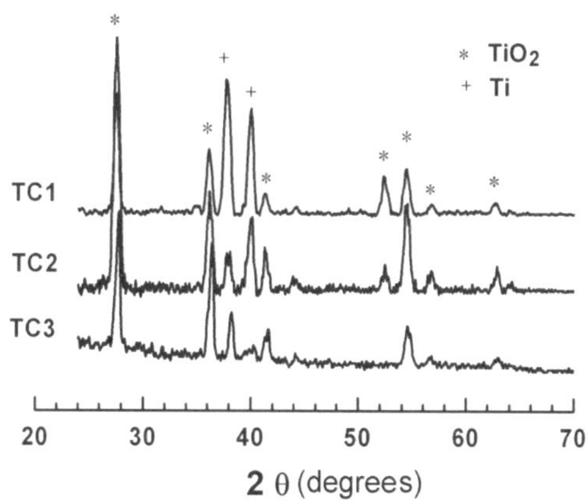

Fig. 18. XRD patterns of nc-TiO_2 layers grown on Ti surfaces at 700°C for 1h (TC1), 1.5h (TC2) and 2h (TC3)

3.2.2 Photoluminescent and electrical properties of hybrid junctions

The results of PL measurements of all the samples excited at a short wavelength (ca. 325nm) and at a standard one (ca. 470 nm) are presented. Figure 19 shows plots of the PL spectra measured on MEH-PPV, PON1, PON2 and PON3 samples, using the FL3-2 spectrophotometer with an He-Ne laser as an excitation source (λ = 325 nm). It is seen that all the samples have broad photoemission at two peaks; one higher at 645 nm and another lower at 605 nm. In a work on MEH-PPV+nc-TiO_2 composite (Carter et al., 1997) the author reported that two electroluminescence peaks at 580 nm and 640 nm occurred, where the first peak was higher than the second one. This negligible difference in wavelength values and intensity of the emission peaks can be explained due to electroluminescence. The emission peaks are shifted to longer wavelengths with respect to the main absorbance band. This red-shift is explained due to emission of the most extensively conjugated segments of the polymer (Kersting et al., 1993). From figure 19, it is seen that photoemission of all the hybrid samples exhibit higher luminescence intensity than that of the pure MEH-PPV. However, PL strongest enhancement occurred in PON2 film while for PON1 and PON3 films PL the intensities were not much increased. In these hybrid films no blue shift was observed, as it was obtained for MEH-PPV + nc- TiO_2 (see figure 13) or for PPV+nc-SiO_2, (Yang et al., 2005), as NIP composites. The blue shift was explained by the reduction of the polymer conjugation chain length. Although PL enhancement has been rarely mentioned, one can suggest that the increase PL intensity for such a PON2 thin film can be explained by the large absorption coefficient for TiO_2 nanorods. This similar the effect observed for the MEHPPV-NIP films, which explained due to the non-radiative Förster resonant energy transfer (Heliotis et. al., 2006) from TiO_2 nanorods to polymer with excitation of wavelength less 350 nm.

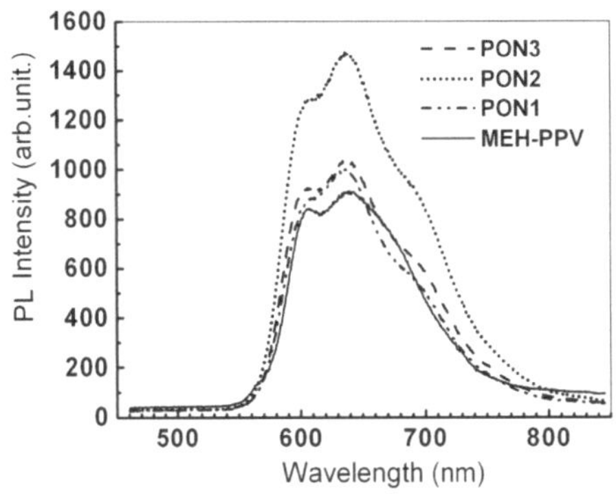

Fig. 19. PL spectra of MEH-PPV and nanohybrid films by using a He-Ne laser excitation at 325 nm. The best PL enhacement is obtained for PON2 sample

In Figure 20 the PL spectra of MEH-PPV and hybrid film samples with excitation wavelength of 470 nm (on FL3-22 using Xe lamp) are plotted. In this case, the MEH-PPV luminescence quenching occurred clearest in the PON2 sample. These spectra exhibited quite similarly to the spectra obtained for the MEH-PPV+nc-TiO_2 (NIP) samples (see figure 14).

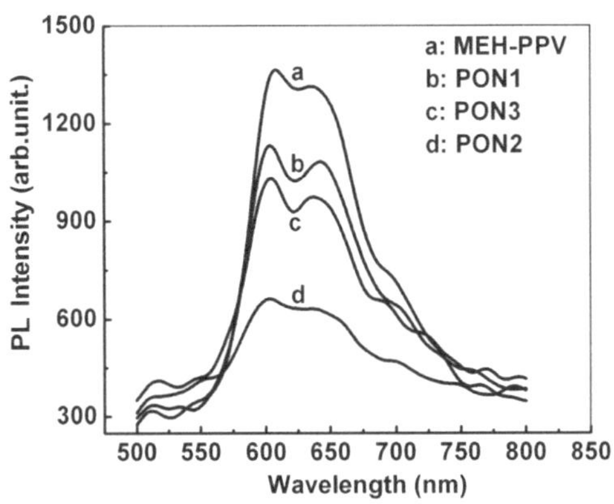

Fig. 20. PL spectra of MEH-PPV and nanohybrid films by using a Xe lamp excitation at 470 nm. The strongest MEH-PPV fluorescence quenching is obtained for PON2 sample

For all the samples the photoemission has two broad peaks at 605 nm and 645 nm as in the case of short wavelength excitation. Moreover, from figure 19 and figure 20 one can see that in these samples the larger enhancement in PL intensity (under short wavelength excitation), the stronger fluorescence quenching (under normal excitation) has occurred. The fact that the peak at 605 nm is larger than the peak at 645 nm is similar to the electroluminescence spectra plotted in a work of Carter et al (1997). As seen in a work of Petrella et al (2004), for a NIP composite, in presence of rod-like TiO_2 nanocrystals, PPV quenching of fluorescence is significantly high. This phenomenon has been explained due to the transfer of the photogenerated electrons to the TiO_2. In our case, among three hybrids films the PON2 sample is the most porous, and the rods are well separated from each other. Thus this sample is likely to be a NIP composite. Perhaps, this is the reason why PON2 exhibited the strongest quenching effect.

3.2.3 Current-voltage characteristics

Figure 21 shows the I-V curves of a pure MEH-PPV based diode and three hybrid diodes denoted as PON1, PON2 and PON3. It is seen that such a diode of Ag/Ti/MEHPPV/AlAg does not have both the transparent anode and hole transport layer (HTL). Thus, stating from some applied voltage, IV characteristics present a linear dependence of current on voltage as for a resistance (bottom curve, figure 5). For all the nanohybrid devices a turn-on voltage is of around 3 V, ascending from PON2 sample to PON1 and PON3, but the current density is not large (about 5 ÷ 10 mA/cm^2 at 4 V). For PON2 device although the turn-on voltage is smaller, the current began increasing with voltage right from 0. For PON1 and PON3 devices, it grew up from 2 V. This means that in the PON2 the reverse current of the device appeared from starting switch-on voltage, it can cause the device to be heated up. The PON1 and PON3 devices have a rather low turn-on voltage and no reverse current was observed up to an applied voltage of 2 V. It is known (Carter et al., 1997) that the fluorescence quenching of MEH-PPV results in charge-separation at interfaces of TiO_2/MEH-PPV, consequently reducing the barrier height at the last. This indicates that the PON2 film will be a better candidate for a photovoltaic solar cell than for the OLED.

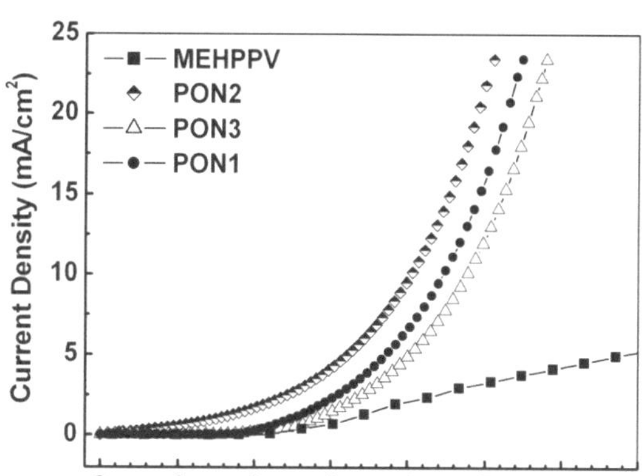

Fig. 21. I-V characteristics measurement of AgTi/MEH-PPV/Al-Ag (MEHPPV) and three devices of Ag/Ti/MEH-PPV+nc-TiO_2/Al-Ag (PON1, PON2 and PON3)

The fact, that PON1 and PON3 samples have very weak fluorescence quenching means that under the light illumination an inconsiderable electron/hole generation may occur at the TiO_2/MEH-PPV interfaces. Therefore, the PON1 and PON3 are not suitable for the photocurrent conversion. However, the improvement in I-V of the PON1 and PON3 devices can be attributed to a thin TiO_2 layer sandwiched between the polymer and Ti substrate. In this case the nc-TiO_2 layer played the role of HTL in OLEDs. Thus, contrarily to the PON2, such a laminar device as Ag-Al/PON/Ti/Ag is preferable to be used for OLEDs rather than for polymeric solar cells. However, to make a reverse OLED, instead of AgAl thin film, it is necessary to deposit a transparent cathode onto the emitting layer.

4. Conclusion and remarks

We have given an overview of the recent works on nanocomposites used for optoelectronic devices. From the review it is seen that a very rich publication has been issued regarding the nanostructured composites and nano-hybrid layers or heterojunctions which can be applied for different practical purposes. Among them there are organic light emitting diodes (OLED) and excitonic or organic solar cells (OSC).

Our recent achievements on the use of nanocomposites for OLEDs were also presented. There are two types of the nanocomposite materials, such as nanostructured composites with a structure of nanoparticles embedded in polymers (abbreviated to NIP) and nanocomposites with a structure of polymers deposited on nanoporous thin films (called as PON). Embedding TiO_2 nanoparticles in PEDOT, one can obtain the enhancement of both the contact of hole transport layer with ITO and the working function of PEDOT films. The improvement was attributed to the enhancement of the hole current intensity flowing through the devices. The influence of nanooxides on the photoelectric properties of the NIPs is explained with regard to the fact that TiO_2 particles usually form a type-II heterojunction with a polymer matrix, which essentially results in the separation of nonequilibrium electrons and holes. NIPs with the TiO_2 nanoparticles in MEH-PPV have been studied as photoactive material. MEH-PPV luminescence quenching is strongly dependent on the nature of nanostructral particles embedded in polymer matrix. Actually, the higher quenching of the polymer fluorescence observed in presence of titania nanoparticles proves that transfer of the photogenerated electrons to TiO_2 is more efficient for rods. Characterization of the nanocomposite films showed that both the current-voltage (I-V) characteristics and the photoluminescent properties of the NIP nanocomposite materials were significantly enhanced in comparison with the standard polymers. OLEDs made from these layers can exhibit a large photonic efficiency. For a PON-like hybrid layer of MEH-PPV/nc-TiO_2, the photoluminescence enhancement has also been observed. Thin nanostructured TiO_2 layers were grown by thermal annealing, then they were spin-coated by MEH-PPV films. Study of PL spectra of pure MEH-PPV and MEHPPV-PON films has shown that with excitation by a 331.1 nm wavelength laser lead to the largest enhancement in photoluminescent intensity as observed in the PON samples, and with an excitation of a 470 nm wavelength laser, the strongest fluorescence quenching occurred in this sample too. Current-voltage characteristics of laminar layer devices with a structure of Ti/PON/Al-Ag in comparison with that of Ti/MEH-PPV/Al-Ag showed that the turn-on voltage of the devices was lowered considerably. Combining I-V with SEM and PL, it is seen that PON are suitable for an reverse OLED, where the light goes out through the transparent or semi-transparent cathode, moreover to do Ohmic contact to the metallic Ti electrode is much easier.

However, to realize making reverse OLEDs, it is necessary to carry-out both the theoretical and technological researches to find out appropriate materials which can be used for the transparent cathode.

Acknowledgement

This work was supported by the Vietnam National Foundation for Science and Technology Development (NAFOSTED) in the period 2010 - 2011 (Project Code: 103.02.88.09).

5. References

Burlakov, V. M.; Kawata, K.; Assender, H. E.; Briggs, G. A. D.; Ruseckas, A. & Samuel, I. D. W. (2005). Discrete hopping model of exciton transport in disordered media. *Physical Review* 72, pp. 075206-1 ÷ 075206-5.

Carter, S. A.; Scott, J. C. & Brock, J. (1997). Enhanced luminance in polymer composite light emitting diodes. *J. Appl. Phys.* **71(9)**, pp. 1145 - 1147.

Cullity, B. D. (1978). *Elements of X-Ray diffraction*, 2nd ed., p.102. Addison, Wesley Publishing Company, Inc., Reading, MA.

Dinh, N. N.; Chi L. H., Thuy, T.T.C; Trung T.Q. & Vo, Van Truong. (2009). Enhancement of current, voltage characteristics of multilayer organic light emitting diodes by using nanostructured composite films, *J. Appl. Phys.* 105, pp. 093518-1÷ 093518-7.

Dinh, N. N.; Chi, L. H.; Thuy, T. T. C.; Thanh, D. V. & Nguyen, T. P. (2008). Study of nanostructured polymeric composites and hybrid layers used for Light Emitting Diodes. *J. Korean Phys. Soc.* **53**, pp. 802-805.

Dinh, N. N.; Trung, T. Q.; Le H. M.; Long P. D. & Nguyen T., P. (2003). Multiplayer Organic Light Emmiting Diodes: Thin films preparation and Device characterization, *Communications in Physics* 13, pp. 165-170.

Dittmer, J. J.; Marseglia, E. A. & Friend, R. H. (2000). Electron Trapping in Dye/Polymer Blend Photovoltaic Cells. *Adv. Mater.* 12, pp.1270-1274.

Haugeneder, A.; Neges, M.; Kallinger, C.; Spirkl, W.; Lemmer, U. & Felmann, J. (1999). Exciton diffusion and dissociation in conjugated polymer/fullerene blends and heterostructures. *Phys. Rev. B, 59*, pp. 15346–15351.

Heliotis, G.; Itskos, G.; Murray, R.; Dawson, M. D.; Watson, I. M. & Bradley, D. D. C. (2006). Hybrid inorganic/organic semiconductor heterostructures with efficient non, radiative Förster energy transfer. *Adv. Mater.* **18**, pp. 334-341.

Huynh, W. U.; Dittmer, J. J. & Alivisatos, A. P. (2002). Hybrid Nanorod, Polymer Solar Cells. *Science* 295, pp. 2425 - 2427.

Kersting, R.; Lemmer, U.; Marht, R. F.; Leo, K.; Kurz, H.; Bassler, H. & Gobel, E. O. (1993). Femtosecond energy relaxation in π, conjugated polymers. *Phys. Rev. Lett.* **70**, pp. 3820 - 3823.

Klabunde, K. J. (2001). *Nanoscale Materials in Chemistry*, John Wiley & Sons.

Lin, Yu, Ting.; Zeng, Tsung, Wei.; Lai, Wei, Zong.; Chen, Chun, Wei.; Lin, Yun, Yue.; Chang, Yu, Sheng. & Su, Wei, Fang. (2006). Efficient photoinduced charge transfer in TiO_2 nanorod/conjugated polymer hybrid materials. *Nanotechnology* 17, pp. 5781–5785.

Ma, W.; Yang, C.; Gong, X.; Lee, K. & Heeger, A. J. (2005). Thermally Stable, Efficient Polymer Solar Cells with Nanoscale Control of the Interpenetrating Network Morphology. *Adv. Func. Mater.* 15, pp.1617 - 1622.

Petrella, T. M.; Cozzoli, P. D.; Curri, M. L.; Striccoli, M.; Cosma, P.; Farinola, G. M.; Babudri, F.; Naso, F. & Agostiano, A. (2004). TiO_2 nanocrystals - MEH, PPV composite thin films as photoactive material. *Thin Solid Films* 451/452, pp. 64–68.

Quyang, J.; Chu, C., W.; Chen, F., C.; Xu, Q. & Yang, Y. (2005). High, Conductivity Poly(3,4, ethylenedioxythiophene): Poly(styrene sulfonate) Film and Its Application in Polymer Optoelectronic Devices. *Advanced Functional Materials* **15,** pp. 203 - 208.
Quyang, J.; Xu, Q.; Chu, C., W.; Yang, Y.; Li, G. & Shinar, J. (2004). On the mechanism of conductivity enhancement in poly(3,4, ethylenedioxythiophene):poly(styrene sufonate) film through solvent treatment. *Polymer* **45**, pp. 8443 - 8450.
Ravirajan, P.; Bradley, D. D. C.; Nelson, J.; Haque, S. A.; Durrant, J. R.; Smit, H. J. P. & Kroon, J. M. (2005). Efficient charge collection in hybrid polymer/TiO_2 solar cells using poly(ethylenedioxythiophene)/polystyrene sulphonate as hole collector. *Appl. Phys. Lett.* 86, pp. 143101 - 143113.
Salafsky, J. S. (1999). Exciton dissociation, charge transport, and recombination in ultrathin, conjugated polymer, TiO2 nanocrystal intermixed composites. *Physical Review B* 59, pp. 10885 – 10894.
Scott, J. C.; Kaufman, J.; Brock, P. J.; DiPietro, R.; Salem, J. & Goitia, J. A. (1996). MEH, PPV Light, Emitting Diodes: Mechanisms of Failure. *J. Appl. Phys.* 79, pp. 2745 – 2753.
Tehrani, P.; Kanciurzewska, A.; Crispin, X.; Robinson, N. D.; Fahlman, M. & Berggren, M. (2007). The effect of pH on the electrochemical over, oxidation in PEDOT:PSS films. *Solid State Ionics* 177, pp. 3521 – 3528.
Thuy, T. T.C.; Chi, L. H. & Dinh, N. N. (2009). Study of Photoluminescent and Electrical Properties of Nanostructured MEH, PPV/ TiO_2 hybrid films, *JKPS* 54, pp. 291 - 295.
Yang, S. H.; Nguyen, T. P.; Le Rendu, P. & Hsu, C. S. (2005). Optical and electrical properties of PPV/SiO_2 and PPV/TiO_2 composite materials. *Composites Part A: Appl. Sci..Manufact.* 36, pp. 509 - 513.

5

Carrier Transport and Recombination Dynamics in Disordered Organic Light Emitting Diodes

Shih-Wei Feng[1] and Hsiang-Chen Wang[2]
[1]*Department of Applied Physics, National University of Kaohsiung, Taiwan, R.O.C.*
[2]*Graduate Institute of Opto-Mechatronics, National Chung Cheng University, Chia-Yi, Taiwan, R.O.C.*

1. Introduction

Organic light emitting diode (OLED) displays are forecast to be the promising display technology. They are thin, flexible, energy conserving, and suitable for large screen displays. For the developments of high-performance devices, high efficiency and good color purity are necessary. The emission wavelengths can be modified by blending dopants into the polymers light emitting diodes or by the incorporation of fluorescent dyes into the emissive layers for small molecule devices. The incorporation of fluorescent dyes into host materials has the advantages of efficient color tuning, good device efficiency, and narrow emission spectrum width [1-4].

In OLEDs, carriers are localized in molecules and charge transport is a hopping process [2]. Carrier mobility is determined by charge transport between neighboring hopping sites. The mobility usually shows the Poole-Frenkel characteristic [5]. By controlling the distance between hopping sites, carrier mobility can be adjusted [6]. At thermodynamic equilibrium, charge carriers mostly occupy the deep tail states of the density-of-states (DOS) distribution [7]. Carrier hopping occurs mostly via shallower states [8,9]. This shows that carrier density could affect mobility. Furthermore, dopants in OLEDs act as shallow trapping centers, which trap carriers and change the carrier density. Carrier trapping is the main emission mechanism in doped organic systems [10]. This also shows the dependence of the mobility on the dopant concentration. Although the efficiency of doped OLEDs has been improved, the carrier dynamics have not been well discussed [1-4]. To further improve the efficiency and lifetimes of OLEDs, the carrier transport as well as recombination dynamics of doped OLEDs should be well explored.

In this study, the dependences of carrier transport behavior and luminescence mechanism on dopant concentration of OLEDs were studied. In the lightly-doped sample, higher carrier mobility and better device performance were observed. This shows that dopants create additional hopping sites and shorten the hopping distance. At a higher dopant concentration, dopants tend to aggregate and the aggregations degrade the device performance. In addition, the observed decay rates and luminescence efficiencies of the doped samples can be used to calculate the radiative and nonradiative decay rates. The trend suggests that the lightly-doped sample can exhibit better luminescence efficiency at higher applied voltage while the highly-doped sample shows poorer luminescence efficiency even operated at lower applied voltage. The resulting recombination dynamics can be used to explain the device characteristics and performance of the doped samples.

2. Sample Structures and Experimental Procedures

The OLEDs are fabricated by vacuum deposition of the organic materials onto an indium-tin-oxide (ITO)-coated glass at a deposition rate of 1-2Å s^{-1} at 10^{-6} Torr. The device structures are ITO/*N*, *N*'-bis(naphthalen-1-yl)-*N*, *N*'-bis(phenyl) benzidine (NPB : 55nm) /Tris(8-quinolinolato)-aluminum($A1q_3$) : 10-(2-benzothiazolyl)-1, 1, 7, 7-tetramethyl-2, 3, 6, 7-tetrahydro-lI I, 5H, 11H-benzo[*l*]pyrano[6, 7, 8-*i j*] quinolizin-11-one (C545T : 40nm)/Alq_3 (40nm)/LiF(1nm)/Al(200nm). NPB and Alq_3 are used as the hole transport layer (HTL) and electron transporting layer (ETL), respectively. The dopant concentrations of C545T in $A1q_3$ are 1%, 3%, and 7%. The active areas of each device were 9 mm^2. A blank sample (no doping) was also prepared for comparison. Figure 1 shows the sample structures of OLEDs.

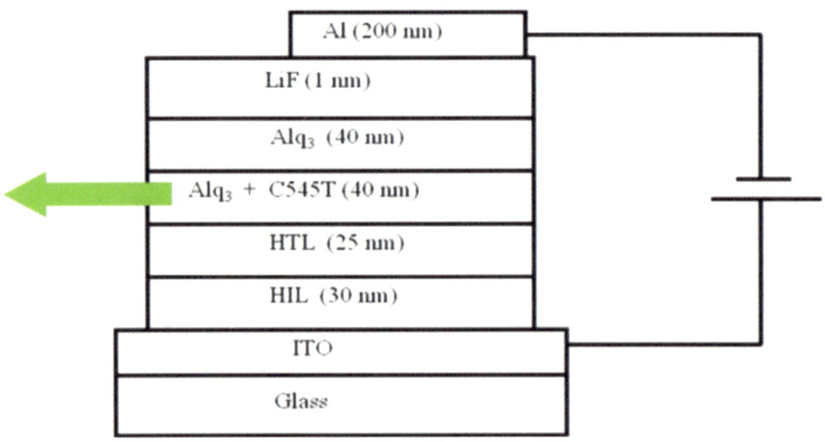

Fig. 1. Sample structures of OLEDs.

The morphological study was done by a scanning electron microscopy (SEM) (Hitachi Model S-4300N) with the excitation 5kV electrons. The electroluminescence (EL) spectra were measured by a Hitachi (model 4500) fluorescence spectrometer together with a power supply. Current-voltage (*I-V*) and capacitance-voltage (*C-V*) characteristics were measured with a semiconductor parameter analyzer (Agilent 4145B) and a LCR meter (Agilent 4284), respectively.

For transient electroluminescence measurements, a pulse generator (Agilent 8114A 100 V/2) was used to generate rectangular voltage pulses to the devices. The repetition rate and width of the pulse were 1 kHz and 5 μs, respectively. The light output was detected by a fast-biased silicon photodiode (Electro-Optics Technology Inc., model : ET-2020) operating directly on the surface of the devices. The transit time is a function of both the time required to charge the device (a function of the *RC* time constant of the circuit) and mobility [11]. In order to reduce the contribution of the time to charge the device, attention was paid to the

RC time constant of the EL cells. The maximum measured capacitance, *C*, of the EL cells was about 6 nF. The series resistance of our cells was estimated to be about 10 Ω. Therefore, the *RC* time constant was estimated to be less than 60 ns and the selected pulse width was greater than the charging time of the devices [4,12]. The temporal evolutions of EL signals were recorded by the average mode of a 50Ω input resistance of a digital oscilloscope (Agilent Model DSO 6052A, 500 MHz/4Gs/s). The oscilloscope was triggered by the pulse generator. The two coaxial cables for measuring transit EL and voltage pulse were equal in length, so that the time delay, except for the intrinsic delay, was negligible. All the measurements were carried out at room temperature (RT).

3. Experimental Results

3.1 SEM Images and EL Spectra (9 pt, bold)

Figure 2 (a) and (b) shows the SEM images of 1% and 3% C545T-doped Alq_3 films, respectively. The morphology of 1% C545T-doped Alq_3 film shows a homogeneous and flat image while that of 3% C545T-doped Alq_3 shows aggregations. This shows that dopants tend to aggregate as the dopant concentration becomes higher.

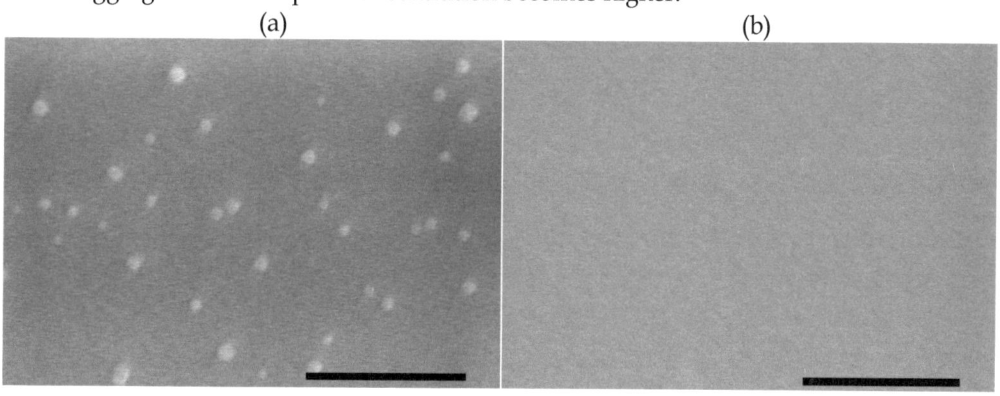

Fig. 2. SEM images of (a) 1% and (b) 3% C545T-doped Alq_3 films.

Figure 3 shows the EL spectra of 1%, 3%, and 7% C545T-doped Alq_3 samples and the undoped one. The EL spectra of the doped samples are significantly narrower than that of the undoped one. This is a tremendous advantage in the color mixing of red, green, and blue light for full-color applications. In order to create saturated colors, it is important for the individual red, green, and blue to be as pure as possible. Furthermore, as the dopant concentration increases, the peak position was slightly red-shifted and a shoulder in the long-wavelength side becomes apparent. Similar phenomena were also observed in Alq_3 films with 4-(dicyanomethylene)-2-methyl-6-(*p*-dimethylaminostyryl)-4H-pyran doapnt (DCM) aggregations [1,13,14]. The aggregations not only represent spatially distributed potential minimums but also broaden the effective DOS distribution. Hence, the broader spectrum width and the long-wavelength shoulder in EL spectra imply a larger degree of disorder.

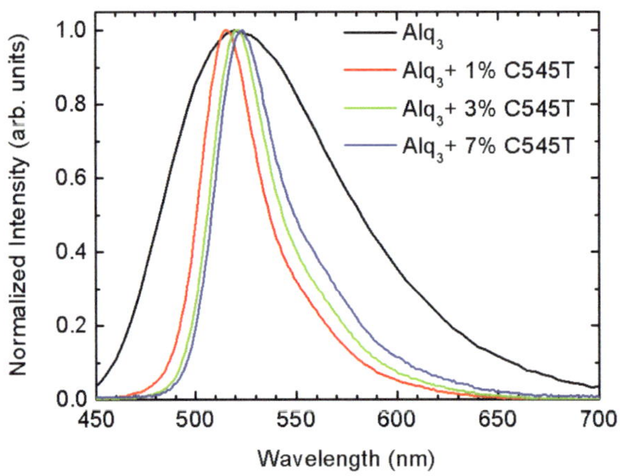

Fig. 3. EL spectra of the undoped and 1%, 3%, and 7% C545T-doped Alq_3 samples at RT.

3.2 *I-V* and *C-V* Characteristics

Figure 4(a) shows the current density versus applied voltage (*I-V*) characteristic of the four samples. Compared with the doped samples, the undoped sample shows a higher operational threshold and a shallow slope of current density versus applied voltage. This shows that the incorporation of dopants into host materials can improve device performance. In addition, with a higher dopant concentration, the driving voltage is higher and the current density is lower. This suggests that the aggregations tend to degrade the device performance.

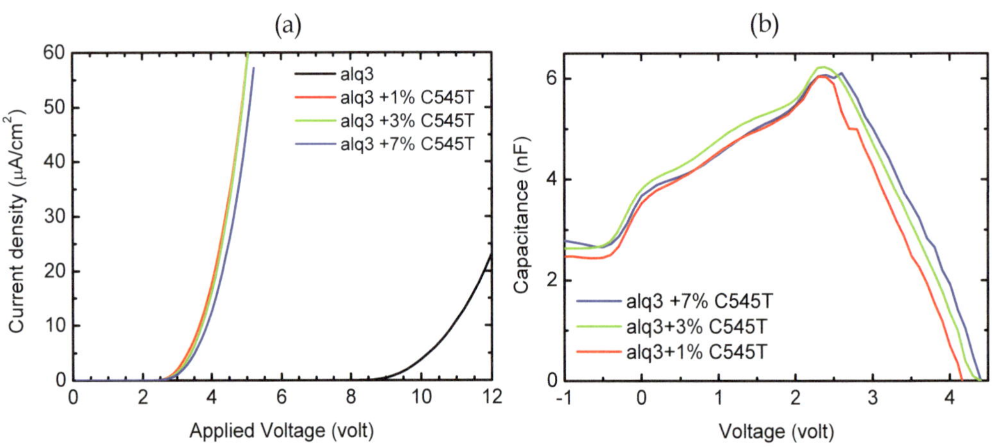

Fig. 4. (a) Current density versus applied voltage (*I-V*) characteristics of the undoped and three doped samples；(b) the differential capacitance as a function of bias (*C-V*) at a fixed frequency of 10 Hz of the three doped samples.

Furthermore, Figure 4(b) shows the differential capacitance $C = dQ/dV$ as a function of bias for a fixed frequency 10 Hz. No apparent difference was observed for negative bias. For the positive bias, the capacitance increases significantly and reaches a maximum at the "transition voltage V_0". V_0 is regarded as the built-in voltage V_{bi}, ie. the difference in work function between the two contacts [15]. The transition voltages V_0 for 1%, 3% and 7% C545T-doped Alq_3 samples are 2.3, 2.38 and 2.6 volts, respectively. Aggregations can trap carriers for self-quenching and luminescence losses, which leads to a higher turn-on voltage in the highly-doped sample. This argument is consistent with the long-wavelength shoulder in the EL spectrum. Furthermore, as the applied voltage is beyond V_0, the electrons and holes start to recombine and the capacitance decreases. The negative slope is related to the recombination efficiency. The lower the capacitance, the better the recombination efficiency. The slower decreasing trend of the highly-doped samples implies a low recombination efficiency.

3.3 Carrier Transport and Recombination Dynamics

The dynamic behavior of EL under electrical fast-pulse excitation provides important insights into the carrier transport behaviors and internal operation mechanisms of OLEDs. The response time is determined by the time delay, t_d, between addressing the device with a short, rectangular voltage pulse and the first appearance of EL [16,17]. The EL onset is identified as the time at which the two leading fronts of injected holes and electrons meet in the device. The time after the EL tends to saturate is the time at which electron and hole distributions have interpenetrated. The temporal decay of the EL at the end of the applied voltage pulse reflects the depletion of the carrier reservoir established during the preceding on-phase. Because the doped samples performed better than the blank one, the discussions in this section were focused on the three doped samples.

Figure 5(a) shows the transient EL as a function of time for different applied voltages for 1% C545T-doped Alq_3 sample. With increasing applied voltage, a shorter time delay (i.e. an earlier EL onset) and a steeper rise of the transient EL were observed. This shows a faster response time and more carrier mobility.

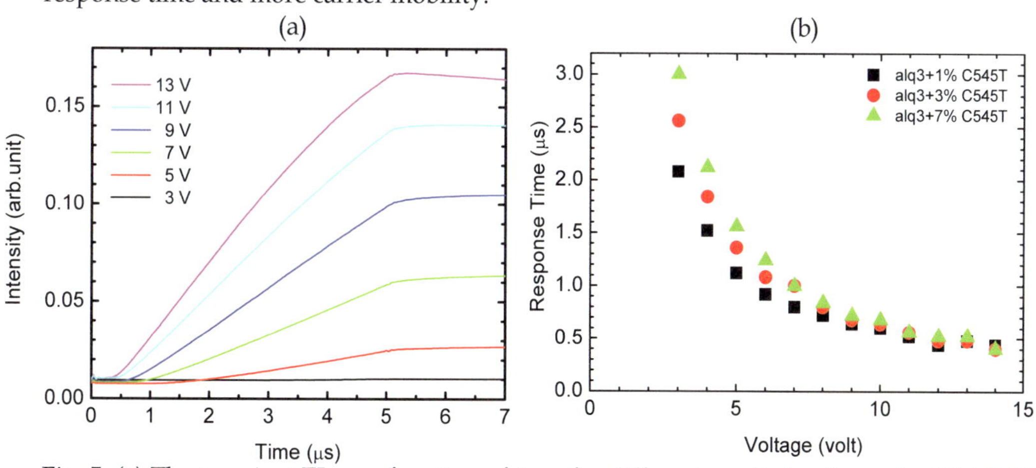

Fig. 5. (a) The transient EL as a function of time for different applied voltages for 1% C545T-doped Alq_3 sample. (b) Response time as a function of applied voltage for three doped samples.

The response time as a function of the applied voltage for the three samples are shown in Figure 5(b). At low applied voltages ($V_{applied} \leq 8$ volts), the response time increases with dopant concentration. In the highly-doped sample, some carriers are trapped and then quenched in aggregations. This slows down carrier mobility and decreases the overlap integral of electron-hole wavefunctions. Hence, the response time is longer. On the other hand, with high applied voltages ($V_{applied} > 8$ volts), carriers have more mobility among the hopping sites so that carriers may not be quenched in aggregations. This leads to the response times nearly independent of dopant concentration. The constant response time within the large bias range ($V_{applied} \geq 11$ volts) implies a saturation of carrier mobility.

The transient EL decay as a function of time for different applied voltages for 1% C545T-doped Alq_3 sample is shown in Figure 6. The EL decay can be fitted with a single exponential to obtain decay time. Figure 7(a) shows the decay time as a function of applied voltage for the three doped samples. The decay rate ($\kappa = 1/\tau$), the reciprocal of the decay time (τ), is shown in Figure 7(b). The decay rate shows an increasing and then decreasing trend with increasing applied voltage. It is noted that the measured decay rate is the sum of the radiative decay rate and nonradiative decay rate by the following equation [18] :

$$\kappa = \kappa_r + \kappa_{nr} = \frac{1}{\tau} \tag{1}$$

where κ_r, κ_{nr}, and κ are the radiative decay rate, nonradiative decay rate, and total decay rate, respectively. At somewhat high applied voltages ($V_{applied} \geq 5$ volts), the slower decay rate may imply an enhanced nonradiative decay rate. The details will be discussed later.

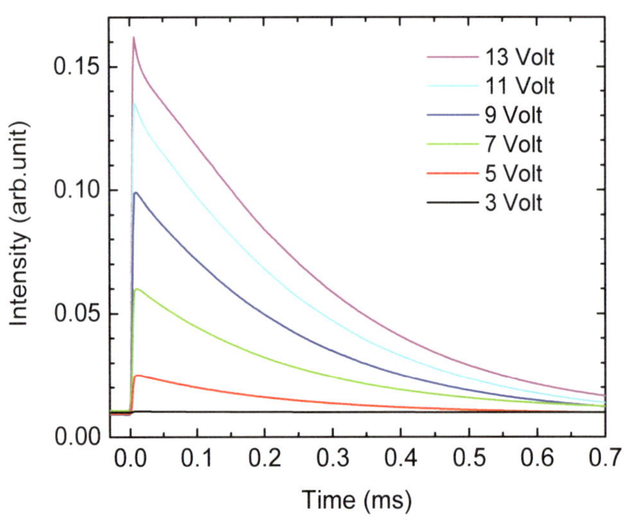

Fig. 6. The transient EL decay as a function of time for different applied voltages for 1% C545T-doped Alq_3 sample at RT.

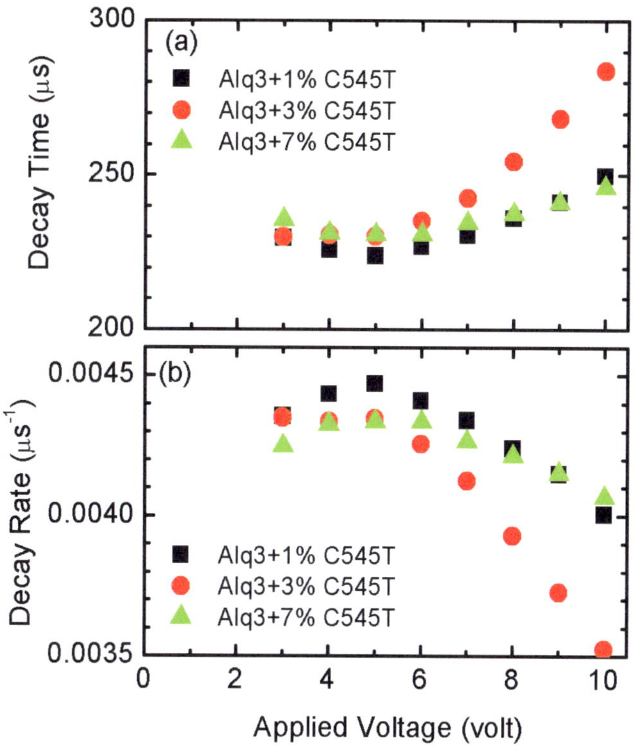

Fig. 7. (a) Decay times as well as (b) decay rates as a function of applied voltage for the three doped samples.

Figure 8 shows the luminescence efficiency as a function of applied voltage for the three doped samples. The luminescence efficiency exhibits a steep rise, then a substantial decrease with increasing current density. This phenomenon, called 'efficiency roll off' [19,20], was often observed in OLEDs and can be explained with the following mechanisms : (i) singlet-singlet and singlet-heat annihilations [21], (ii) exciton-exciton annihilation, (iii) excitons quenching by charge carriers, and (iv) field-assisted exciton-dissociation into an electron-hole pair [22]. In addition, the 1% C545T-doped Alq_3 sample has the best luminescence efficiency among the three samples. This shows that a small amount dopant improves the quantum efficiency. As the dopant concentration goes beyond a certain value, the dopants tend to aggregate. This degrades the device performance. Also, the response time seems to be related to the luminescence efficiency. Shorter response time correlates with luminescence efficiency. The shorter response time suggests higher carrier mobility and larger overlap integral of electron-hole wavefunctions. These factors improve the luminescence efficiency.

As shown in Figure 8, we normalize the luminescence efficiency at the maximum efficiency (at 3 volts) of 1% C545T-doped Alq_3 sample to get the normalized quantum efficiency. Because they have the same device structures, the extraction efficiencies of these samples are assumed to be the same and the normalized quantum efficiency can be regarded as the internal quantum efficiency. The internal quantum efficiency, η, is defined as the ratio of the number of light quanta emitted inside the device to the number of charge quanta undergoing recombination. η is given by the radiative decay rate over the total decay rate of recombination [18,23]. The decay rate is the reciprocal of decay time $(\kappa = 1/\tau)$. Hence, η can be expressed as

$$\eta = \frac{\kappa_r}{\kappa_r + \kappa_{nr}} = \frac{\kappa_r}{\kappa} \tag{2}$$

where κ_r, κ_{nr}, and κ are the radiative decay rate, nonradiative decay rate, and total decay rate, respectively. η can be improved when the radiative decay rate, κ_r, is enhanced. Radiative recombination requires spatial overlap of the electron-hole wavefunctions and κ_r is expected to decrease when carrier separation occurs. κ_r is in the μs^{-1} to the ns^{-1} range when electron-hole pairs are located on a single conjugated polymer chain. It is difficult to give an order of magnitude to the nonradiative process, since it depends on the defect density. In order to quantitatively study the recombination dynamics, the observed decay rate (κ) and internal quantum efficiency (η) can be used to trace out the radiative decay rate and nonradiative decay rate by solving equations (1) and (2).

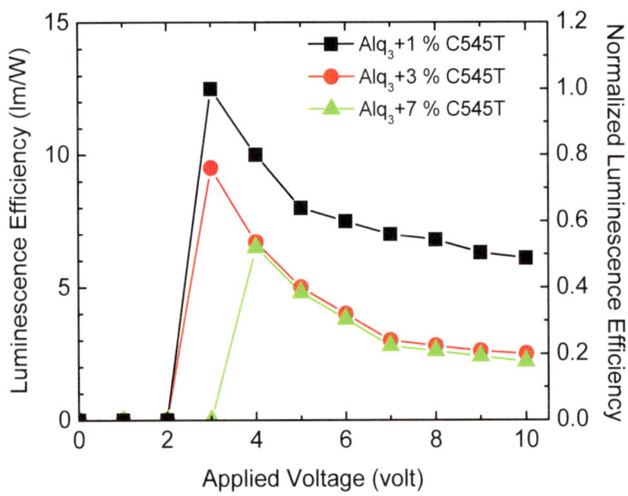

Fig. 8. Luminescence efficiency and normalized quantum efficiency as a function of applied voltage for the three doped samples.

Figure 9 shows the calculated results of κ_r and κ_{nr}. Some phenomena associated with recombination dynamics are shown in this figure. For the 1% C545T-doped Alq_3 sample, κ_r exhibits a decreasing trend with increasing the applied voltage while κ_{nr} does the opposite. Around the applied voltage $V_{1\%}$~7.5 volts, κ_r and κ_{nr} are equal, at about 0.0022 μs^{-1}. With the applied voltage lower than 7.5 volts, the larger κ_r implies better luminescence efficiency. At larger forward bias, the lower radiative decay rate and higher nonradiative decay rate are responsible for the lower luminescence efficiency of the OLED devices. The trends of κ_r and κ_{nr} can explain the luminescence efficiency as a function of applied voltage for the 1% C545T-doped Alq_3 sample, as shown in Figure 8. In addition, for the 3% and 7% C545T-doped Alq_3 samples, κ_r becomes lower while κ_{nr} is enhanced. At larger forward bias, the radiative decay rates, κ_r, are larger than nonradiative decay rates, κ_{nr}. These trends, due to the existence of aggregations, lead to the lower luminescence efficiency. For the 7% C545T-doped Alq_3 sample, the largest κ_{nr} and the smallest κ_r suggests the strongest nonradiative recombination and poorest luminescence efficiency among the three doped samples. In addition, it was found that κ_r and κ_{nr} are equal, at about 0.0022 μs^{-1} for the three doped samples. The applied voltages $V_{3\%}$ and $V_{7\%}$, corresponding to equal κ_r and κ_{nr}, are ~4.3 and ~4.0 volts for the 3% and 7% C545T-doped Alq_3 samples, respectively. The applied voltages corresponding to equal κ_r and κ_{nr} decrease with increasing dopant concentration. These demonstrate that the lightly-doped sample exhibits better luminescence efficiency than the highly-doped samples at all applied voltages and that all the doped samples exhibit peak luminescence efficiency at relatively low applied voltage, with luminescence efficiency decreasing for all the doped samples as the applied voltage is increased. The resulting recombination dynamics are correlated with the device characteristics and performance of the doped samples.

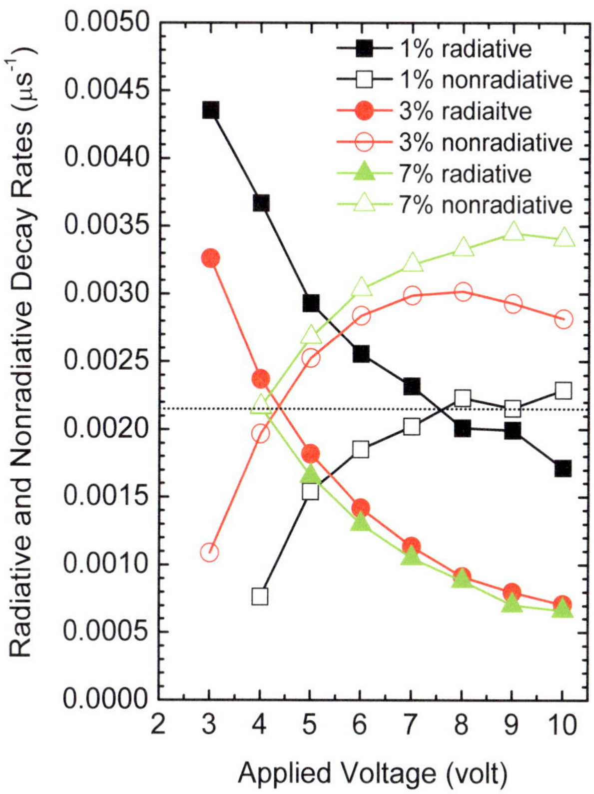

Fig. 9. Radiative decay rate (filled symbol) and nonradiative decay rate (empty symbol) as a function of applied voltage for the three doped samples.

4. Conclusion

In summary, the dependence of recombination dynamics and carrier transport on dopant concentration of OLEDs studied. In the lightly-doped sample, a higher carrier mobility and better device performance were observed. Due to the aggregations in the highly-doped samples, carrier quenching as well as nonradiative recombination degrade the device performance. In addition, the radiative decay rate and nonradiative decay rate were calculated to explain the recombination dynamics. It was found that the lightly-doped sample exhibits better luminescence efficiency than the highly-doped samples at all applied voltages and that all the doped samples exhibit peak luminescence efficiency at relatively low applied voltage, with luminescence efficiency decreasing for all the doped samples as the applied voltage is increased. The results for recombination dynamics are correlated with the device characteristics and performance of the doped samples. The investigations are beneficial for discovering recombination dynamics, investigating quantum efficiency, and developing device applications.

5. References

1. A.A. Shoustikov, Y. You, and M.E. Thompson, IEEE Journal of Selected Topics in Quantum Electronics, 4 (1998) 3.
2. H. Bässler, Phys. Status Solidi B 175 (1993) 15.
3. J. Chen and D. Ma, J. Appl. Phys., 95 (2004) 5778.
4. D.G. Moon, R.B. Pode, C.J. Lee, and J.I. Han, Appl. Phys. Lett., 85 (2004) 4771.
5. A.B. Walker, A. Kambili, and S.J. Martin, J. Physics : Condens. Matter 14, (2002) 9825.
6. Y. Zhang, Y. Hu, J. Chen, Q. Zhou, and D. Ma, J. Phys. D 36 (2003) 2006.
7. V.I. Arkhipov, P. Heremons, E.V. Emelianova, G.J. Adriaenssens, and H. Bässler, Appl. Phys. Lett. 82 (2003) 3245.
8. D. Monroe, Phys. Rev. Lett. 54 (1985) 146.
9. V.I. Arkhipov, E.V. Emelianova, and G.J. Adriaenssens, Phys. Rev. B 64, (2003) 125.
10. M. Uchida, C. Adachi, T. Koyama, and Y. Taniguchi, J. Appl. Phys. 86 (1999) 1680.
11. C. Hosokawa, H. Tokailin, H. Higashi, and T. Kusumoto, Appl. Phys. Lett., 60 (1992) 1220.
12. R.A. Klenkler, G. Xu, H. Aziz, and Z.D. Popovic, Appl. Phys. Lett., 88 (2006) 242101.
13. K.H. Weinfurtner, H. Fujikawa, S. Tokito, and Y. Taga, Appl. Phys. Lett., 76 (2000) 2502.
14. E.M. Han, L.M. Do, M. Fujihira, H. Inada, and Y. Shirota, J. Appl. Phys., 80 (1996) 3297.
15. D.Y. Kondakov, J.R. Sandifer, C.W. Tang, and R.H. Young, J. Appl. Phys. 98 (2003) 1108.
16. W. Brütting, H. Riel, T. Beierlein, and W. Riess, J. Appl. Phys. 89 (2002) 1704.
17. M. Ichikawa, J. Amagai, Y. Horiba, T. Koyama, and Y. Taniguchi, J. Appl. Phys., 94, (2003) 7796.
18. E.F. Schubert, Light-Emitting Diodes, Cambridge University Press, 2006.
19. F.X. Zang, T.C. Sum, A.C. H. Huan, T.L. Li, W.L. Li, and F. Zhu, Appl. Phys. Lett., 93 (2008) 2339.
20. M. Cocchi, V. Fattori, D. Virgili, C. Sabatini, P. Di Marco, M. Maestri, and J. Kalinowski, Appl. Phys. Lett., 84 (2004) 1052.
21. H. Nakanotani, H. Sasabe, and C. Adachi, Appl. Phys. Lett., 86 (2005) 213506.
22. J. Szmytkowski, W. Stampor, J. Kalinowski, and Z.H. Kafafi, Appl. Phys. Lett., 80 (2002) 1465.
23. Q. Huang, S. Reineke, K. Walzer, M. Pfeiffer, and K. Leo, Appl. Phys. Lett., 89 (2006) 263512.

6

Solution Processable Ionic *p-i-n* Organic Light-Emitting Diodes

Byoungchoo Park
Department of Electrophysics, Kwangwoon University, Seoul
Korea

Strong and efficient light emission from organic light emitting devices (OLEDs) fabricated by simple solution process is potentially useful for the manipulation of light in information display and lighting technologies. Recently, intensive research has been conducted into the development of OLEDs for realizing strong light emission from a simple OLED structure. In this chapter, highly efficient and enhanced light emission will be described for solution-processed phosphorescent OLEDs (PHOLEDs) doped with ionic salt, treated with the simultaneous electrical and thermal annealing. Because the simultaneous annealing causes the adsorption of charged salt ions at the electrode surfaces, the separated charges act as ionic doping in organic semiconductor layer and the electronic energy levels of the organic molecules are bended by the electric fields due to the adsorbed charged ions at the electrode interfaces, i.e., the simultaneous annealing can induce the proper formation of ionic *p-i-n* structure.

1. Introduction

Recent research has focused on the development of organic materials and device structures for use in organic light-emitting devices (OLEDs), with the aim of realizing cost-efficient, lightweight, and large-area flat panel displays (Tang & VanSlyke, 1987, Burroughes et al., 1990, Baldo et al., 1998). In order to achieve this aim, the scientific developments of the greatest interest to researchers are the improved efficiency, stability, and simplicity of the device fabrication process. In respect of the efficiency of these devices, for example, their internal quantum efficiency has been improved significantly of late, and is currently typically near 100%, as a result of incorporating phosphorescent dopant into the electroluminescent (EL) layer. This innovation has resulted in strong spin-orbit coupling, which leads to a rapid intersystem crossing and a radiative transition from triplet states to a ground state, thus promoting enhanced EL emissions (Baldo et al., 1998, Baldo et al., 1999, Adachi et al., 2002, He et al., 2004). By making use of the electro-phosphorescent Ir complex, it has been possible to create phosphorescent OLEDs (PHOLEDs) with an increased peak luminescence of up to ~ 50,000 cd/m^2 (Baldo et al., 1998, Baldo et al., 1999, Adachi et al., 2002, He et al., 2004). In contrast, relatively little progress has been made to date in designing a reliable and simple fabrication process that ensures the formation of a flat and uniform EL layer over a large area, which is particularly important for achieving the highly efficient and reliable device performance that is required for OLEDs. During the fabrication of OLEDs, the organic layers used are typically prepared using physical vapor deposition (Tang & VanSlyke, 1987, Baldo et al., 1998, Baldo et al., 1999, Adachi et al., 2002, He et al., 2004) or wet solution-coating processes (Friend et al., 1999, Pardo et al., 2000, Jabbour et al., 2001, Ouyang et al., 2002, de Gans et al., 2004, So et al., 2007). To date, OLEDs manufactured using vapor-deposited organic multi-layers of small molecular materials have the best performance record. However, the vapor deposition process is quite complex and expensive. Solution-processed devices made of polymeric or small molecular materials are also of interest, because these techniques make possible a simple production technique that uses a non-vacuum process such as continuous coating, screen printing, and Ink-jet printing (Pardo et al., 2000, Jabbour et al., 2001, Ouyang et al., 2002, de Gans et al., 2004, So et al., 2007). In such solution-processed devices, it is of critical importance to achieve strong light emission from a simple OLED structure. For that purpose, several trials have been made. One of the simplest devices tested is a solution-processed PHOLED (Yang & Neher, 2004, Liu et al., 2005, Niu et al., 2005, Suzuki et al., 2005). Usually, solution-processed PHOLEDs are prepared by doping a low molecular weight phosphorescent dye, such as an iridium complex, into a proper polymer matrix that contains a large-band-gap polymer, such as poly(vinylcarbazole) (PVK) (Yang & Neher, 2004, Liu et al., 2005, Niu et al., 2005). To achieve a better balance of charge transport in these devices, an interfacial layer, such as CsF, LiF, and/or surfactant layer, was also introduced between the phosphor-doped emissive layer and the metal (Al) cathode (Yang & Neher, 2004, Liu et al., 2005, Niu et al., 2005, Park et al., 2007). For a representative example, a power conversion efficiency (η_P) of 24 lm/W at current efficiency (η_C) of 30 cd/A was reported for PHOLEDs with the 1 nm CsF interfacial layer and the Al cathode (Yang & Neher, 2004). Another method is to use a bilayer cathode that consists of an electron-injection layer, such as Ca, Ba, or Cs, and Al evaporated onto the light-emitting polymer layer (Suzuki et al., 2005). A peak η_P of 38.6 lm/W was reported when the bilayer cathode was used (Suzuki et al., 2005). Although such efficient single-layered PHOLEDs with the interfacial layer have been demonstrated, the processes by which PHOLEDs are fabricated remain inadequate; the device structure is still complex and thus the fabrication process is complicated. In order to realize strong light emission from simple OLED devices, another possible method of preparation is to use an organic light emitting layer doped with organic salt (Sakuratani et al., 2001, Lee et al., 2002, Xu et al., 2003). By using fluorescent OLEDs with the doped organic layer and a simple Al cathode, enhanced EL emission was observed after treatment with a high electric field at room temperature (Sakuratani et al., 2001). This device is simple, and hence potentially inexpensive to make. However, there still remain problems regarding inhomogeneous emission, low reproducibility, and undesirable electric field treatment at high field strength over 20 V. Thus, the effect of doping with salt on performance in OLEDs has not yet been fully investigated (Yim et al., 2006). Hence, a new work was initiated to improve the device's high efficiency and brightness further by doping with organic salt and simultaneous treatments of electrical and thermal annealing. It will be shown that the improved device yields homogeneous emission and an increase in EL emission with high reproducibility.

The structure of the single-layered PHOLED used is shown in Figure 1. On a transparent substrate, a transparent indium-tin-oxide (ITO) layer was formed as an anode, over which a single electrophosphorescent EL organic layer was formed and upon which a metal cathode Al layer was then deposited. After the PHOLED was fabricated, it was treated thermally and electrically. It was heated to *T*°C by a hot plate and then an electric field of *V* was applied between the anode and the cathode of the device. When the device began to emit EL light, the field was terminated and then the device was cooled to room temperature. In this process, *T*°C was set to be below the glass transition temperature of the used organic materials and *V* to be below 15 V, to prevent deformation of the EL layer. Electric field treatment at elevated temperatures can induce the charge separation of organic salt towards the electrodes efficiently and homogeneously, even at relatively low voltages. That is to say, at the electrode surfaces, the adsorbed ions (positive ions for the n-doped region and negative ions for the p-doped region) can bend the heights of the lowest unoccupied molecular orbital (LUMO) level and the highest occupied molecular orbital (HOMO) level of the organic molecules near the electrodes. Thus, the electrons and holes can be injected easily into the organic layer through the reduced tunneling barriers from the electrodes. Note that this situation is quite similar to the ion implantation in inorganic semiconductors (Chason et al., 1997, Gerstner et al., 2001). Thus, one may also expect that the simultaneous treatments will be useful for further increasing EL emission from a homogeneous emitting area with high reproducibility.

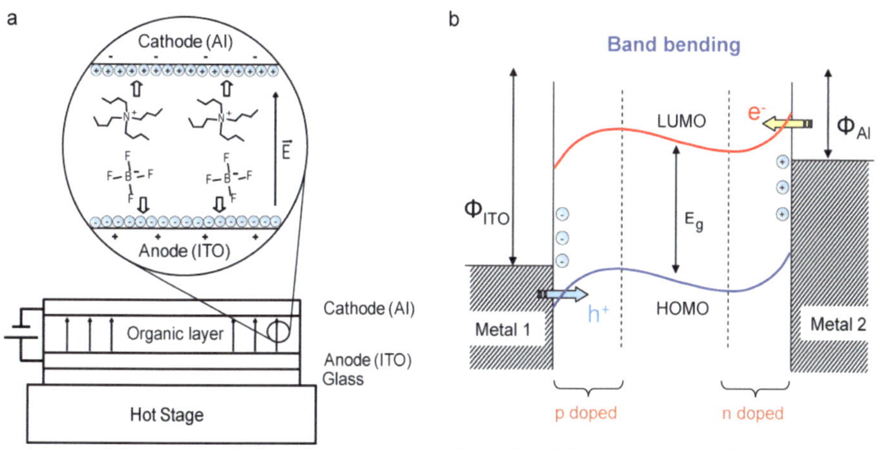

Fig. 1. (a) Architecture of the devices showing the fabrication of the device with the simultaneous electrical and thermal treatments. (b) Schematic energy levels of the device after the simultaneous treatments.

2. Fabrication of solution processable ionic *p-i-n* OLEDs

For the fabrication of devices, glass substrates (0.7 mm) coated with ITO (80 nm, 10-20 ohm/square sheet resistance) were used. After routine cleaning procedures for the substrate with wet (acetone and isopropyl alcohol) and dry (UV-ozone) processes, a blended solution of organic materials was spin coated (700 rpm) on top of ITO, precoated with a poly(3,4-ethylenedioxythiophene) : poly(4-styrenesulphonate) (PEDOT:PSS) hole-injecting buffer layer. The basic organic solution consisted of a hole transporting material of N,N'-diphenyl-N,N'-bis(3-methylphenyl)-1,1'biphenyl-4,4'-diamine (TPD: 0.08 wt%), an electron transporting material of 2-(4-biphenylyl)-5-(4-tert-butylphenyl)-1,3,4-oxadiazole (Bu-PBD: 0.32 wt%), a green emitting material of tris(2-phenylpyridinato) iridium ($Ir(ppy)_3$: 0.06 wt%), and a hole-transporting host material of PVK (0.34 wt%) into mixed solvents of 1,2-dichloroethane and chloroform (mixing weight ratio 3:1), which have different volatilities.

The organic salt of tetrabutylammonium tetrafluoroborate (Bu_4NBF_4) was further dissolved into the basic organic solution at an appropriate concentration. The thickness of the spin-coated organic layer was about 80 nm. Then, an Al cathode layer (100 nm) was formed on the top of the organic layers via thermal deposition at a rate of 0.7 nm/s under a base pressure of 2×10^{-6} Torr. In this experiment, phosphorescent OLEDs were fabricated and compared: one with Bu_4NBF_4 (0.0050 wt%) annealed electrically at V = +7 V (forward bias) at T = 65°C; the other for reference with Bu_4NBF_4 (0.0050 wt%) annealed electrically at V = +20 V (forward bias) at T = 25°C. It should be noted that, except for the emissive layer, the device structure of the reference device was identical to that of the sample device. The structures of the devices and materials used were identical. The devices were prepared in inert Ar gas environments; this preparation included electrical and thermal treatments.

3. Ionic *p-i-n* PHOLEDs made by spin-coating

3.1 Performance of ionic *p-i-n* PHOLEDs with a structure of [ITO / PEDOT:PSS / EL layer / Al]

First, the current flows were observed for sample and reference devices doped with Bu_4NBF_4 during the electric field treatment under given temperature. Figure 2 shows the plots of the current flows of the devices for the applied voltage profiles, which are shown in the inset of the figure. As shown in the figure, the current flow of the sample device was 2.1 mA/cm^2 at t = 0 s (before annealing) and increased to 12.4 mA/cm^2 at t = 33 s (after annealing), while that of the reference device was 63 mA/cm^2 at t = 0 s (before annealing) and increased abruptly to 730 mA/cm^2 at t = 4.5 s (after annealing). The figure shows clearly the smooth increase in current flow of the sample device with time during the simultaneous treatments. This smooth increase means that the organic salt ions that have been separated by the electric field move slowly towards the interfaces of the electrodes at elevated temperature and the accumulated ions near the interface reduce the interfacial barriers for carrier injection between the electrode and the organic layer, which may lead to improved electrical characteristics of the sample device. By comparing this smooth increase with the abrupt current increase of the reference device during the electrical treatment, one can see clearly that the increase in current flow of the sample device was smaller than that of the reference device. Thus, the deterioration of the reference device due to the abrupt increase in current flow through the organic layer under high voltage will be prevented by application of the simultaneous treatments. Thus one can easily fabricate sample devices with high reproducibility by applying the simultaneous treatments under low voltage. Note that during the treatment, one can see the current spikes at the rising edges of squared waveform for the sample device. These spikes may be due to the reduction of capacitance by the separated ions near the electrode surface.

Next, we observed the EL operation of the sample and reference devices with the naked eye. Figure 3 shows a photograph of the operating sample and reference devices. For clear comparison, we took a photograph of the device operation under different bias. The figure clearly demonstrates that all across the active areas, there is bright and homogeneous EL emission from the 3×3 mm^2 active area in the sample device, while relatively small and inhomogeneous emission was observed in the reference device even under higher bias. Similar to the current flows, it is also noted that the sample device exhibited much greater reproducibility than the reference device. We also observed the EL spectra for sample and reference devices (Fig. 3 (b)). As shown in the figure, the spectra are almost identical to those for the multilayered PHOLEDs reported previously (Baldo et al., 1999); there is an emission peak wavelength of 516 nm with full width at half maximum (FWHM) of ~70 nm. These results for the spectra indicate clearly that the doped organic salt does not disturb the EL spectrum, i.e., the energy levels of the organic material emitting layer. Note that the Commission Internationale de L'Eclairage (CIE) coordinates of (x=0.296, y=0.631) for the device and the EL spectrum are independent of current density.

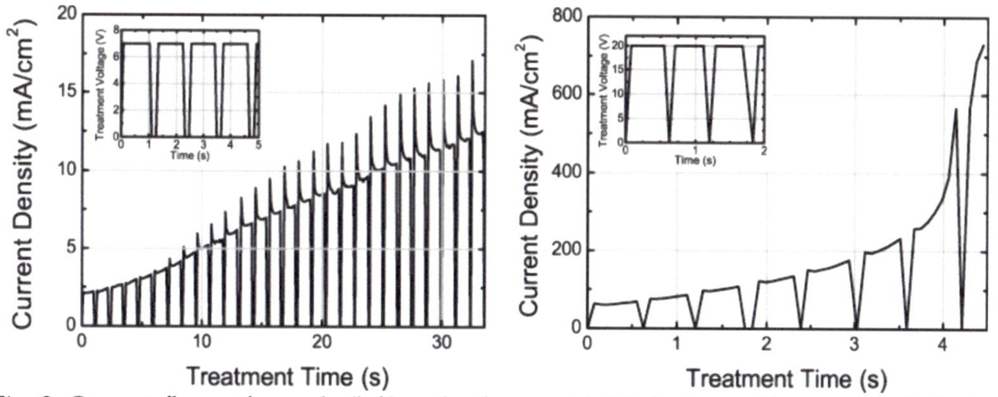

Fig. 2. Current flows of sample (left) and reference (right) devices with organic salt during electrical and thermal treatments for given temperatures. The inset shows the voltage profile applied to the devices. (Oh et al., 2007)

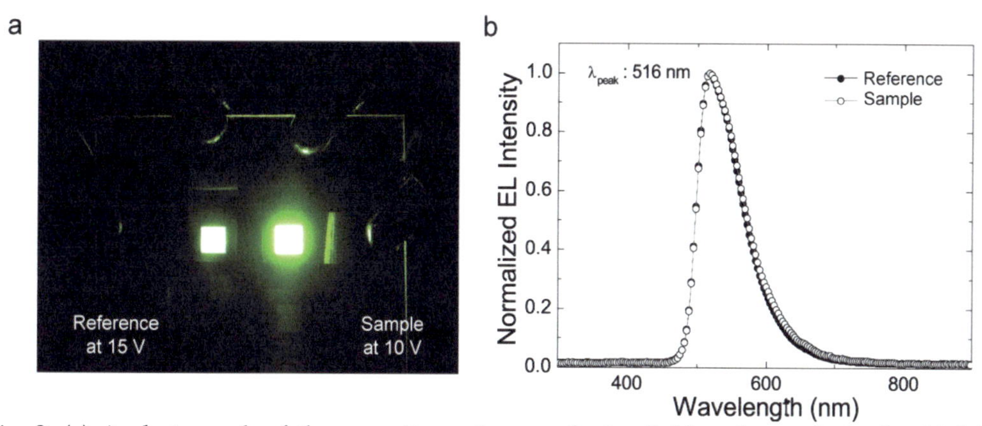

Fig. 3. (a) A photograph of the operating reference device (left) and sample device (right). The active area of each PHOLED is 3 × 3 mm². (b) The normalized EL spectra for the operating sample device (open circles) and reference device (closed circles). (Oh et al., 2007)

Next, in order to understand the effect of the organic salt, we observed the dependence of the EL characteristics on the doping concentration of the salt for the sample device, treated at V = +7 V (forward bias) and T = 65°C. Figure 4 (a) shows the current density-voltage (J-V) characteristics of the treated devices for various concentrations of Bu_4NBF_4. The rate of increase of the current density increases as the doping concentration increases. For the doped PHOLEDs, the current density behaves according to the power law equation of J =

$k \cdot V^{m+1}$ (Burrows & Forrest, 1994, Brutting et al., 2001) (where k is a proportional constant and m is 4 for doped devices, while m is 3 for an undoped device). This indicates that the space charge limit current (SCLC) due to the traps increases as the doping concentration increases. The luminance-voltage (*L-V*) characteristics of the devices are shown in Fig. 4 (b). As the doping concentration increases from zero, the EL luminescence also increases greatly. As shown in the figure, for the sample device with a doping concentration of 0.0050 wt%, the low turn-on voltage (2.5 V for 1 cd/m^2) and a steep increase in the *L-V* curve under low *J-V* characteristics suggest that both holes and electrons can easily be injected into the organic layer. The operating voltage for a typical display application is 4.2 V to obtain a brightness of 100 cd/m^2 (0.48 mA/cm^2) and 6 V for 1,000 cd/m^2 (3.47 mA/cm^2); the luminescence reached ~51,000 cd/m^2 (at 13.5 V), which is two orders of magnitude higher than that (~426 cd/m^2) of the device with 0 wt% concentration. Note that the EL properties of devices without any thermal/electrical annealing were exactly the same as those of the 0 wt% sample. Note also that the luminance of the device was comparable to that of the previously reported EL device (22,100 cd/m^2 at 10 V) fabricated with a CsF interfacial layer and Al cathode (Yang & Neher, 2004).

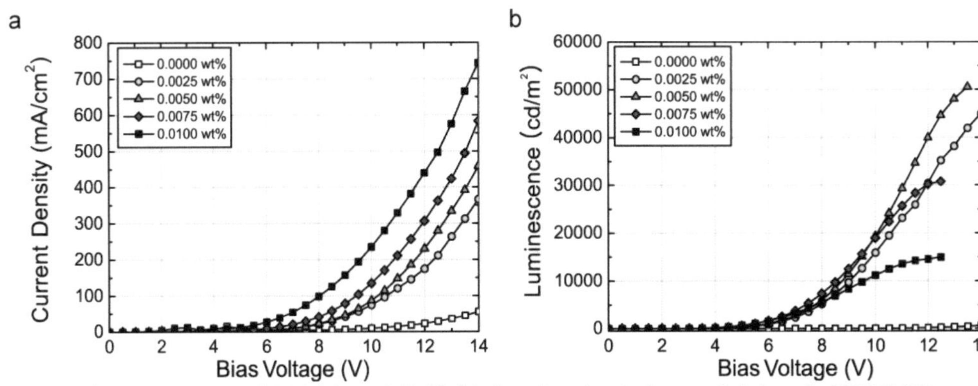

Fig. 4. Characteristics of *J-V* (a) and *L-V* (b) for the single-layered $Ir(ppy)_3$ PHOLEDs with various concentrations of Bu_4NBF_4. (Oh et al., 2007)

Here, it is noted that the enhanced EL emission caused by the adsorption of ions at the electrode surface for the salt-doped PHOLED treated by the thermal and electrical annealing. At the ITO anode contact, the accumulation of separated negative BF_4^- ions can assist hole injection from ITO to the organic layer and an Ohmic contact is achieved at the ITO interface (Sakuratani et al., 2001). At the same time, the accumulation of the positive Bu_4N^+ ions near the cathode aids the injection of electrons from the metal cathode into the organic layer by reducing the tunneling barrier of the cathode interface. Thus, proper adsorption of ions at the interfaces between the organic layer and the electrodes can enhance the injection of charge carriers into the organic layer, which results in the enhancement of current flow (SCLC) and EL luminance.

Next, the sample ionic *p-i-n* PHOLEDs was compared with typical (frozen) light-emitting cells (LECs) (Pei et al., 1995, Gao et al., 1997, de Mello et al., 1998). In LECs, the organic salt is also doped into the light-emitting polymer layer to improve the device performance. However, there are five main differences between typical LECs and the sample devices presented in this study. 1) In typical LECs, an ionic conducting material, such as poly-(ethylene oxide) (PEO), is required. However, in the sample devices presented in this study, no ionic conducting material is necessary, even though the active organic layer used in the sample device has very low ionic mobility. 2) The characteristics of the sample devices depend strongly on the thickness of active layer, in contrast to the LECs (Pei et al., 1995, Gao et al., 1997, deMello et al., 1998). 3) The forward- and reverse-biased *J-V* and *L-V* curves of the typical LECs were almost symmetric about zero bias, in contrast to the diode-like behaviors exhibited in the sample devices. 4) The polarity of the *p-i-n* junction of the sample device cannot be switched by reversing the polarity of the electric field at elevated temperature, in contrast to the polarity of the frozen junction LECs, which is dictated by the polarity of the prebias at high temperature (Gao et al., 1997). These results mean that it is only for forward simultaneous treatments (Al connected as the cathode) that electrons and holes are injected efficiently from the electrodes to recombine in the organic EL layer to generate photons, in contrast to the typical frozen LECs. 5) The dynamic response of the typical LECs was determined by the ionic mobility. Thus, the response time depended on whether the PEO ion-transport polymer was added (~1 s) or not (~60 s) to the polymer blend. However, the dynamic response (~10 μs) of the sample device is determined by mobility of the charge carrier (electron and hole). It will thus be evident that our device is quite different from typical LECs. The fast response time of the sample device also clearly indicates that the ions separated by the simultaneous treatments remain at the contacts in a stable fashion and the ions at the interfaces enhance the charge injection into the organic layer.

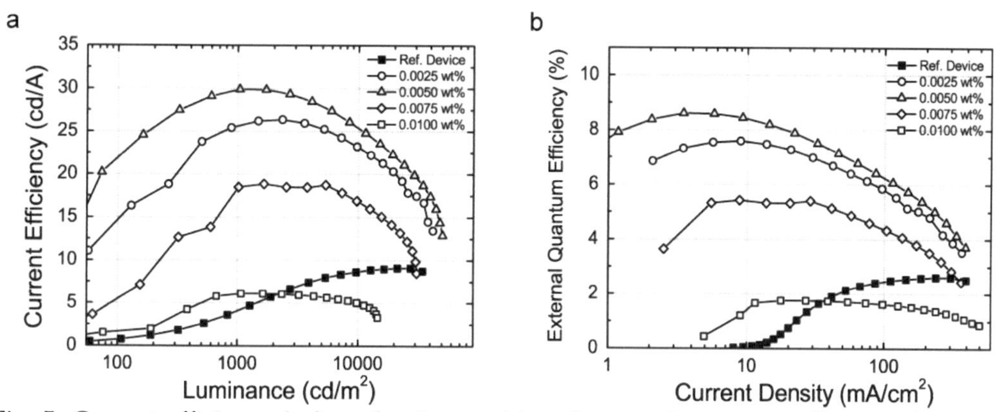

Fig. 5. Current efficiency (η_C) vs. luminance (a) and external quantum efficiency (η_{ext}) vs. current density (b) of the single-layered $Ir(ppy)_3$ sample devices (open symbols) with various concentrations of Bu_4NBF_4 and reference device (closed square). (Oh et al., 2007)

Next, in order to confirm the effect of the simultaneous annealing, we also investigated the efficiency characteristics of the sample devices. Fig. 5 (a) shows the current efficiency-luminance (η_C-*L*) of the PHOLEDs. As shown in the figure, PHOLEDs (0.0050 wt%) after the annealing treatments are more efficient than the reference device: for the sample device, η_C of 22 cd/A was obtained at 100 cd/m^2, reaching η_C = 30 cd/A at 1000 cd/m^2, while for the reference device, η_C of 0.7 cd/A at 100 cd/m^2, and η_C = 4 cd/A at 1,000 cd/m^2. We also observed the external quantum efficiency η_{ext} of the sample devices. Here, η_{ext} was determined from the conventional luminance-current characteristics of the EL spectrum

(Okamoto et al., 2001). As shown in Fig. 5 (b), η_{ext} of the sample device (0.0050 wt%) is much higher than that of the reference device: for the sample device, η_{ext} increases, reaches a maximum of 8.6 %, and then slowly decreases with increasing current density, while for the reference device, η_{ext} reaches only a maximum of 2.6 %. For another comparison, η_{ext} s of 8 ~ 12 % of the hetero-structured PHOLEDs, reported in References (Baldo et al., 1999, Adachi et al., 2002), are comparable to that of the sample device. These results indicate clearly that the charge balance in the charge injection was improved significantly by controlled adsorption of ions at the interfaces.

Therefore, by applying electric and thermal treatments simultaneously, homogeneous and enhanced EL emission was obtained from the active area of the devices with high reproducibility. Moreover, the efficiency of the devices was also observed to improve. As a result, an ionic *p-i-n* PHOLED with a peak external quantum efficiency of 8.6 % was achieved in the sample device. On the basis of these results, it is demonstrated that simultaneous annealing can lead to more efficient electroluminescence through increased and balanced carrier injection. This improvement can be attributed to the excellent balancing of holes and electrons.

3.2 Performance of PHOLEDs with a structure of [ITO / EL layer / CsF / Al]

Although efficient solution-processed PHOLEDs have been demonstrated, the process of their fabrication is still complex because a hole injecting buffer layer of PEDOT:PSS has been introduced between the emissive layer and the transparent ITO anode. Given the state of research, a new work was initiated to achieve high efficiency and brightness from the simplest PHOLED structure that it is possible to achieve. In order to realize strong light emission from a real single-organic-layered PHOLED, a modified PHOLED was proposed by including an ionic salt-doped emissive layer, treated by appropriate simultaneous electrical annealing at elevated temperature.

The proposed device structure is very simple; on a glass substrate, an ITO layer was formed as an anode, over which an electro-phosphorescent EL layer doped with organic salt was formed by a solution-process and upon which a metal cathode was deposited. After fabrication, the PHOLED was treated by simultaneous annealing by applying an electric field of V at elevated temperature of T. Then, the ions that separate and accumulate at the electrode surfaces as a result of the treatments induce the electric fields, which can bend the electronic energy levels of organic molecules, enhancing the charge injection into the organic layer from the electrodes across the whole area of the device.

For the experiments, an organic solution consisted of a hole injecting material of 4,4',4"-Tris(N-3-methylphenyl-N-phenyl-amino)-triphenylamine (m-MTDATA), a hole transporting material of TPD, an electron transporting material of Bu-PBD, a green emitting material of $Ir(ppy)_3$, and a hole-transporting host material of PVK at an appropriate concentration into mixed solvent of 1,2-dichloroethane and chloroform. The organic salt, Bu_4NBF_4, was also dissolved into the organic solution. The spin-coated organic layer was about 80 nm thick. Then, a cathode layer of CsF (1 nm) / Al (100 nm) was formed on the top of the organic layers via thermal deposition at a rate of 0.7 nm/s under a base pressure of 2 × 10^{-6} Torr. In this experiment, the following PHOLEDs were fabricated and compared: annealed sample devices with Bu_4NBF_4 (0.0050 wt%) at V = +9 V (forward bias) at T = 65°C, and, as references, other annealed devices that were not doped.

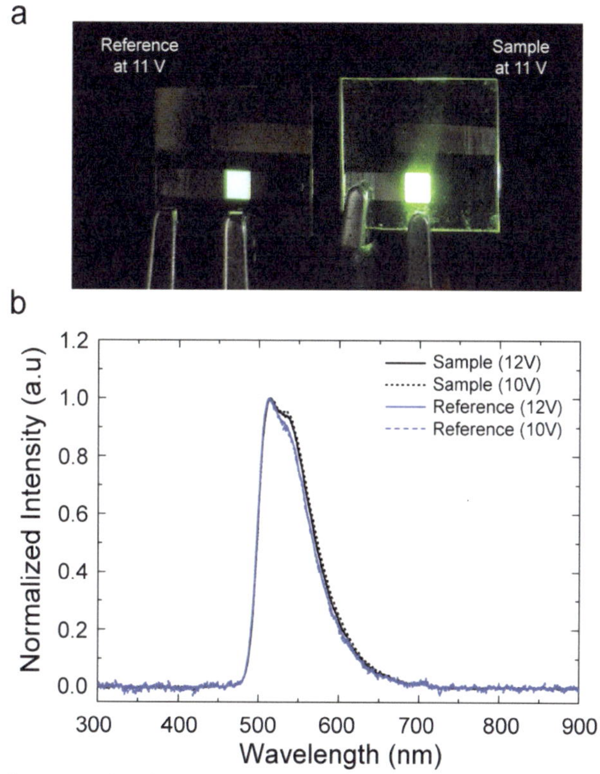

Fig. 6. Upper: A photograph of the operating sample device (right) and reference device (left). Lower: Normalized EL spectra for the operating sample device and reference device. (Park et al., 2008)

The EL operation was observed for the fabricated devices with the naked eye. Figure 6 (a) shows a photograph of the operating devices. For clear comparison, we took a photograph of the device operation under the same bias of 11 V. The figure clearly demonstrates that there is bright and homogeneous EL emission from the active area in the sample device, while there is relatively weak emission from the active area in the reference device. Next, we observed the EL spectra of the devices. As shown in Fig. 6(b), the emission spectra of the sample devices are almost identical to those of the reference devices; there is a spectral peak wavelength of 516 nm with FWHM of ~70 nm. These results for the EL spectra indicate that the doped organic salt does not interfere with the energy band gap of the organic materials. It is likely that the slight increase in the shoulder luminescence near 530 nm for the sample device arises from the more phosphorescent sites at the excited energy level (Xie et al., 2001). It is noted that the CIE coordinate of (x=0.32, y=0.62) for the sample devices is independent of current density. Next, we observed the dynamic responses of the operating devices. As shown in Fig. 7, the observed dynamic responses of the sample device are nearly identical to those of the reference device; the rising time and falling time are about 10 μs and 9 μs, respectively. These fast responses of the sample device indicate that the separated ions remain at the contacts in a stable fashion and the ions at the interfaces can increase the charge injection into the organic layer. In other words, at the ITO anode contact, the accumulated negative BF_4^- ions can bend the energy level and reduce the tunneling barrier of the anode interface, while the positive Bu_4N^+ ions at the cathode aids the injection of electrons into the organic layer by reducing the tunneling barrier of the cathode interface.

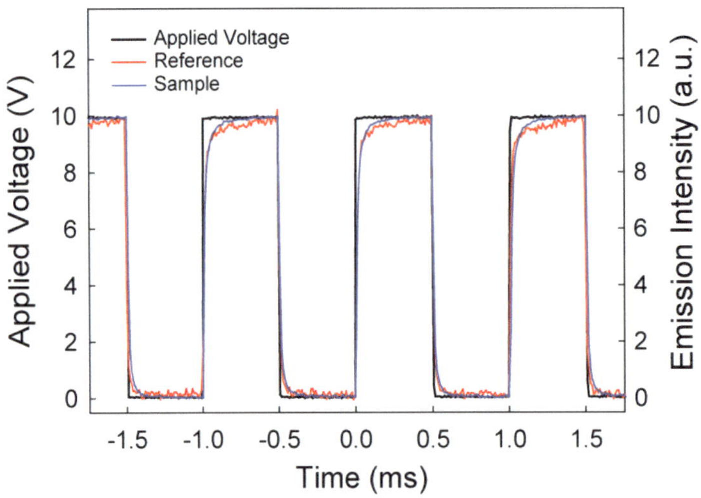

Fig. 7. The dynamic responses for the operating sample device and reference device. (Park et al., 2008)

Next, in order to determine the optimal conditions of operation of the sample device, the dependence of the device characteristics on the concentration of the doping salt has been observed. Figure 8(a) shows The *J-V* characteristics of the sample devices for several concentrations of the salt. As shown in the figure, the rate of increase of the current density increases as the doping concentration increases. The *L-V* characteristics of the devices are shown in Fig. 8(b). As the doping concentration increases from zero, the EL luminescence also increases. The low turn-on voltage (2.5 V for 1 cd/m^2) with a sharp increase in the *L-V* curve even under low current density suggests that both holes and electrons can easily be injected into the annealed organic layer for a doping concentration of 0.0075 wt%, giving ionic concentration of about 10^{19} ~ 10^{20} cm^{-3}. The operating voltage is about 4 V to obtain a brightness of 100 cd/m^2 and about 7 V for 1,000 cd/m^2; the luminescence reached ~35,400 cd/m^2 (at 17.5 V), which is almost three times higher than that (~13,000 cd/m^2) of the reference device. Note that the luminance of the sample device was also comparable to that of a previously reported PHOLED device (Yang & Neher, 2004) that was fabricated with a hole-injecting PEDOT:PSS layer. Thus, it is clear that the proper adsorption of ions at the electrode surface can result in the formation of the ionic *p-i-n* structure and enhance the injection of charge carriers into the organic layer, which results in the enhancement of current flow and EL luminance. To confirm the high performance of the devices, the efficiency was deduced for the studied devices. For the sample device (0.0025 wt%), η_C of 37 cd/A was obtained at 100 cd/m^2, reaching η_C = 34 cd/A at 1,000 cd/m^2, while for the reference device, η_C of 6.5 cd/A at 100 cd/m^2, and η_C = 10.4 cd/A at 1,000 cd/m^2. η_P was also obtained for the sample device; η_P reaches a maximum of 42 lm/W, while for the reference device, η_P reaches only a maximum of 2.8 lm/W. For another comparison, η_Ps of ~30 lm/W of the hetero-structured PHOLEDs, reported in References (Baldo et al., 1999,

Adachi et al., 2002), are comparable to that of the sample device. These results indicate clearly that the balance in the charge injection due to the ionic *p-i-n* structure was improved significantly by controlled adsorption of ions at the interfaces.

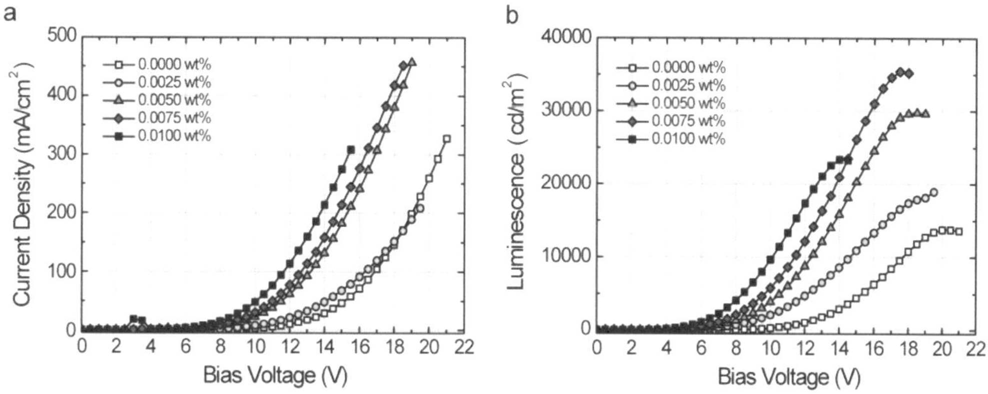

Fig. 8. Characteristics of *J-V* (a) and *L-V* (b) for the single-layered ionic *p-i-n* PHOLEDs with various concentrations of doped salt of Bu_4NBF_4. (Park et al., 2008)

Therefore, herein, a solution-processed single-layered ionic *p-i-n* $Ir(ppy)_3$ PHOLED has been demonstrated. By applying the simultaneous thermal and electrical annealing, homogeneous and enhanced EL emission with increased brightness and high efficiency can be obtained from the devices in a simple fashion. As a result, an ionic *p-i-n* PHOLED with peak power efficiency over 40 lm/W was achieved by means of increased and balanced carrier injections. Given that this device can be fabricated by a simple wet process, combining simultaneous annealing with the use of highly luminous organic materials will surely lead to highly efficient ionic *p-i-n* OLED devices.

4. Ionic *p-i-n* PHOLEDs made by Horizontal-dipping

4.1 Horizontal-dipping method for large area PHOLEDs

For solution-processed devices, spin-coating has until now been the most popular method of forming organic layers. This method is convenient, but has several disadvantages, such as the high stress caused by the spinning motion, the poor uniformity at the edges of large areas, and the large amounts of wasted solution (Pardo et al., 2000, Jabbour et al., 2001, de Gans et al., 2004, So et al., 2007, Tseng et al., 2008). These factors make spin-coating unsuitable for application to large active areas. An alternative method for depositing the solution is to use such techniques as screen printing (Pardo et al., 2000, Jabbour et al., 2001), ink-jet printing (de Gans et al., 2004), or blade coating (Tseng et al., 2008). By using these techniques, organic or polymeric layers may be formed on substrates in a controlled fashion. However, despite the recent developments in such solution-processed devices, an alternative solution-coating process is nevertheless required, because of the continued difficulty of controlling the uniformity of the organic semiconducting layers resulting from the conventional coating methods that have so far been proposed. Hence, further research on solution deposition techniques is required in order to achieve simpler and more reliable fabrication of efficient OLEDs.

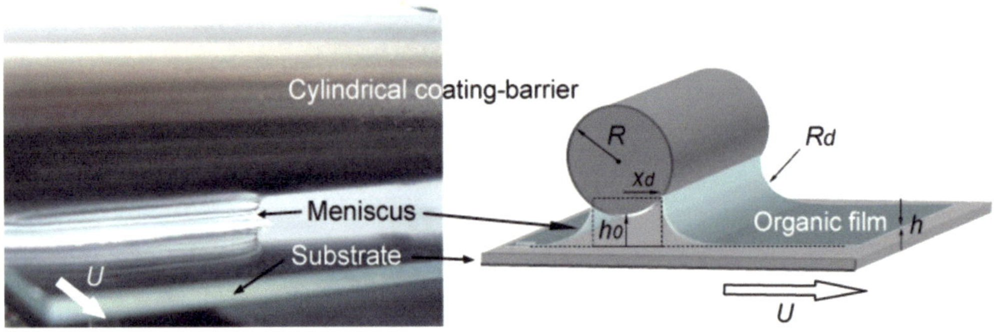

Fig. 9. A photograph (left) with schematic illustration (right) of the premetered horizontal-dip (H-dip) coating process described herein: a cylindrical coating barrier (SUS steel) with a diameter *R*, a gap height h_0, and a carrying speed *U*. (Park & Han, 2009)

The advantage of using premetered coating is that the coating thickness is predetermined, in contrast to more typical metered methods such as fixed-gap blade, knife, or wire-bar coatings. Recently, there was a report describing a solution-processed, highly efficient polymer solar cell that was fabricated using a premetered solution-process (Park & Han, 2009). In the study described herein, this premetered process was used to demonstrate the fabrication of efficient, high-performance solution-processed ionic *p-i-n* PHOLEDs.

4.2 Operating principle of H-dipping

It is well-known that coating flows can be divided into two categories, metered or premetered, according to whether the thickness of the coated film is determined by the process or imposed externally. The thickness of the film coated by the metered process is independent of the capillary number, while that produced by premetered process usually increases with increasing capillary number ($C_a=(\mu U/\sigma)$), where μ and σ represent the viscosity and surface tension of the coating solution, respectively, and *U* is the carrying (coating) speed. Examples of premetered coating flow include meniscus and dip coatings. A photograph and schematic illustration of the premetered solution coating process under investigation are shown in Figure 9. This figure shows a cylindrical coating barrier hanging at a specific height (h_0) above a rigid substrate laid on a carrying stage that transports the substrate in a horizontal direction. The coating process occurs in the following sequence. (1) The substrate is attached to the carrying stage, and the coating barrier is placed at the front edge of the substrate. A blended organic semiconducting molecular solution is then introduced into the empty space between the barrier and the substrate by capillary action, so that a uniform meniscus of the solution may be formed on the substrate by attraction to the barrier (i.e. by surface tension). (2) The substrate is then transported horizontally at constant velocity whilst maintaining the shape of the downstream meniscus. A thin solution layer of the downstream meniscus is then spread evenly on the substrate. While the substrate is being transported, the blended organic solution may be supplied into the gap space at an appropriate injection rate. (3) Having been spread on the substrate, the wet film is dried, and a heater may be used to assist the evaporation of the residual solvent in the wet film on the substrate. Following this process, it is possible to obtain a substrate coated with a solid organic film of uniform thickness. The transport of the substrate through the meniscus of the solution is similar to that which occurs in the typical dip-coating method (Landau & Levich, 1942, Krozel et al., 2000). In that method, the substrate is immersed in the coating solution and a wet layer is then formed by withdrawing the substrate vertically through the meniscus of the coating solution. Our proposed coating method differs from the conventional dip-coating method, however, because the wet film is formed by withdrawing the substrate horizontally. Therefore, we call the proposed process horizontal dipping (H-dipping) (Park & Han, 2009). It is noted that the apparatus of H-dipping is similar to that of the zone casting (Yabu & Shimomura 2005, Miskiewicz, 2006, Duffy, 2008), that has been used in fabrication of organic transistors, but the coating speed of the zone casting is extremely low (several μm/s) compared with the premetered H-dipping (several cm/s).

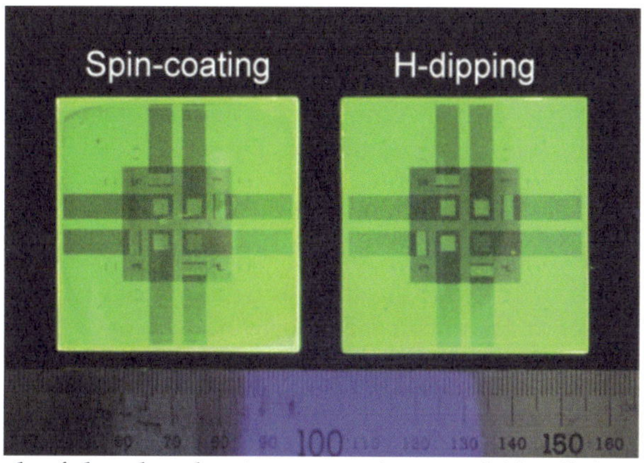

Fig. 10. A photograph of the photoluminescent spin-coated film (left, at 1000 rpm) and the H-dip-coated film (right) on pre-patterned 2'' substrates. (Park & Han, 2009)

Photoluminescence images obtained from spin- and H-dip-coated organic films on 2-inch glass substrates (pre-patterned ITO) using a UV light source of 365 nm is shown in Figure 10. It may be seen from the figure that the luminescent intensity and thickness of the spin-coated films vary near the edges of the substrates due to the Bernoulli effect (Luurtsema, 1997). In order to produce a fully smooth spin-coated film, the speed of rotation must be adjusted and the amount of spreading solution increased. On the other hand, it may be seen that the H-dip-coated film is very smooth and uniform. Variation in the thickness of the film was observed only at the very rear edge of the substrate. The surface morphology investigation using AFM showed clearly that the topography was fairly uniform, the root mean square roughness for the H-dip-coated film being only ~0.9 nm, which was comparable to that (~1.0 nm) of the spin-coated films. Moreover, the surface roughness of the H-dip-coated films was identical at different positions. This uniformity was achieved because no external centrifugal force was applied during the formation of the film. Thus, compared to spin-coating, it is possible to achieve uniformity in the film thickness of the EL layer in a reliable way when using H-dipping, even on a large-area substrate. This is a result of the control of the undesirable free-surface flow that occurs at the top organic solution-air interface, via the surface tension effect between the solution and the coating barrier.

The film thickness that results from the H-dipping process may be explained by the description of the associated drag-out problem suggested by Landau and Levich (Landau & Levich, 1942). Based on their description, for a small capillary number ($C_a << 1$), a useful relationship may be obtained that relates the thickness of the film emerging from a coating bead to the radius of the associated meniscus and carrying speed, U (Landau & Levich, 1942, Park & Han, 2009):

$$h = 1.34\left(\frac{\mu U}{\sigma}\right)^{2/3} \cdot R_d, \quad n \cdot R_d = \left(\frac{x_d^2}{2R} + 2h_0\right) - h, \tag{1}$$

where R_d represents the radius of curvature of the downstream meniscus. Here, R and h_0 represent the radius of the cylindrical coating barrier and the minimum gap height, respectively, and n is 1 for a contact angle of 90° or 2 for a contact angle of 0° measured on the contact line at the interface between the solution and the coating barrier. In our study, n was assumed to be 2, as shown in the photograph (Fig. 9).

It is worthy of note that the thickness of the H-dip-coated film is much less than the gap height. This is characteristic of the main way in which the premetered H-dipping process differs from the conventional metered doctor-blade (or wire-bar) coating (Kuo et al., 2004). In the conventional approach, the doctor-blade (or wire-bar) coating process produces a film thickness of the order of the gap size whose thickness is independent of the carrying speed of the substrate. In the H-dipping method, the premetered process allows the critical control of the thickness and can produce superior quality and extremely thin films at line speeds of the order of a few meters per minute.

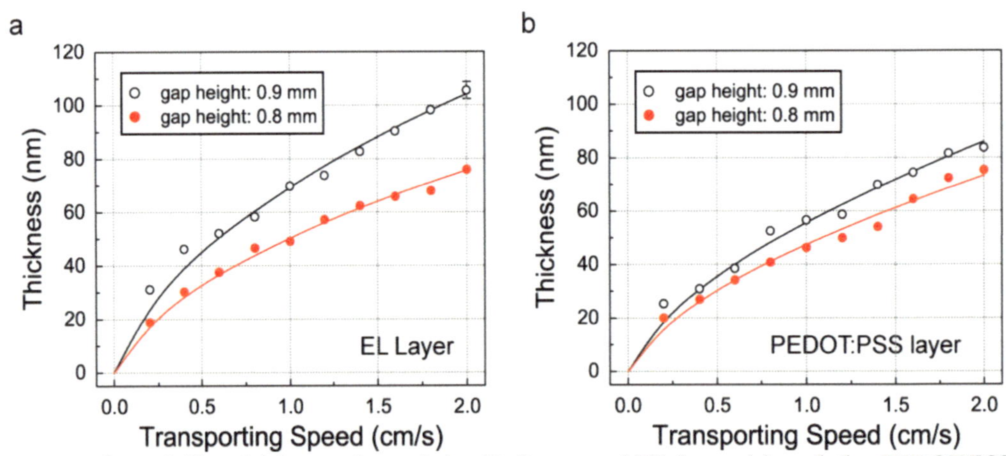

Fig. 11. Coated film thickness data of the H-dip-coated EL layer (a) and the PEDOT:PSS layer (b) as a function of carrying speed for two gap heights (0.9 and 0.8 mm). The solid curves show the theoretical predictions of the Landau & Levich equation. (Park & Han, 2009)

4.3 Fabrication of *p-i-n* PHOLEDs made by H-dipping

For the fabrication of devices, the PEDOT:PSS and the organic EL layers were successively deposited by H-dipping on an ITO-coated glass substrate. The PEDOT:PSS solution used was a mixture of 1 % PEDOT:PSS solution (CLEVIOS™ P VP AI 4083, H.C. Starck) and isopropyl alcohol with a weight ratio of 2:1. The viscosity of the mixed PEDOT:PSS solution, measured by viscometer (RVDV II +, Brookfield Inc.), was about 11.6 cp. For the blended EL solution, we used TPD, Bu-PBD, $Ir(ppy)_3$, and PVK without further purification, in mixed solvents of 1,2-dichloroethane and chloroform (3:1). The organic salt, Bu_4NBF_4, was also dissolved into the EL solution. The viscosity of the EL solution was about ~1.0 cp at a temperature of 25°C. The apparatus used for H-dipping had a maximum work space of 15 × 15 cm^2. A small volume of the solution (~ 6 μl) per unit coating area (1 × 1 cm^2) was fed into the gap between the cylindrical barrier (SUS steel, R = 6.35 mm) and the glass substrate using a syringe pump (Pump Systems Inc. NE-1000). The height of the gap, h_0 was adjusted vertically using two micrometer positioners, and the carrying speed U was controlled using a computer-controlled translation stage (SGSP26-200, Sigma Koki Co., Ltd). After a meniscus had formed on the solution, the substrate was transported horizontally, so that the barrier spread the solution on the transporting substrate. The transporting speed U was 1.5 cm/s. It took 2 seconds to prepare a complete film on a substrate with an area of 1.8 × 2.0 cm^2. The H-dip-coated PEDOT:PSS layer and electrophosphorescent EL layer doped with Bu_4NBF_4 were then dried using a heating plate at 110°C for 60 minutes and at 60°C for 5 minutes, respectively, in order to remove the remaining solvents. 1 nm CsF and 60 nm of Al were evaporated sequentially on the EL layer via thermal deposition (0.5 nm/s) at a base pressure below 2×10^{-6} Torr. The PHOLED fabricated thus had a device configuration of ITO/ PEDOT:PSS/ EL layer/ CsF/ Al. In the experiment, the sample PHOLEDs with Bu_4NBF_4 (0.0050 wt%) were annealed at V = +8 V (forward bias) at T = 75°C.

4.4 Performance of *p-i-n* PHOLEDs made by H-dipping

By using the AFM, we investigated the dependence of the film thickness, h of the H-dip-coated organic/polymer layer on the transporting speed U and the gap height h_0. The results obtained are shown in Figure 11. As shown in the figure, for a gap height, h_0 of 0.8 mm, the thickness of the H-dip-coated layer increases continuously as the speed U increases in the observed region (filled circles). Furthermore, when h_0 was increased from 0.8 mm to 0.9 mm, the thickness of the H-dip-coated layer also increased with increasing speed U. These results may be explained by the description of the associated drag-out problem, using Equation (1). The theoretical curves resulting from Equation (1) are shown in the figure as solid lines. The observed data fitted the theoretical values predicted by Equation (1) rather well, indicating that the thickness of the H-dip-coated organic film may be controlled by adjusting the gap height h_0 and the carrying speed U. These results indicate that the H-dipping process can be used to produce an organic layer at least as well as spin-coating can. It is further evident that the thickness of the H-dip-coated layer follows nearly the same trends as those shown in previous results using the H-dip-coated photovoltaic layer (Park & Han, 2009).

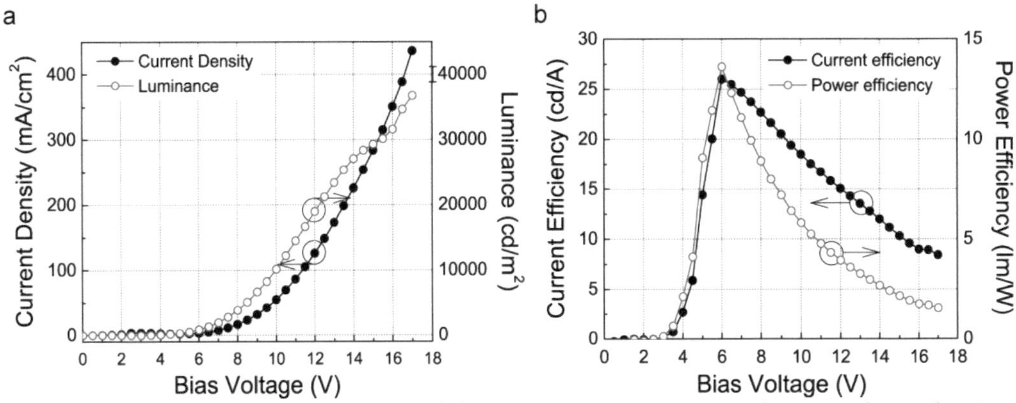

Fig. 12. (a) *J-V* and *L-V* characteristics of the ionic *p-i-n* PHOLED made using the H-dipping process. (b) η_C-*V* and η_P-*V* characteristics of the studied PHOLED.

We then investigated the EL characteristics of the ionic *p-i-n* PHOLEDs produced by the H-dipping process. In the device, the thicknesses of the PEDOT:PSS and the EL layers were adjusted to about 40 nm and 80 nm, respectively. Figure 12(a) shows the observed *J-L-V* characteristics of the fabricated ionic *p-i-n* PHOLED after the simultaneous treatments at T = 75°C and V = +8.0 V. The slope of the *J-V* curve between 0 and 18 V shows the excellent diode behavior of the fabricated OLED and thus indicates good coverage of the H-dip-coated PEDOT:PSS buffer layer and the EL layer. It is clear from the *J-L-V* curves that both the charge injection and turn-on voltages are below 2.7 V, with sharp increases in the *J-L-V* curves occurring at higher applied voltages. An operating voltage of about 4.3 V yields a brightness of 100 cd/m^2, 6.3 V yields 1,000 cd/m^2, and 9.8 V yields 10,000 cd/m^2. The luminescence reached ca. 36,700 cd/m^2 (at 17.0 V), which is comparable to that of a previously reported PHOLED device (Yang & Neher, 2004) made by spin-coating. Thus, it is clear that the proper adsorption of ions at the electrode surface can result in the formation of the ionic *p-i-n* structure and enhance the injection of charge carriers into the H-dip-coated organic layer, which results in the enhancement of current flow and EL luminance. In order to confirm the high performance of the sample devices, we also calculated the efficiency of the devices studied, as shown in Figure 12(b). For the H-dip-coated ionic *p-i-n* PHOLED, η_C of 3.0 cd/A was obtained at 100 cd/m^2, reaching η_C = 26.0 cd/A at 800 cd/m^2. We also calculated η_P of the H-dip-coated device, which reached a maximum of 13.6 lm/W. These results clearly indicated that the EL layer manufactured by H-dip-coating possesses bright and efficient EL characteristics due to the formation of a uniform layer with the appropriate ionic *p-i-n* structure.

Next, in order to check the processing ability of large-area ionic *p-i-n* PHOLEDs, we also fabricated a 10 × 10 cm^2 ionic *p-i-n* PHOLED device using the H-dipping process on an ITO-coated glass substrate. A photographic image of the fabricated device is shown in Figure 13. A PEDOT:PSS layer and an EL layer were deposited on a strip-patterned 10 × 10 cm^2 ITO-coated glass substrate by H-dipping, in order to fabricate a passive-matrix display device. The pixel array was 10 × 10 and the pixel size was 9 × 9 mm^2. It may be seen from the figure that the fabricated ionic *p-i-n* PHOLEDs were fairly luminous. The EL spectra were collected from each of the 100 individual pixels on the substrate, and were almost identical for each pixel, the emission peak wavelength being ~510 nm with a FWHM of about 70 nm. The variation of the emitting intensity at different pixels was quite low. This result implies that the variation in the thickness of the organic thin film was small, because the EL intensity from a PHOLED is sensitive to the layer thickness. The low variation of EL intensity is quite acceptable for large-scale fabrication. These results confirm that the H-dipping method shows considerable promise for use in simple fabrication techniques that may easily be scaled up to a larger size at a lower cost than other processes. It should be noted that we were not able to form a homogeneous and uniformly thin EL layer by spin-coating for EL solutions on a 10 × 10 cm^2 substrate. From the results reported above, it is clear that the H-dipping process for solution coating shows considerable promise for the fabrication of bright and large-area ionic *p-i-n* PHOLEDs. It is worth noting that the performance of ionic *p-i-n* PHOLEDs may be further enhanced by, for example, the selection of more suitable materials, solvents, solution concentrations and viscosities, and by optimizing the gap height between the barrier and the substrate.

Fig. 13. A photograph of the operating 10 × 10 pixels of *p-i-n* PHOLEDs made by the H-dipping method at 15 V on a glass substrate (10 × 10 cm^2).

A simple premetered H-dipping process has been investigated as a promising organic thin-film coating process for the manufacture of cost-efficient and large-area ionic *p-i-n* PHOLEDs. Organic semiconducting thin films were fabricated successfully on a 10 × 10 cm^2 substrate with a high uniformity using H-dipping in a solution whose meniscus was controlled by adjusting the gap height and coating speed. It was also shown that bright and efficient ionic *p-i-n* PHOLEDs were produced. Experimental results indicate that the H-dipping method also shows great potential for applications involving large-area ionic *p-i-n* PHOLEDs. This novel process for depositing the solution on the substrate can be expanded to slot-die and slit-die coatings, and will provide a solid foundation for extending the fabrication of large-area solution processed PHOLEDs.

5. Summary

This chapter presented the fabrication and operation of the solution processed ionic *p-i-n* PHOLEDs. By applying the simultaneous electric and thermal treatments, homogeneous and enhanced EL emission with increased efficiency can be obtained from the devices in a simple fashion. Combining the simultaneous annealing process presented here with luminous organic materials will surely lead to the development of highly luminous large-area ionic *p-i-n* PHOLEDs, which will render the use of such devices possible for many applications, such as lighting, displays, and/or optoelectronic devices.

Acknowledgments

This research was supported by Basic Science Research Program through the National Research Foundation of Korea (NRF) funded by the Ministry of Education, Science and Technology (2010-0005557 and 2010-0016549). BP thanks Ms. M. Han and Mr. H. G. Jeon for their useful discussion at the early stage.

6. References

Adachi. C.; Thompson. M. E. & Forrest. S. R. (2002). Architectures for Efficient Electrophosphorescent Organic Light-Emitting Devices. *IEEE Journal on Selected Topics in Quantum Electronics*, vol. 8, no. 2, 372-7.

Baldo. M. A.; O'Brien. D. F.; You. Y.; Shoustikov. A.; Sibley. S.; Thompson. M. E. & Forrest. S. R. (1998). Highly efficient phosphorescent emission from organic electroluminescent devices. *Nature(London)*, vol. 395, no. 6698, 151-4.

Baldo. M. A.; Lamansky. S.; Burrows. P. E.; Thompson. M. E. & Forrest. S. R. (1999). Very high-efficiency green organic light-emitting devices based on electro-phosphorescence. *Applied Physics Letters*, vol. 75, no. 1, 4-6.

Brütting. W.; Berleb S. & Mückl. A. G. (2001). Device physics of organic light-emitting diodes based on molecular materials. *Organic Electronics*, vol. 2, no. 1, 1-36.

Burroughes. J. H.; Bradley. D. D. C.; Brown. A. R.; Marks. R. N.; Mackay. K.; Friend. R. H.; Burns. P. L. & Holmes. A.B. (1990). Light-Emitting diodes based in conjugated polymer. *Nature*, vol. 347, no. 6301, 539-541.

Burrows. P. E. & Forrest. S. R. (1994). Electroluminescence from trap-limited current transport in vacuum deposited organic light emitting devices. *Applied Physics Letters*, vol. 64, no. 17, 2285-7.

Chason. E.; Picraux. S. T.; Poate. J. M. & Borland. O. (1997). Ion beams in silicon processing and characterization. *Journal of Applied Physics*, vol. 81, no. 10. 6513-6562.

de Gans. B.-J.; Duineveld. P. C. & Schubert. U. S. (2004). Inkjet Printing of Polymers: State of the Art and Future Developments. *Advanced Materials*, vol. 16, no. 3, 203-213.

de Mello. J. C.; Tessler. N.; Graham. S. C. & Friend. H. (1998). Ionic space-charge effects in polymer light-emitting diodes. *Physical Review B*, vol. 57, no. 20, 12951-12963.

Duffy. C. M.; Andreasen. J. W.; Breiby. D. W.; Nielsen. M. M.; Ando. M.; Minakata. T. & Sirringhaus. H. (2008). High-Mobility Aligned Pentacene Films Grown by Zone-Casting. *Chemistry of Materials*, vol. 20, no. 23, 7252–9.

Friend. R. H.; Gymer. R. W.; Holmes. A. B.; Brroughes. J. H.; Marks. R. N.; Taliani. C.; Bradly. D. D. C.; Dos Santos. D. A.; Bredas. J. L.; Logdlund. M. & Salaneck. W. R. (1999). Electroluminescence in conjugated
polymers. *Nature,* vol. 397, no. 6715, 121-8.

Gao. J.; Yu. G. & Heeger. A. J. (1997). Polymer light-emitting electrochemical cells with frozen *p-i-n* junction. *Applied Physics Letters,* vol. 71, no. 10, 1293-5.

Gerstner. E. G.; Cheong. T. W. D. & Shannon. J. M. (2001). Formation of bulk unipolar diodes in hydrogenated amorphous silicon by ion implantation. *IEEE Electron Device Letters,* vol. 22, no. 11, 536-8.

He. G.; Pfeiffer. M.; Leo. K.; Hofmann. M.; Birnstock. J.; Pudzich. R. & Salbeck. J. (2004). High-efficiency and low-voltage *p-i-n* electrophosphorescent organic light-emitting diodes with double-emission layers. *Applied Physics Letters,* vol. 85, no. 17, 3911-3.

Jabbour. G. E.; Radspinner. R. & Peyghambarian. N. (2001). Screen Printing for the Fabrication of Organic Light-Emitting Devices. *IEEE Journal on Selected Topics in Quantum Electronics,* vol. 7, no. 5, 769-773.

Krozel. J. W.; Palazoglu. A. N. & Powell. R. L. (2000). Experimental observation of dip-coating phenomena and the prospect of using motion control to minimize fluid retention. *Chemical Engineering Science,* vol. 55, no. 18, 3639-3650.

Kuo. C.-C.; Payne. M. M.; Anthony. J. E. & Jackson. T. N. (2004). TES Anthradithiophene Solution-Processed OTFTs with 1 cm^2/V-s Mobility, *2004 International Electron Device Meeting Technical Digest,* 373-6.

Landau. L. D. & Levich. V. G. (1942). Dragging of a liquid by a moving plate. *Acta Physicochimica URSS,* vol. 17, 42-54.

Lee. T. W.; Lee. H. C. & Park. O. O. (2002). High-efficiency polymer light-emitting devices using organic salts: A multilayer structure to improve light-emitting electrochemical cells. *Applied Physics Letters,* vol. 81, no. 2, 214-7.

Liu. H.-M.; He. J.; Wang. P.-F.; Xie. H.-Z.; Zhang. X.-H.; Lee.C.-S. & Xia.Y.-J. (2005). High-efficiency polymer electrophosphorescent diodes based on an Ir (III) complex. *Applied Physics Letters,* vol. 87, no. 22, 221103-5.

Luurtsema. G. A. (1997). Spin coating for rectangular substrates. U.Califonia, Berkely, [Online]. Available: http://bcam.berkeley.edu/ARCHIVE/theses/gluurtsMS.pdf.

Miskiewicz. P.; Mas-Torrent. M.; Jung J.; Kotarba. S.; Glowacki. I.; Gomar-Nadal. E.; Amabilino. D. B.; Veciana. J.; Krause. B.; Carbone. D.; Rovira. C. & Ulanski. J. (2006). Efficient High Area OFETs by Solution Based Processing of a pi-Electron Rich Donor. *Chemistry of Materials,* vol. 18, no. 20, 4724-9.

Niu. Y. -H.; Ma. H.; Xu. Q. & Jen. K.-Y. (2005). High-efficiency light-emitting diodes using neutral surfactants and aluminum cathode. *Applied Physics Letters,* vol. 86, no. 8, 083504-6.

Okamoto. S.; Tanaka. K.; Izumi. Y.; Adachi. H.; Yamaji. T. & Suzuki. T. (2001). Simple Measurement of Quantum Efficiency in Organic Electroluminescent Devices. *Japanese Journal of Applied Physics,* vol. 40, no. 7B, L783-4.

Ouyang. J.; Guo. T. -F.; Yang. Y.; Higuchi. H.; Yoshioka. M. & Nagatsuka. T. (2002). High-Performance, Flexible Polymer Light-Emitting Diodes Fabricated by a Continuous Polymer Coating Process. *Advanced Material,* vol. 14, no. 12 , 915-918.

Pardo. D.A.; Jabbour. G. E. & Peyghambarian. N. (2000). Application of Screen Printing in the Fabrication of Organic Light-Emitting Devices. *Advanced Material,* vol. 12, no. 17, 1249-1252.

Park. B. & Han M. (2009). Photovoltaic characteristics of polymer solar cells fabricated by pre-metered coating process. *Optics Express,* vol. 17, no. 16, 13830-13840.

Park. J. H.; Oh. S. S.; Kim. S. W.; Choi. E. H.; Hong. B. H.; Seo. Y. H.; Cho. G. S. & Park. B. (2007). Double interfacial layers for highly efficient organic light-emitting devices. *Applied Physics Letters,* vol. 90, no. 15, 153508-1-3.

Pei. Q.; Yu. G.; Zhang. C.; Yang. Y. & Heeger. A. J. (1995). Polymer Light-Emitting Electrochemical Cells. *Science,* vol. 269, no. 5227, 1086-8.

Sakuratani. Y.; Asai. M.; Tokita. M. & Miyata. S. (2001). Enhanced electron injection and electroluminescence in poly(*N*-vinyl carbazole) film doped with ammonium salt. *Synthetic. Metals,* vol. 123, no. 2, 207-210.

Suzuki. M.; Tokito. S.; Sato. F.; Igarashi. T.; Kondo. K.; Koyama. T. & Yamaguchi. T. (2005). Highly efficient polymer light-emitting devices using ambipolar phosphorescent polymers. *Applied Physics Letters,* vol. 86, no. 10, 103507-9.

So. F.; Krummacher. B.; Mathai. M. K.; Poplavskyy. D.; Choulis. S. A. & Choong. V. -E. (2007). Recent progress in solution processable organic light emitting devices. *Journal of Applied Physics,* vol. 102, no. 9, 091101-1-21.

Tang. C. W. & Van Slyke. S.A. (1987). Organic electroluminescent diodes. *Applied Physics Letters,* vol. 51, no. 12, 913-5.

Tseng. S. -R.; Meng. H. -F. & Lee. S. -F. (2008). Multilayer polymer light-emitting diodes by blade coating method. *Applied Physics Letter,* vol. 93, no. 15, 153308-1-3.

Xie. H. Z.; Liu. M. W.; Wang. O. Y.; Zhang. X. H.; Lee. C. S.; Hung. L. S.; Lee. S. T.; Teng. P. F.; Kwong. H. L.; Zheng. H. & Che. C. M. (2001). Reduction of Self-Quenching Effect in Organic Electrophosphorescence Emitting Devices via the Use of Sterically Hindered Spacers in Phosphorescence Molecules. *Advanced Materials,* vol. 13, no. 16, 1245-1248.

Xu. H.; Meng. R.; Xu. C.; Zhang. J.; Hee. G. & Cui. Y. (2003). A monolayer organic light-emitting diode using an organic dye salt. *Applied Physics Letters,* vol. 83, no. 5, 1020-2.

Yabu. H. & Shimomura. M. (2005). Preparation of self-organized mesoscale polymer patterns on a solid substrate: Continuous pattern formation from a receding meniscus. *Advanced Functional Materials,* vol. 15, no. 4, 575-581.

Yang. X. H. & Neher. D. (2004). Polymer electrophosphorescence devices with high power conversion efficiencies. *Applied Physics Letter,* vol. 84, no. 14, 2476-8.

Yim. Y. C.; Park. J. H.; Kim. S. W.; Choi. E. H.; Gho. G. S.; Seo. Y. H.; Kang. S. O.; Park. B.; Cho. S. H.; Kim. I. T.; Han. S. H.; Lim. J. & Takezoe. H. (2006). Enhanced light emission from one-layered organic light-emitting devices doped with organic salt by simultaneous thermal and electrical annealing. *Applied Physics Letters,* vol. 89, no. 10, 103507-9.

7

High-Contrast OLEDs with High-Efficiency

Daniel Poitras,† Christophe Py† and Chien-Cheng Kuo‡
†*Institute for Microstructural Sciences, National Research Council of Canada*
1200 Montreal Road, Ottawa K1A 0R6 Canada
‡*Thin Films Technology Center, Department of Optics and Photonics,*
National Central University, 32201 Chung-Li, Taiwan

1. Introduction

As more electronic devices with display are targeted for both indoor and outdoor uses (e.g. cameras, telephones, music players), it becomes increasingly important to solve the problem of contrast of the display under strong external lighting, more particularly under sunlight. In such conditions, the eye has difficulty discriminating the light emitted by the display from the light reflected from the device and surrounding. Increasing the contrast thus consists basically in making sure that the light emitted by the display dominates any other surrounding light reaching the observer. Undesired light from the display itself could be residual light emitted from "*off*" (or dark) pixels, or ambient light reflected on or within the display.

Numerically, the contrast can be expressed as a ratio of the brightest and the darkest elements of a display, taking into account the ambient light reflected by it. In the case of liquid crystal displays, generally with a white backlight source, this contrast is related to the transmittance values of "*on*" and "*off*" pixels (Bahadur, 1991). In the case of light emitting devices, such as organic light emitting displays (OLEDs), the transmittance is replaced by the luminance of the brightest and darkest pixels, and the contrast ratio (CR) is expressed as (Dobrowolski et al., 1992):

$$CR = \frac{L_{on} + R_D L_{ambient}}{L_{off} + R_D L_{ambient}}, \tag{1}$$

where L_{on} and L_{off} are the luminance values of "on" and "off" pixels on the display, respectively, $L_{ambient}$ is the ambient luminance, and R_D is the luminous reflectance of the display, given by

$$R_D = \frac{\int_{\lambda_1}^{\lambda_2} V(\lambda)\cdot R(\lambda)\cdot S(\lambda)\, d\lambda}{\int_{\lambda_1}^{\lambda_2} V(\lambda)\cdot S(\lambda)\, d\lambda}, \quad (2)$$

$V(\lambda)$ being the photopic curve (an eye sensitivity spectrum standard defined by CIE 1931), $R(\lambda)$ is the reflectance of the pixel (*on* or *off*), and $S(\lambda)$ is the source of ambient light [for calculation, CIE standards such as D65 are used (Wyszecki, 1968)]. A value of 20 for CR is usual for a cathode ray tube television in a living room, while a cinema typically has a *CR* of 80 (Poynton, 2003). Care should be taken when comparing CR values as a few different expressions are used to calculate them; for example L_{on}/L_{off} is often used as an expression for *CR*, but should be valid only when the ambient light is sufficiently low, which excludes the cases studied here. Representative luminance values for ambient light and display devices are given in Table I. Without ambient light (i.e. $L_{ambient}$=0 in Eq. 1), *CR* is limited by the darkness of the *off* pixel, which is not as dark for liquid-crystal displays (due to imperfect blocking of its back illumination) as it is for emitting devices (see Table 1). When ambient light is considered, the viewer is seeing the light reflected on the pixels and the only way to prevent it from affecting too much *CR* is to increase the ratio $L_{on}/R_D L_{ambient}$ by (i) increasing L_{on} and (ii) reducing R_D to 1% or less (see Table 2). Thus an ideal display should have a high L_{on}/L_{off} ratio and $L_{on} >> R_D L_{ambient}$.

SOURCE	TYPICAL LUMINANCE L [cd/m^2]
Clear day	10^4
Heavily Overcast day	10^2
Bright moonlight	10^{-2}
Moonless overcast night	10^{-4}
CRT	90 – 150
CRT, "off" pixel	0.01
LCD	400 – 500
LCD, "off" pixel	0.72
OLED	70 – 600
OLED, "off" pixel	0

Table 1. Typical values of luminance for different ambient light conditions and display devices (Boff et al., 1988; Anderson, 2005).

R_D [%]	Contrast Ratio *CR*	
	$L_{ambient}$=10^4 cd/m^2	$L_{ambient}$=10^2 cd/m^2
50	1.1	11
10	1.5	**51**
1.0	6.0	**501**
0.1	**51**	**5001**

Table 2. Values of Contrast Ratio (Eq. 1) corresponding to different values of R_D and $L_{ambient}$ (assuming L_D=500 cd/m^2).

In organic light emitting displays (OLEDs), electrons and holes are injected from the cathode and the anode, respectively, to one or several organic layers between them in which they recombine radiatively, resulting in light emission. We distinguish bottom- and top-emission OLEDs, for which emission occurs through a transparent anode/substrate or a semi-transparent cathode, respectively. In most OLEDs, a thick metal layer is encountered as the electrode material on the non-emitting side; the light reflection from such an electrode is high and this results in a low *CR* value. Replacing the metal electrode by a transparent conductor (such as ITO) can contribute to lower the OLED reflectance, but this generally results in a lower carrier injection into the organic layers. For efficient injection, the cathode requires a material with a low work function (such as Ca, Mg:Ag, or Al/LiF), which are all metallic and possess high reflectance. The anode material should have a high work function, and transparent conductors such as ITO are usually the preferred choice for bottom-emitting devices –they obviously don't have high reflectance.

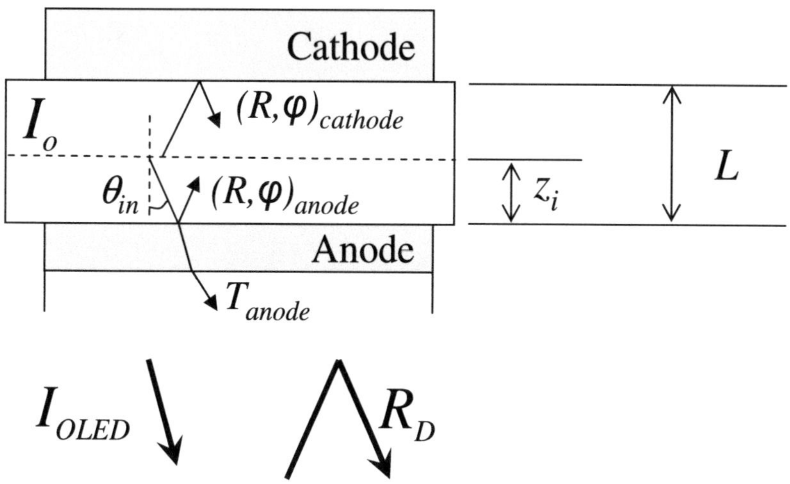

Fig. 1. Schematic view of an OLED showing its Fabry-Perot-like structure and the parameters used in Eq. 3.

2. Theory

2.1 Theory of emission

Several comprehensive models for the emission of dipoles in a multilayer structure have been presented in the literature, which take into account the orientation of dipoles in the emitting layer (Björk, 1991). Less elaborated expressions for the emission of a thin-film structure with an emitting layer can also be developed using an approach similar to the one presented by Smith for describing the transmittance of Fabry-Perot structures, using the concept of effective interfaces (Smith, 1958). We used this approach to obtain the following expression for bottom-emission OLEDs (similar to other expressions that can be found in the literature, for example Lee et al., 2002):

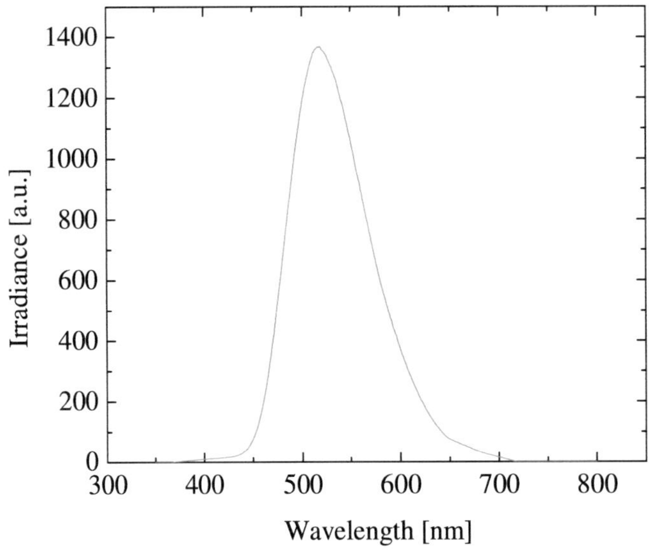

Fig. 2. Emission spectrum of Alq_3. The curve was taken as representing I_0 inside the OLED emitting layer.

$$I_{OLED}(\lambda) = \frac{T_{anode} \frac{1}{N} \sum_{i=1}^{N} \left[1 + R_{cathode} + 2\sqrt{R_{cathode}} \cos\left(\frac{4\pi z_i \cos\theta_{in}}{\lambda} + \phi_{cathode} \right) \right]}{1 + R_{cathode} R_{anode} - 2\sqrt{R_{cathode} R_{anode}} \cos\left(\phi_{cathode} + \phi_{anode} + \frac{4\pi L \cos\theta_{in}}{\lambda} \right)} I_0(\lambda), \tag{3}$$

where R_{anode} and $R_{cathode}$ are the internal reflectance values of the two electrodes, φ_{anode} and $\varphi_{cathode}$ are the phase changes on internal reflection from the mirrors surrounding the cavity layers, T_{anode} is the transmittance of the exit anode, L is the total optical thickness of the cavity layer, $I_0(\lambda)$ is the irradiance of the emitter, $I_{OLED}(\lambda)$ is the irradiance emitted in the glass substrate, z_i is the optical distance between the emitting sublayer i and its interface with the cathode, and θ_{in} is the angle of the emitted beam when measured from inside the emitting material. As shown in Eq. 3, the emitting layer can be divided into N sublayers and their contribution summed up (this step is not essential when the electric field intensity does not change significantly over the emitting layer, as with thin emitting layer, or weak microcavity effect). This equation can include the absorption and the dispersion of the optical constants of the materials. Luminance $L(\lambda)$ spectra can be obtained from Eq. 3 simply by modulating $I_{OLED}(\lambda)$ with the photopic curve. Assuming that the phase conditions in Eq. 3 are optimal, the maximum of emission is obtained approximately when $R_{anode}/(R_{anode}+T_{anode})=R_{cathode}$, which reduces to $R_{anode}=R_{cathode}$ when there is no absorption.

We see that Eq. 3 depends on the internal irradiance $I_0(\lambda)$, which is difficult to determine exactly. In this work, we approximated $I_0(\lambda)$ with the photoluminescence spectra of a thick Alq_3 layer, having a green emission peak (as shown in Figure 2) (Tang, 1987).

As mention above, Eq. 3 is similar to the equation describing the transmittance of a Fabry-Perot, except for the cosine at the numerator. As in Fabry-Perot filters, the multiple internal reflections in OLEDs induce, at some specific wavelengths, a resonance of the light electric-field intensity (or more accurately, the irradiance) distribution inside the OLED.

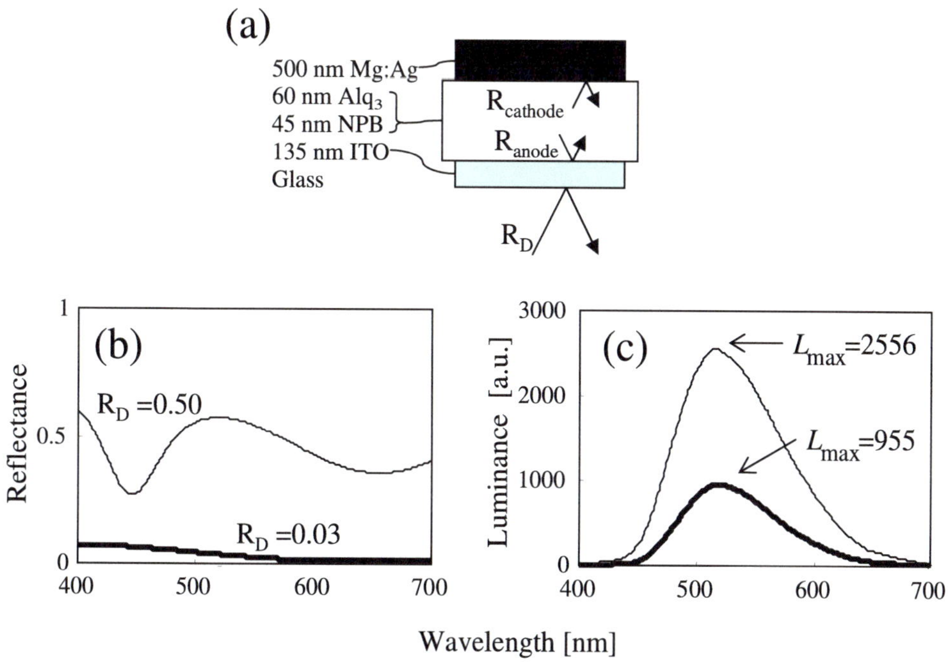

Fig. 3. (a) Schematic representation of a bottom-emitting OLED, (b) Example of reflectance and emission of a conventional OLED (thin line), and one with $R_{cathode}$=0 (thick line). (R_L is the luminous reflectance, given by Eq. 2)

The phenomenon known as "microcavity effect" refers to the enhancement or annihilation of the emitted irradiance related to the position of the emitting material relative to this resonance peak of the irradiance. A weak microcavity effect is usually present in conventional OLEDs because internal reflections are caused by the higher refractive index of the ITO anode compared to most organic layers, and the cathode is highly reflective (Bulovic, 1998). This is usually considered a nuisance, but has been exploited in microcavity OLEDs (Jordan, 1996). With Fabry-Perot filters, the phase condition for the appearance of resonance peaks is given by the following equation:

$$\frac{\phi_{anode} + \phi_{cathode}}{2} - \frac{2\pi \cos\theta}{\lambda} = m\pi. \tag{4}$$

For OLEDs, this condition is slightly shifted due to the top cosine term in Eq. 3. When all-dielectric mirrors are used, the phase terms φ_{anode} and $\varphi_{cathode}$ can be set to zero; however, when absorbing materials (such as a metal) are used in at least one of the mirrors, the phase terms have to be considered.

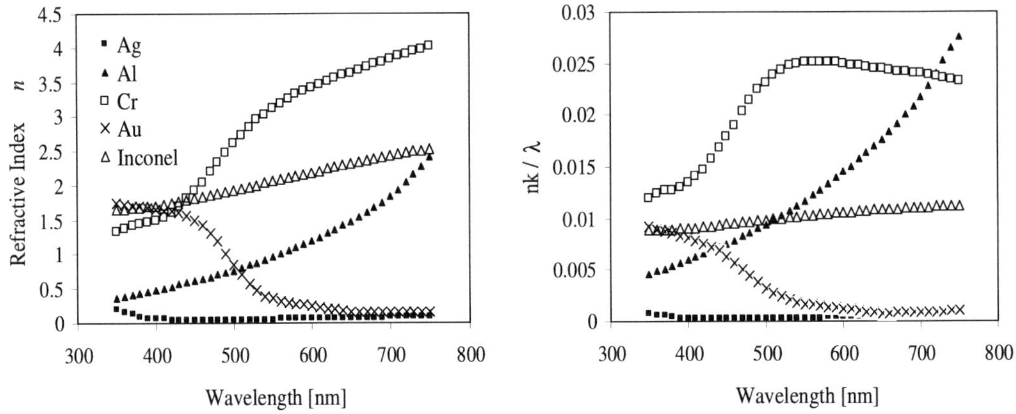

Fig. 4. Refractive index and nk/λ dispersion curves for a few metals.

2.2 Calculation of the reflectance and design

The reflectance of the OLED, R_{OLED}, is calculated using well-know iterative approach for multilayer optics (Macleod, 2001). It is interesting to note that this approach can also be used not only to optimize (minimize) the reflectance of the OLED, but also to simultaneously optimize (maximize) its emission through the optimization of its transmittance (from the substrate towards the cathode), instead of using an exact expression for the emission, such as Eq. 3. It is the comparison of Eq. 3 with the Fabry-Perot equation that makes it possible; the position of the resonance peak in emission is first approximated by Eq. 4 from the Fabry-Perot transmittance peak, and then refined using Eq. 3 to maximize the emission.

3. Review of current approaches for reducing R_D

As seen in Eq. 1 and Table 1, significantly reducing the reflectance of OLEDs is crucial to increase the contrast ratio. Several approaches have been proposed for reaching that goal, which are described briefly in the following paragraphs.

3.1 Use of a polarizer

OLED display manufacturers have so far used circular polarizers borrowed from the LCD technology to improve contrast (Trapani et al., 2003). This approach does not require introduction of new layers in the OLED structure and results in reflectance similar to that of glass. However, polarizers are expensive, generally not flexible, and absorb a substantial amount of the light (up to 40%) (Wu, 2005).

3.2 All-dielectric antireflection coating

Using an all-dielectric antireflection (AR) coatings is the proper way to remove the reflection from the front glass surface when the light is emitted through a glass substrate (bottom-emission).

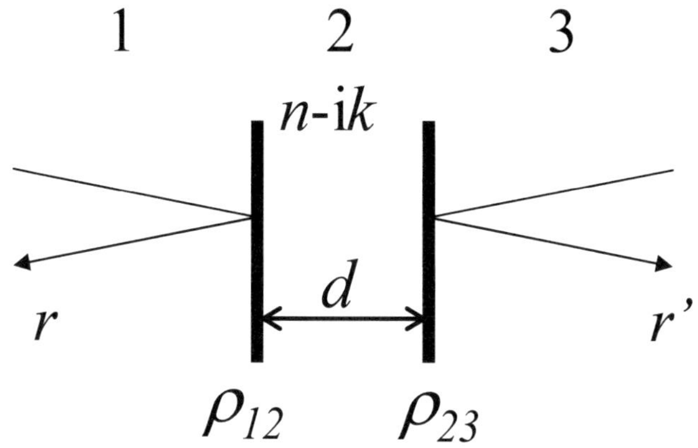

Fig. 5. Schematic view of a metal layer, surrounded by arbitrary materials (ρ_{12} and ρ_{23} can represent the reflection coefficient of multilayers, media 1 and 3 can be different).

However, this AR coating does not remove the reflectance from the OLED structure, on the other side of the glass substrate.

One can also introduce an all-dielectric AR coating in the OLED structure, i.e. between the anode of a bottom-emitting OLED and a glass substrate. Such a coating can have a limited efficiency in reducing R_D when the cathode is a highly reflecting metal (i.e. Al or Mg:Ag). In such cases, metal-dielectric AR coatings can lead to a better performance.

3.3 Scattering Anti-glare surface

One can reduce the specular reflectance of the display by making its top-surface non-planar, an approach that can also enhance the extraction of light from the device. However, the fact that the emitted light is scattered by the surface structure may have a detrimental effect on the resolution of the display (Nuijs & Horikx, 1994).

3.4 Black electrode

In the last few years, many attempts have been made to reduce the reflectance of metal-based electrodes, mainly by making the cathode black (Renault et al., 2000; Krasnov, 2002; Aziz, 2003; Dobrowolski, 1981; Lemarquis & Marchand, 1999) by using absorbing materials in it, or covering it with a conductive black layer coating (similar to the metal-dielectric AR coating described below) (Krasnov, 2002). The result is a reduced reflectance of the OLED sometimes below 1% at the detriment of the emission; or a relatively good emission, but with a higher luminous reflectance (Wu, 2005). Using Eq. 3, we can show that a completely dark cathode (or anode in the case of top emission) is not desired because (i) it does not take into account the contribution to R_D from the other interfaces in the OLED structure, and more important (ii) it destroys any beneficial microcavity effect and reduce the emission, as shown in Fig. 3. In typical OLEDs, a weak microcavity effect is generally present and can be exploited to improve contrast.

3.5 Absorbing Pigments

Absorbing pigments in front of the OLEDs can be used in the fabrication of displays to create red, green and blue (RGB) pixels when combined with a wide band emission OLEDs. These pigments can be used to absorb the light with a wavelength that does not correspond to that of the light emitted by the pixel, thus contributing to reduce the ambient light reflection (Urabe et al., 2004). When combined, these RGB pigments can darken the surfaces surrounding the pixels. The only remaining ambient reflected light is the one corresponding to the wavelength of the '*off*' pixel (e.g. red pixel will reflect red light even when '*off*').

3.6 Metal-dielectric antireflection coating

It has been known for some time that for efficiently reducing the reflectance of highly reflective substrate with a complex admittance (i.e. metals, or coated metals, such as an OLED), it is convenient to use simple metal-dielectric AR coatings similar to those used in black absorbers (Dobrowolski, 1981; Lemarquis & Marchand, 1999) or heat-reflector in solar-cells applications (Macleod, 1978). This type of coatings has been demonstrated for the contrast-enhancement of electroluminescent (EL) displays (Dobrowolski et al., 1992) and on the cathode side of bottom-emitting OLED (see above) (Krasnov, 2002).

4. Our design approach

We mentioned in Secs. 2.1 and 3.4 that keeping a weak microcavity effect is important for maintaining a relatively high emission. When designing the high-contrast OLEDs, our goals are thus (i) to minimize the external R_D of the OLED, and (ii) to maintain R_{anode} and $R_{cathode}$ large enough to keep the emission high. Many of the approaches mentioned above concentrate on darkening the electrode on the non-emitting side of the OLED, neglecting the reflections on emitting side of the OLED and the contribution of the electrodes' reflectance to the efficiency of the OLED. In order to take these aspects into consideration and achieve the goals mentioned above, our approach combines in the OLED structure three types of optical coatings phenomena: antireflection with a metal-dielectric coating on the anode side, microcavity effect at the emitting layer, and an asymmetric reflectance of the anode.

A small microcavity effect, as seen in Sec. 2, is necessary for maintaining a good emission of the device. For that purpose, internal reflections R_{anode} and R_{cathode} must not be reduced to zero, and the organic layers inside the OLED act as cavity layers, so that the position of the emitting layer (the thin recombination layer) must be at a resonance peak of the electric field.

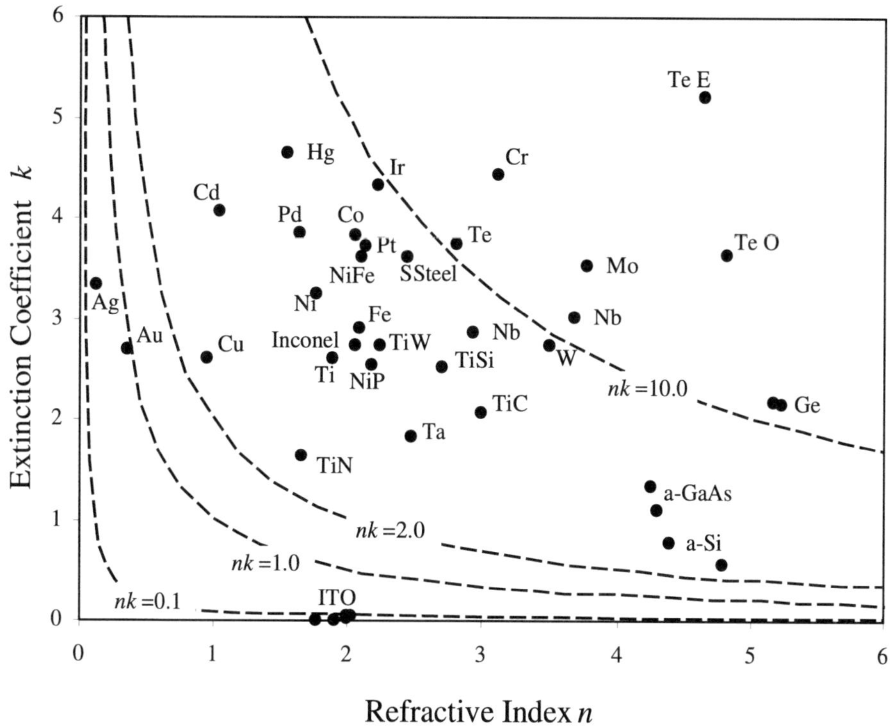

Fig. 6. Refractive indices and extinction coefficients (both given at a wavelength of 550 nm) of several metals and semiconductor materials, as found in the literature. Some isovalue-curves of *nk* product are shown (most optical constants values are extracted from Palik, 1985, and from J.A. Woollam WVASE software, 2009).

As shown in Fig. 1, the combination of good AR coating and small microcavity effect apparently lead to a contradiction of the anode's role: it must have simultaneously a low external reflectance when seen from the substrate and a relatively large internal reflectance when seen from the cavity layers. It has been observed for a long time in thin-film optics that a thin layer of a material with a large extinction coefficient k can lead to the kind of asymmetric reflectance (Goos, 1937). In our design, such a layer has thus to be introduced on the anode side of the OLED structure.

Of course, a compromise must still be made between low reflectance and high emission. Also, a too-high microcavity effect is usually not desirable in display application, since it leads to a large dependence of the emission on the angle of view. The key to our design is the asymmetry of the anode internal and external reflection.

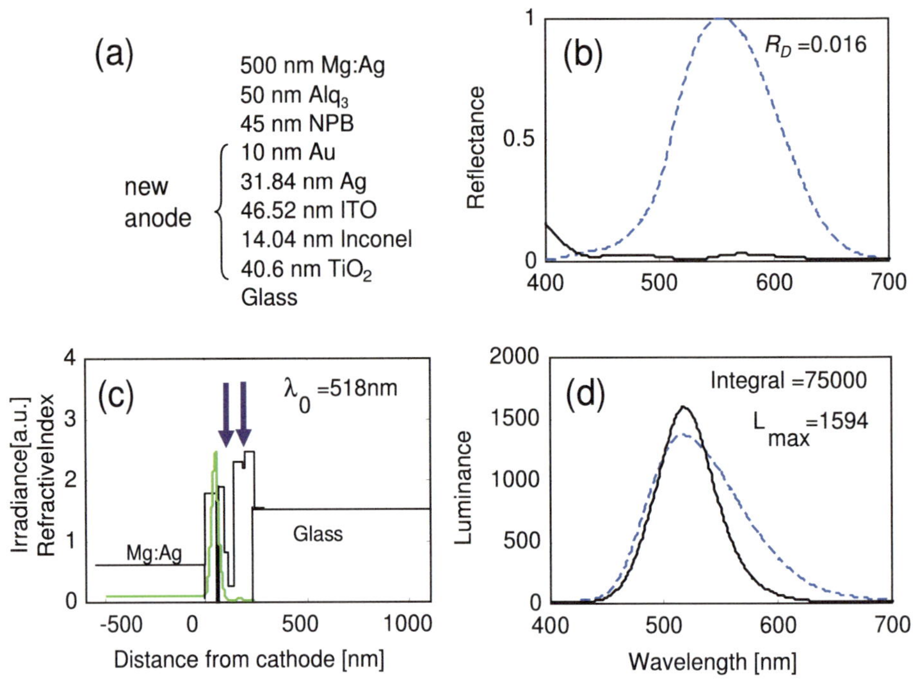

Fig. 7. (a) OLED design; (b) calculated reflectance (solid line) with the photopic curve (dash line) and the value of the luminous reflectance RD; (c) refractive index profile (step) and irradiance profile inside the OLED, with the arrows showing the metal layers, and the emitting layer marked in black.

5. Choice of Materials

5.1 Diode consideration

The selection of materials composing the OLED is important from an electronic point of view. For example, electrode materials must be adequate for carrier injection in organic materials, and they must, along with the organic materials, act as good carrier transport materials. In particular, the cathode must be selected with care, and requires a material with a low work-function to promote injection to an organic layer.

In the present work, we choose well-known materials for the OLED "core" layers: Mg:Ag as a cathode (electron injection) material, Alq3 for the electron transport and emitting layer, NPB simultaneously as a electron-blocking and hole-transport layer (to ensure that electrons and holes recombine in Alq3). Given the low mobility of organic materials, it is also important that their thickness be close to the diffusion length of the charges they transport; this usually constraints the optical design since the resulting thickness of the organic stack is somewhat less than a half-wavelength. In some cases, ITO was used for the anode (hole injection) material. The other materials included in the design are mentioned in the following sub-sections.

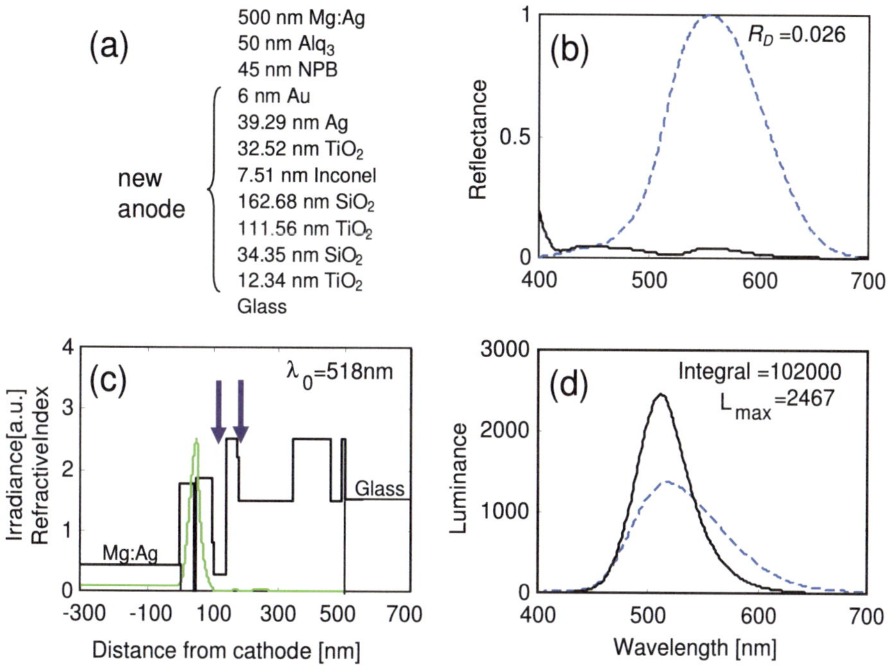

Fig. 8. (a) OLED design; (b) calculated reflectance (solid line) with the photopic curve (dash line) and the value of the luminous reflectance R_D; (c) refractive index profile (step) and irradiance profile inside the OLED, with the arrows showing the metal layers, and the emitting layer marked in black.

5.2 Optical consideration, metal-dielectric antireflection coating

In metal-dielectric AR coatings, the main role of the metal layers is not to absorb the light but to benefit from its complex admittance *n-ik* in order to more efficiently reach to AR condition (Lemarquis & Marchand, 1999). For that reason, metals with relatively large *k* are required for this type of coatings (see Fig. 6). Metals that are highly reflecting, such as metals with $n < 1$, are usually avoided. In addition, metals with n that decreases with decreasing λ (often called "abnormal dispersion") are needed to compensate for the increase of optical thickness in the dielectrics at shorter wavelengths. This type of optical constants dispersion is also needed so that the metal does not introduce chromatic absorption in the device, which requires a constant nk/λ for all the wavelengths of interest. Figure 4 shows *n* and nk/λ dispersion curves for several metals. Chromium is often used for metal-dielectric black absorbers, but our preferred choice is Inconel (an alloy of Cr:Ni:Fe), which is less absorbing and has a very flat nk/λ curve.

5.3 Optical consideration, electrode with asymmetric reflection

As mentioned in Sec. 4, a material with $k > 0$ is required at the anode to maintain a cavity effect in the OLED while reducing its external reflectance, i.e. introducing an asymmetry of the internal and external reflectance of the anode. The optical constants required for that purpose can be found by looking at the reflection coefficients r and r' from both side of an arbitrary layer, with arbitrary interfaces (they could include multilayer), as shown in Fig. 5:

$$r = \frac{\rho_{12} + \rho_{23} exp(-2i\beta)}{1 + \rho_{12}\rho_{23} exp(-2i\beta)},$$
$$r' = \frac{\rho_{23} + \rho_{12} exp(-2i\beta)}{1 + \rho_{12}\rho_{23} exp(-2i\beta)}, \tag{5}$$

with $$\beta = \frac{2\pi}{\lambda}\tilde{n} d cos\theta = \frac{2\pi}{\lambda}(n - ik) d cos\theta.$$

Clearly, the exponential term differentiates r and r'. It can be shown from Eq. 5 that a large k value is essential to increase the asymmetry in reflectance, with a sufficiently large thickness d; a large n value will also increase the asymmetry, but is not essential. In the case of the anode (as in many other cases involving asymmetric reflectance), a reduction of the light absorption in the layer is important. The irradiance absorbed by a layer is given by the following relation (Macleod, 2001):

$$I_{abs} = \frac{2\pi}{\lambda} nkd|E|^2 \gamma, \tag{6}$$

where E is the average amplitude of the electric field in the film considered and γ is the free space admittance (a constant). From this equation, we see that reducing the thickness and the amplitude of the electric field inside the layer will lead to a low absorption. This can be done when refining the design of the OLED. Equation 6 also indicates that materials with a low nk product will lead to lower absorption. Figure 6 shows the optical constants of many absorbing materials, and help to find materials having the required (i) high k value and (ii) low nk value. We see on Figure 6 that semiconductor materials (Ge, GaAs, and Si) have relatively low k values and large nk products, while ITO has a low nk product, but also a low k. Metals, on the other hand, have larger k values, but most of them are too absorbing (large nk product). Only silver (Ag) and gold (Au), two transition metals, have a suitably low n value and large k value; they are the preferred choice for our application.

6. Application to high-contrast OLED

6.1 Examples of design

We used the ideas presented in the previous Section and optimized the layers thicknesses of OLED structures consisting of thick-Mg:Ag | organics | Au/Ag | ITO | metal-dielectric-AR | glass in order to reduce R_D, while keeping $R_{cathode}$ and R_{anode} sufficiently high for maintaining a weak cavity effect.

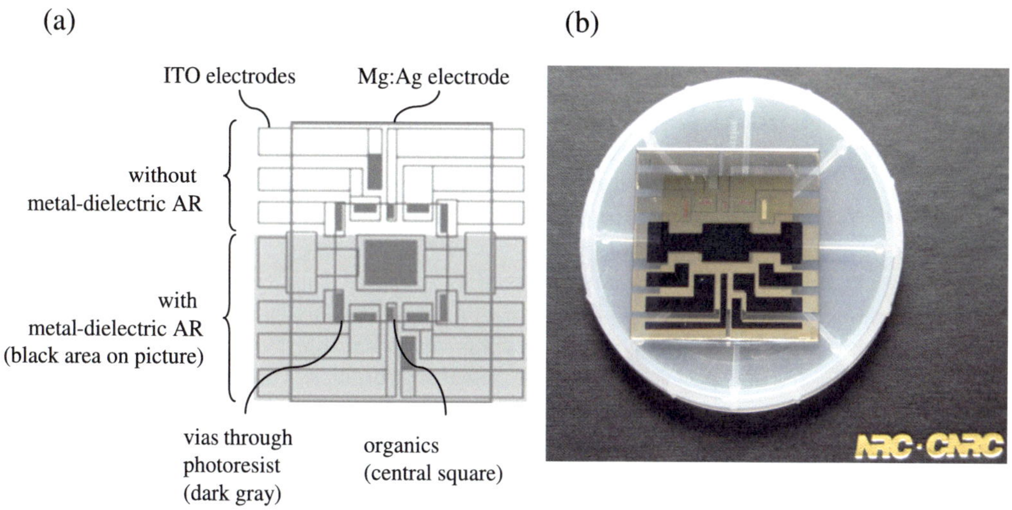

Fig. 9. (a) Schematic bottom view of multi-segment OLED device with and without metal-dielectric AR. (b) Picture of such a device after fabrication. This device corresponds to the design presented in Figure 8.

During theses optimizations, it was important to constrain the thickness values of organic materials to ensure high efficiency OLEDs. For a similar reason, we introduced an Au layer (which has a higher work function than Ag) to facilitate hole injection in the hole transport layer (see Sec. 7). In addition, the thickness of the ITO film had to be large enough to form a low resistivity anode and facilitate the contact with an external lead (although in some cases, we found that the Au/Ag layer was thick enough so that no ITO layer was required).

Figures 7 and 8 show two different designs with a different number of layers in the metal-dielectric AR part of the structure, along with their calculated performances (reflectance and luminance spectra). The complex refractive index of all layers were measured from films deposited in the same conditions as our devices. When compared to the performance of a conventional OLED shown in Fig. 3(b) and (c), we see that the new designs reduce the reflectance to 2% and less, which is 25 times less than that of a typical OLED, and that the emission is of the same order of magnitude.

Figure 7(c) and 8(c) also show the distribution of the irradiance inside the OLEDs at the peak wavelength. The maximum of irradiance at the position of the emitting layer indicates that a microcavity effect occurs in the OLED. Not shown here is the fact that the optimization of such designs with absorbing layers involves the adjustment of phase values φ_{anode} and $\varphi_{cathode}$ in Eq. 3 (Poitras et al., 2003). In addition, the reduced irradiance values at positions corresponding to the metal layers contribute to reduce the absorption of emitted light in these layers (see Eq. 6).

6.2 Example of actual device

SiO_2, TiO_2 and Inconel were deposited in a dual ion-beam sputtering deposition chamber (Spector, Veeco-IonTech), and all other materials were thermally evaporated in a high-vacuum cluster tool (Kurt J. Lesker), in separate chambers for metals and organics to avoid cross-contamination and interface degradation. The complex refractive index spectra of individual films were derived from measurements by ex-situ variable-angle spectroscopic ellipsometer (VASE, J.A. Woollam Co.). These spectra were used to produce the final design described and simulated in Figure 8.

The profile of the calculated irradiance, which is the light radiant flux per unit area, is shown in Figure 1 at the peak wavelength of emission. The cavity is designed so that the irradiance has a maximum in the Alq_3 layer at the NPB interface, where the emission originates, and a minimum in the Ag/Au absorbing bilayer, where light absorption is reduced. High contrast is obtained because the Au/Ag bilayer is highly absorbing seen from the outside. Using published extinction coefficients for evaporated Au and Ag films (AIP, 1972), the transmittance of the Au/Ag bilayer without the cavity effect is calculated to be 0.042.

Actual devices were fabricated with the DBR materials sputtered through a shadow mask on only half of a 2x2 in^2 glass slide to provide direct comparison between filtered and unfiltered sides (see Figure 9). Ag and Au were evaporated through a shadow-mask to define electrode tracks and an electrical separator lithographically patterned to define diode segments (Roth et al., 2001). NPB, Alq_3, Mg:Ag and a Ag capping layer were evaporated with the contacts masked off. The samples were not encapsulated.

Reflectance measurements were performed using a spectrophotometer (Lambda-19, Perkin-Elmer) equipped with a reflectance accessory (with an angle of incidence of 7°). The values obtained (see Figure 10) are in qualitative agreement with our simulation, and show a very clear improvement of the contrast. The spectral shift and discrepancy in values of reflectance between simulated and measured spectra is due to the cumulative error in film thicknesses, most probably from organic materials for which the control is less precise, but also from variations in the optical constants of metallic films, which are critical.

The unfiltered OLED shows a deep absorption peak due to the Fabry-Perot resonance of the naturally-occurring weak microcavity, and the filtered OLED shows oscillations in the reflectance due to the same effect. Lower reflectance filters could be designed with more layers in the DBR, at the expense of added complexity.

7. Conclusion

It is conceivable that future outdoor displays will combine different approaches: intensity control, microstructure for light extraction, or displays based on reflection might be used, but they will certainly include reflection-suppressing designs. As we saw earlier, efficiently suppressing the light reflection from the device requires an integration of the antireflection layers with the entire display device.

We have demonstrated the concept of a multilayer anode comprising an Au/Ag bilayer and a metal-dielectric AR coating that has both a high internal reflectance and a low outside reflectance. The former property is used to maintain a microcavity effect in the OLED that is tuned to maximize light out-coupling, and the latter to improve the OLED contrast ratio.

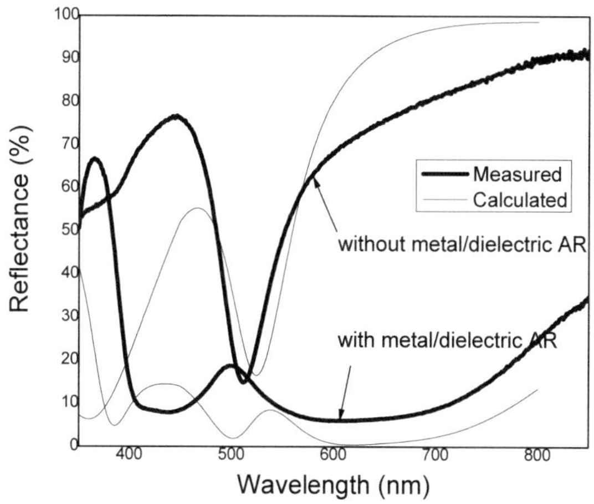

Fig. 10. Theoretical and measured reflectance spectra, for OLED with and without integrated metal-dielectric layers.

Further designs are being considered with varying thicknesses of the Au/Ag layer, and fewer layers in the metal-dielectric coating for a simpler fabrication process.
Although the basic concepts described concerning the microcavity effect have been applied in the present work to bottom-emission OLEDs and specific materials only, they are general and will remain true whatever the materials used in the device (i.e. polymer-based), and for other device structures (such as top-emitting-OLED, tandem-OLED, etc.).
The problem of contrast is complex: the optimum contrast for which a viewer is comfortable depends on the color, and the surrounding light. For outside application, ideal solutions will probably involve not only the reduction of the reflectance of the display, such as explained here, but also the adjustment of display luminance and correction for the gamma parameter (Poynton, 1993; Devlin et al., 2006).

Acknowledgments

The authors wish to thank Hiroshi Fukutani, Eric Estwick and Xiaoshu Tong for their technical assistance. We also are grateful to Dr. Ye Tao for many fruitful discussions, and to Prof. C.C. Lee.
Parts of this work were presented at the OSA 2007 Optical Interference Coating Conference (Tucson, June 2007) and at the 13th Canadian Semiconductor Technology Conference (Montreal, August 2007).

8. References

AIP (1972) *American Institute of Physics Hanbook*, Gray, D.E. ed. McGraw-Hill, 3rd edition, ISBN 978-0070014855, New York.

Anderson, P. (2005). *Advance Display Technologies*, JISC Technology & Standards Watch Report, August 2005. http://www.jisc.ac.uk/whatwedo/services/services_techwatch/techwatch/techwatch_reports_0503.aspx

Aziz, H.; Liew, Y.-F.; Grandin, H. M. & Popovic, Z. D. (2003). Reduced reflectance cathode for organic light-emitting devices using metalorganic mixtures, *Appl. Phys. Lett.* Vol. 83, pp. 186–188.

Bahadur, B. (1991). Display parameters and requirements, In: *Liquid Crystals: Applications and Uses*, B. Bahadur (Ed.), p. 82, World Scientific, ISBN 978-981-02-0111-1, Singapore.

Björk, G. (1991). Modification of spontaneous emission rate in planar dielectric microcavity structures, Physical Review A, Vol. 44, No. 1, pp. 669–681.

Boff, K.R.; Lincoln, J.E. & Armstrong, H.G. (1988). *Engineering Data Compendium. Vol.1. Human Perception and Performance*, Aerospace Medical Research Laboratory, Wright-Patterson Air Force Base, ISBN 978-9992149201, Ohio.

Bulovic, V.; Khalfin, V.B; Gu, G.; Burrows, P.E.; Garbuzov, D.Z. & Forrest, S.R. (1998). Weak microcavity effects in organic light-emitting devices, *Phys. Rev. B.* Vol. 58, No. 7, p. 3730.

Devlin, K.; Chalmers, A. & Reinhard, E. (2006). Visual calibration and correction for ambient illumination, *ACM Transactions on Applied Perception.* Vol. 3, No. 4, pp. 429–452.

Dobrowolski, J.A. (1981). Versatile computer program for absorbing optical thin film systems, *Appl. Opt.* Vol. 20, pp. 74-81.

Dobrowolski, J.A.; Sullivan, B.T. & Bajcar, R.C. (1992). Optical interference, contrast-enhanced electroluminescent device. *Applied Optics*, Vol. 31, No. 28, pp. 5988–5996, ISSN 0003-6935.

Goos, F. (1937). Durchlässigkeit und reflexionsvermögen dünner silberschichten von ultrarot bis ultraviolet, *Zeitschrift für Physik A Hadrons and Nuclei.* Vol. 106, No. 9–10, pp. 606–619.

Jordan, R.H.; Rothberg, L.J.; Dodabalapur, A. & Slusher, R.E. (1996). "Efficiency-enhancement of microcavity organic light-emitting diodes, *Appl. Phys. Lett.* Vol. 69, No. 14, p. 1997.

Krasnov, A. N. (2002). High-contrast organic light-emitting diodes on flexible substrates, *Appl. Phys. Lett.* Vol. 80, pp. 3853–3855.

Lemarquis, F. & Marchand, G. (1999). Analytical achromatic design of metal-dielectric absorbers, *Appl. Opt.* Vol. 38, pp. 4876–4884.

Lee, G.J.; Jung, B. Y.; Hwangbo, C. K. & Yoon, J. S. (2002). Photoluminescence characteristics in metal-distributed feedback-mirror microcavity containing luminescent polymer and filler, *Jpn. J. Appl. Phys.* Vol. 41, p. 5241.

Macleod, H.A. (1978). A new approach in the design of metal-dielectric thin-film optical coatings, *Optica Acta.* Vol. 25, No. 2, pp. 93–106.

Macleod, H.A. (2001). *Thin-Film Optical Filters*, Institute of Physics Publishing, ISBN 0750306882, Bristol.

Nuijs, A. M. & Horikx, J. J. L. (1994). Diffraction and scattering at antiglare structures for display devices, *Appl. Opt.* Vol. 33, No. 18, pp. 4058–4068.

Palik, E. D. (1985). *Handbook of Optical Constants of Solids, Vols. I and II,* Academic Press, ISBN 0125444222, New York.

Poitras, D.; Dalacu, D.; Liu, X.; Lefebvre, J.; Poole, P.J. & Williams, R. L. (2003). Luminescent devices with symmetrical and asymmetrical microcavity structures, *Proceedings of the 46th Annual Tech. Conf. of Society of Vacuum Coaters,* pp. 317—322, Philadelphia, May 2003, ISSN 0737-5921, SVC Publication, Albuquerque.

Poynton, C.A. (1993). 'Gamma' and its Disguises: The Nonlinear Mappings of Intensity in Perception, CRTs, Film and Video, *SMPFTE Journal,* Vol. 102, No. 12, pp. 1099—1108.

Poynton, C.A. (2003). *Digital video and HDTV – algorithms and interfaces,* Morgan Kaufmann Publisher, ISBN 1558607927, San Francisco.

Py, C.; Poitras, D.; Kuo, C.-C. & Fukutani, H. (2008). High-contrast Organic Light Emitting Diodes with a partially absorbing anode, *Opt. Lett.* Vol. 33, No. 10, pp. 1126—1128.

Renault, O.; Salata, O. V.; Etchells, M.; Dobson, P. J. & Christou, V. (2000). A low reflectivity multilayer cathode for organic light-emitting diodes, *Thin Solid Films,* Vol. 379, pp. 195—198.

Roth, D.; Py, C.; Fukutani, H.; Marshall, P.; Popela, M. & Leong, D. (2001). An Organic Digital Integrated Multiplexing Clock Display, *Presented at the 10th Canadian Semiconductor Technology Conference, Ottawa, Canada, Aug 13—17.*

Smith, S.D. (1958). Design of multilayer filters by considering two effective interfaces, *J. Opt. Soc. Am.* Vol. 48, No. 1, pp. 43—50.

Tang, C.W. & VanSlyke, S.A. (1987) Organic electroluminescent diodes, *Appl. Phys. Lett.* Vol. 51, No. 11, pp. 913--915.

Trapani, G.; Pawlak, R.; Carlson, G. R. & Gordon, J. N. (2003). High durability circular polarizer for use with emissive displays, US Patent 6549335.

Uriba, T.; Yamada, J.; Sasaoka, T. (2004) Display and method of manufacturing the same, US Patent 2004/0147200A1.

Wu, C.-C.; Chen, C.-W.; Lin, C.-L. & Yang, C.-J. (2005) Advanced Organic Light-Emitting Devices for Enhancing Display Performances, J. Display Technol. Vol. 1, No. 2, pp. 248—266.

Wyszecki, G. (1968). Recent Agreements Reached by the Colorimetry Committee of the Commission Internationale de l'Eclairage (abstract). , *J. Opt. Soc. Am.* Vol. 58, No. 2, pp. 290—292.

WVASE32 software (J.A. Woollam Co., Lincolrn NE)

8

Optimum Structure Adjustment for Flexible Fluorescent and Phosphorescent Organic Light Emitting Diodes

Fuh-Shyang Juang, Yu-Sheng Tsai, Shun-Hsi Wang, Shin-Yuan Su, Shin-Liang Chen and Shen-Yaur Chen
National Formosa University
Taiwan

1. Introduction

The organic light emitting diodes (OLEDs) [1] is a new-generation flat panel display with the advantages of self-luminescence, wide viewing angle (> 160°), prompt response time (~1 μs), low operating voltage (3~10 V), high luminance efficiency, high color purity, and easy to be made on various substrates. Therefore, it's an important topic that how to improve the luminance efficiency, lifetime and the adhesion characters of ITO/organic interface of flexible OLEDs. Zugang Liu *et al.* reported that the NPB (HTL) is suitable in contact with the emission layer and when they form an energy ladder structure, the driving voltage decreased and the electroluminescent output increased [2]. Thus it can be seen, the hole transport layer [3-6] is very important to balance the injection of hole and electron, to increase the luminance efficiency and lifetime. In recent years, the hole buffer layer of device typically employs LiF [7], CuPc [8], Pani:PSS [9-10] or PEDOT:PSS [9-11] to improve the hole injection efficiency. In addition, a flexible substrate (PET, metal foil, etc.) surface is not completely smooth and will usually have spikes. After the organic thin film evaporates onto the ITO substrate surface the spikes will still exist. When the device is operated under high voltage or high current density, a heavy amount of electric current will concentrate at the spikes and damage the device by causing the device to short circuit, creating Joule heat. The luminance efficiency of the device will therefore be reduced producing shorter device lifetime. Thus, the PEDOT:PSS fabrication process uses spin-coating to obtain a thin film with a smoother surface than that produced by thermal deposition. Spin-coating enhances the organic material adhesion in subsequent processes, thereby directly affecting the performance of flexible OLED. For the above reason, this research dissolved hole transport material N,N'-diphenyl-N,N'-bis(1-naphthyl)- 1,1'biphenyl-4,4''diamine (α-NPD), N,N'-Bis(naphthalene- l-yl) -N,N'-bis(phenyl)-benzidine (NPB) or α-NPD:NPB in tetrahydrfuran (THF) solvent and spin-coated the buffer layer onto ITO surface of flexible OLEDs. Phosphorescent dye gains energy from the radiative recombination of both singlet and triplet excitons [12], improving the internal quantum efficiency of fluorescent OLEDs (FOLEDs) typically 25% at maximum to nearly 100% [13]. Enhancing the luminance efficiency of phosphorescent OLED has attracted the interest of many researchers. Improving device efficiency, the triplet state excitons must be confined in the emitting layer to increase the chance for energy transfer from host to guest. The material that achieves this effect is called the hole blocking layer (HBL), CF-X [14], CF-Y [14], BCP [12], TPBi [15], and BAlq [16]. These materials have higher ionization energy and band gap that can block the diffusion of excitons. When the host-guest orbit overlap is weak, the blocking layer action is particularly important.

2. Experiment

The ITO substrate used in this study was 80Ω/□ PET substrate. Before depositing the patterned ITO substrate was placed in O_2 plasma for surface cleaning. The spin-coating solvents were then prepared by dissolving hole transport materials N,N′-diphenyl-N,N′-bis(1-naphthyl)- 1,1′biphenyl-4,4′′diamine (α-NPD) and N,N′-Bis(naphthalene-l-yl) -N,N′-bis(phenyl)-benzidine (NPB) (α-NPD mixed NPB with 1:1 wt%) in tetrahydrfuran (THF) solvent. The chemicals are vibrated ultrasonically in solution for 60 minutes to facilitate the dissolving process. The coating process is then carried out for 35 seconds at 4500 r.p.m. to deposit the buffer layer onto the ITO surface. After that the substrate was placed in an organic evaporation chamber to deposit the organic layers under 2×10^{-6} torr, α-NPD or NPB was deposited as hole transport layer (HTL), 4,4'-Bis(carbazol-9-yl) biphenyl (CBP) was deposited as the phosphorescent device host, Tris(2-pheny-lpyridine) iridium(III) ($Ir(ppy)_3$) was deposited as the phosphorescent device guest material, 2,9-Dime-thyl-4,7-diphenyl-1,10-phenanhroline (BCP) or 2,2',2''-(1,3,5-Benzinetriyl) -tris(1-phenyl-1-H-benzimidazole) (TPBi) was deposited as the hole blocking layer (HBL), Tris(8-hydroxy-quinolinato)aluminum (Alq3) was deposited as emitting layer (EML) of fluorescent and electron transport layer (ETL), and 1,3-Bis[2-(2,2'-bipyridine-6-yl)-1,3,4-oxadiazo-5-yl]benzene (Bpy-OXD) was deposited as electron transport layer/ hole blocking layer. The chemical structures of all used organic materials are shown in Fig. 1. SpectraScan PR650 and Keithley 2400 equipment were employed to measure the luminance and current-voltage characteristics.

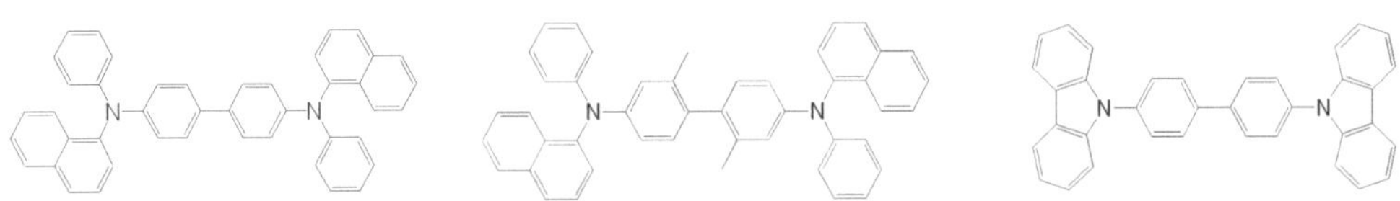

(a) (b) (c)

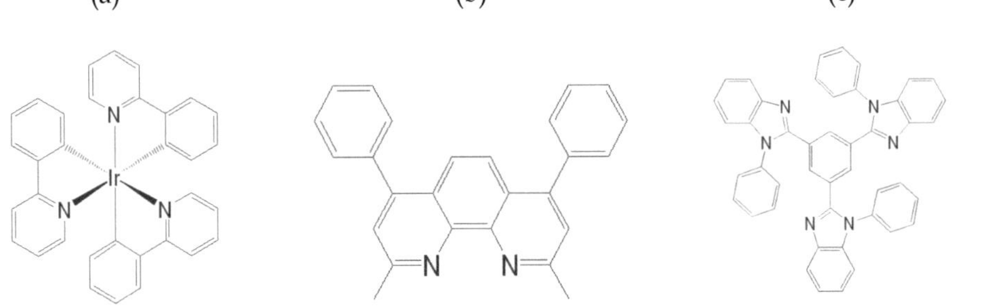

(d) (e) (f)

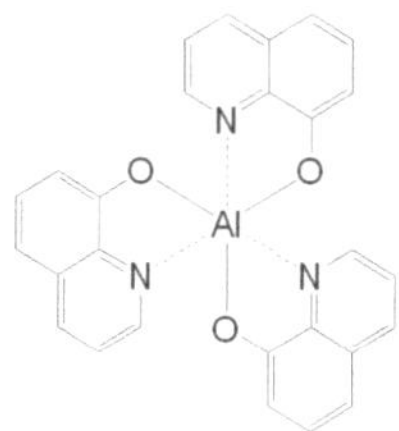

(g)

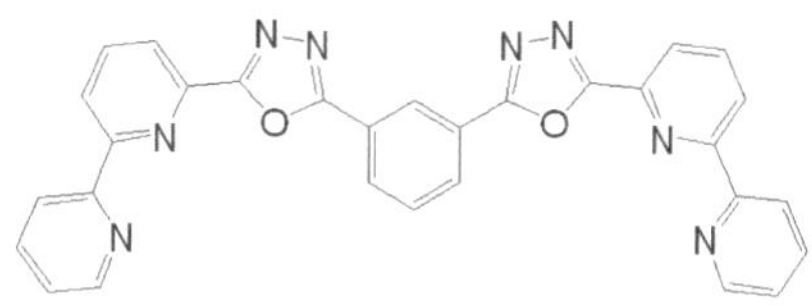

(h)

Fig. 1. The chemical structures of all used organic materials (a) NPB, (b) α-NPD, (c) CBP, (d) $Ir(ppy)_3$, (e) BCP, (f) TPBi, (g) Alq3 and (h) Bpy-OXD

3. Results and discussion

3.1 Optimum Structure Adjustment for Flexible Phosphorescent Organic Light Emitting Diodes

<table>
<tr><th>NO.</th><th>Sub.</th><th>ITO</th><th>NPB</th><th>Dopant 7% Ir(ppy)$_3$ in CBP</th><th>HBL</th><th>Alq3</th><th>LiF</th><th>Al</th></tr>
<tr><td>A1</td><td rowspan="19">Plastic (PET)</td><td rowspan="19">ITO (80Ω/□)</td><td rowspan="4">40</td><td rowspan="4">20</td><td>BCP 0</td><td rowspan="4">40</td><td rowspan="19">0.5</td><td rowspan="19">65</td></tr>
<tr><td>A2</td><td>BCP 5</td></tr>
<tr><td>A3</td><td>BCP 10*</td></tr>
<tr><td>A4</td><td>BCP 15</td></tr>
<tr><td>B1</td><td>30</td><td rowspan="4">20</td><td rowspan="4">BCP 10</td><td rowspan="4">40</td></tr>
<tr><td>A3</td><td>40</td></tr>
<tr><td>B2</td><td>50*</td></tr>
<tr><td>B3</td><td>70</td></tr>
<tr><td>C1</td><td rowspan="4">50</td><td rowspan="4">20</td><td rowspan="4">BCP 10</td><td>30</td></tr>
<tr><td>B2</td><td>40</td></tr>
<tr><td>C2</td><td>50*</td></tr>
<tr><td>C3</td><td>70</td></tr>
<tr><td>D1</td><td rowspan="4">50</td><td>10</td><td rowspan="4">BCP 10</td><td rowspan="4">50</td></tr>
<tr><td>C2</td><td>20</td></tr>
<tr><td>D2</td><td>30</td></tr>
<tr><td>D3</td><td>40*</td></tr>
<tr><td>E1</td><td rowspan="3">50</td><td rowspan="3">40</td><td>TPBi 5</td><td rowspan="3">50</td></tr>
<tr><td>E2</td><td>TPBi 10*</td></tr>
<tr><td>E3</td><td>TPBi 15</td></tr>
</table>

*optimum parameters

Table 1. Adjustment parameters of Phosphorescent organic light emitting diodes (unit: nm)

In this study the device structures are shown in Table 1. First, we inserted a hole blocking layer (HBL) to effectively confine the holes in the emitting layer (EML) for improving the luminance efficiency of the devices. Moreover, varied the thickness of BCP from 0 to 15 nm; it was found that the best hole-blocking result was present at 10 nm of the thickness of BCP (as shown in Fig.1). However, if the thickness of BCP was increased to 15 nm, the hole blocking result was better, but the distance of injecting electrons to EML was increased and caused the brightness decreased. Then, we tried to vary the thickness of NPB to make the amount of the hole injected into EML match with the amount of the electron for increasing the luminance efficiency of the device. From Fig. 2, it was found that the maximum luminance efficiency of the device can be obtained at 50 nm of the thickness of NPB. Furthermore, we varied the thickness of Alq3 to make the amount of the electron injected into EML match with the amount of the hole. From Fig. 3, it was found that the maximum luminance efficiency of the device can be obtained at 50 nm of the thickness of Alq3. However, if the thickness of Alq3 was increased to 70 nm, the distance of the electron injecting to EML was enhanced to decrease the amount of the electron injected into the EML to cause the brightness greatly decreased. At last, we varied the thickness of the EML of CBP:Ir(ppy)3 from 10 nm to 40 nm and hoped that the chance of recombining electron-hole will be increased via varying the thickness of the EML to increase the brightness and luminance efficiency. From the experiment result, it was found that the best luminance efficiency would be obtained at the layer thickness of 40 nm (as shown in Fig. 4); at the moment, the device efficiency was greatly increased to 30.4 cd/A.

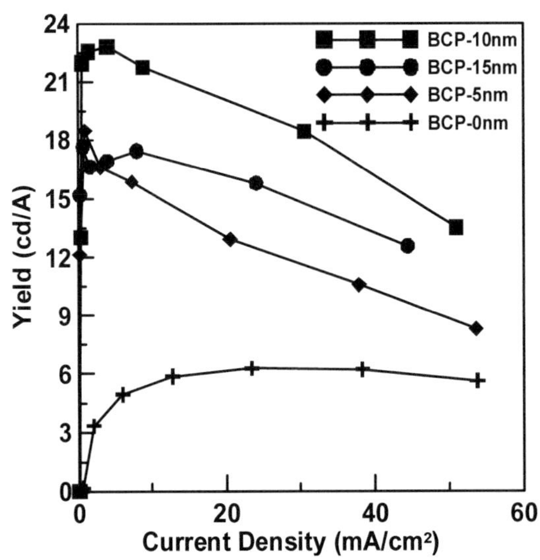

Fig. 1. Luminance efficiency-current density curves for different thicknesses of HBL

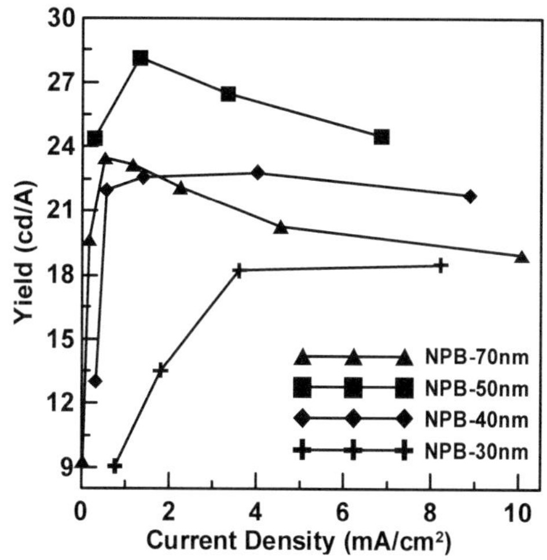

Fig. 2. Luminance efficiency-current density curves for different thicknesses of HTL

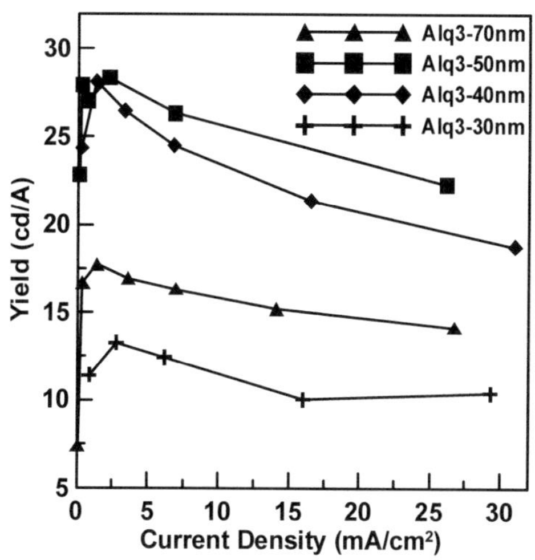

Fig. 3. Luminance efficiency-current density curves for different thicknesses of ETL

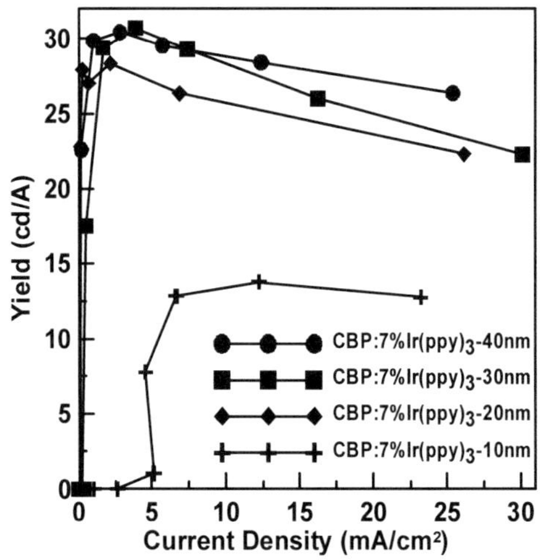

Fig. 4. Luminance efficiency-current density curves for different thicknesses of EML

Finally the doping concentration of ($Ir(ppy)_3$) was adjusted as shown in Table 2. We found in Fig. 5 that when the doping concentration was 7 wt% the device has optimum luminance efficiency. In this study we also changed different hole blocking layers (HBL) BCP and TPBi, whose structures are shown in Table 1, respectively. It is found from luminance efficiency - current density in Fig. 5 that using TPBi as hole blocking layer could enhance the luminance efficiency further to 34.2 cd/A. This is because the HOMO of TPBi (6.7 eV) is higher than the HOMO of BCP (6.3 eV), which confines holes and excitons in the emitting layer more effectively and prevents direct diffusion of holes or excitons to Alq3 layer. After comparing the effects of different HBLs on luminance efficiency as shown in Fig. 5, it is found a suitable material of HBL can improve the luminance efficiency of device effectively.

From the results of these serial experiments, it is found that whenever the thickness of an organic layer has been adjusted, the luminance efficiency of the device has increased (as shown in Fig. 6). It is because the amount of electron-hole of the device becomes more balanced via the optimization of the hole transport layer and the electron transport layer. In addition, extending the charge carrier recombining area and using more suitable HBL material and suitable thickness of its thin film will gradually increase and improve the luminance efficiency of the device.

NO.	Sub.	ITO	NPB	$Ir(ppy)_3$:CBP concen.	thickness	HBL	Alq3	LiF	Al
F1	Plastic (PET)	ITO (80Ω/□)	50	4%	40	BCP 10	50	0.5	65
D3				7%*					
F2				8%					
F3				10%					

*optimum parameters

Table 2. Adjustment parameters of PHOLEDs with different $Ir(ppy)_3$ doping concentrations (unit: nm)

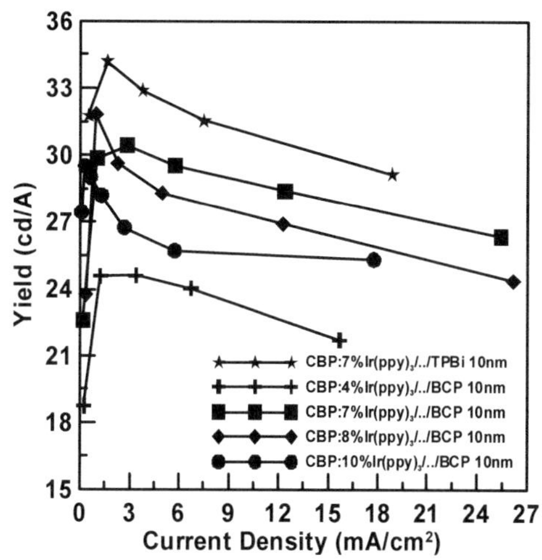

Fig. 5. Luminance efficiency-current density curves of PHOLEDs with different $Ir(ppy)_3$ doping concentrations and HBL materials

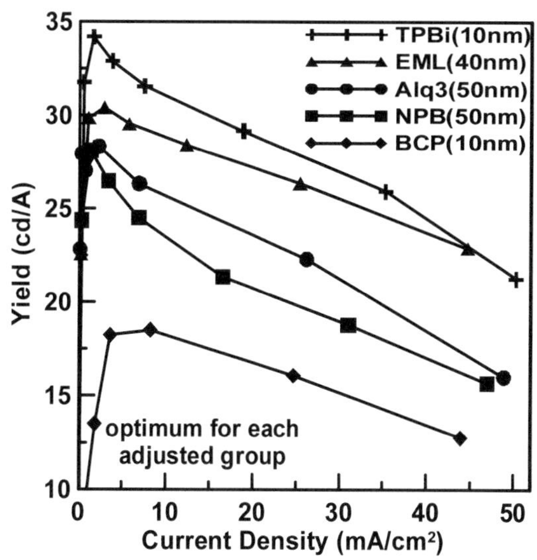

Fig. 6. Luminance efficiency-current density of optimum thickness for each adjusted layer

3.2 Spin-coating (α-NPD:NPB)+THF as the buffer layer and evaporating NPB as the hole transport layer to improve fluorescent OLED characteristics

Because the spin-coating rotation speed directly affects the film thickness, we fixed the spin-coating time at 35 seconds from the beginning of the experiment. The rotation speed of the spinα-NPD:NPB layer (3500, 4000, 4500 and 5000 r.p.m.) was varied to study how the device optoelectronic characteristic was affected by the rotation speed of spin-coating. The device structure was PET/ ITO (160 nm)/spin-coated (NPB:α-NPD)+THF (different rotation speed)/evaporated NPB (41 nm)/Alq3 (52 nm)/Bpy-OXD (15 nm)/LiF (0.5 nm)/Al (135 nm). The device energy band structure was shown in Figs. 7(a). Because the coated film thickness will become thick under lower rotation speed, from Fig. 7(b), we can see the lowest current density flowing through the device at the lowest rotation speed (3500 r.p.m.). According to the Mott-Gurney rule [17], we know that the current density presents an inverse-proportion relation with the thickness, and the greatest NPB spin-coating layer thickness can be estimated to appear at a rotation speed of 3500 r.p.m. However, the layer thickness will become thinner as the rotation speed increases and the current density will increase accordingly. However, at 5000 r.p.m., we found that the device current density decreased. The major reason results from too high rotation speed which causes the centrifugal force to become too strong, preventing the NPB+THF solution from attaching onto the ITO surface; hence, the film thickness becomes uneven. From Fig. 8 luminance-current density characteristic curve, we can observe the same trend in the current density-voltage curve. Hence, we determined that 4500 r.p.m. is the optimal spin-coating parameter to get the best optoelectronic characteristic.

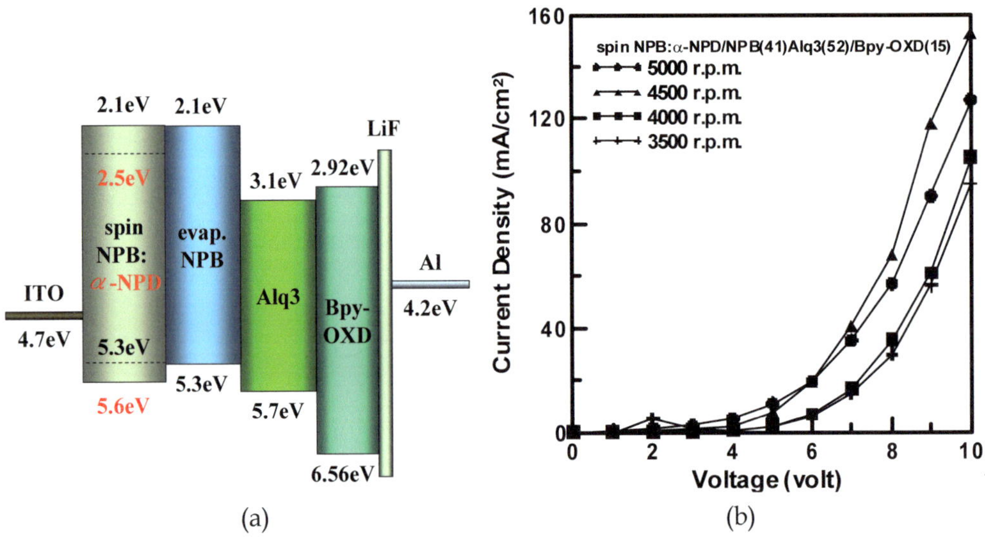

Fig. 7. (a) The energy band structure of the device with spin-coating (NPB:α-NPD)+THF buffer layer; (b) J-V characteristics for different rotation speeds

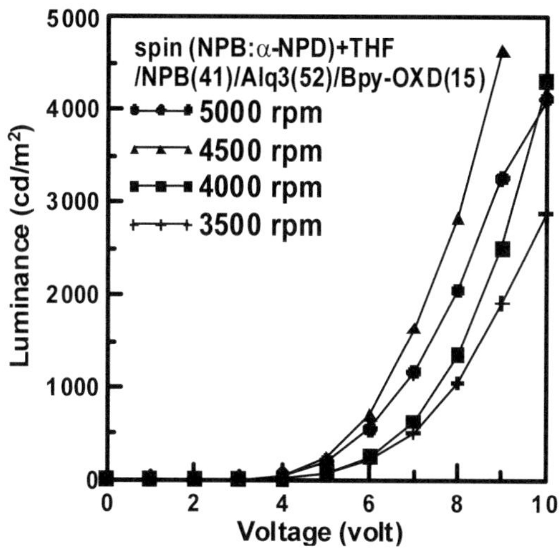

Fig. 8. L–V characteristics for spino-coating (NPB:α-NPD)+THF with different rotation speeds

To further verify the first HTL (NPB:α-NPD)+THF manufacturing process versus improving the device optoelectronic characteristics, the following experiment was conducted. Devices with PET/ITO(160 nm)/ HTL/Alq3(52 nm)/Bpy-OXD(15 nm)/LiF(0.5 nm)/Al(135 nm) structures wherein the HTL was produced via the spin-coating or thermal-evaporation process or both, are shown in Table 3.

Under same current density, the optoelectronic characteristics of Devices G1 and G2 (employing both spin-coated and evaporated double HTLs structures) were higher than that of traditional devices made with single HTL using only thermal-evaporation (Device G3), as shown in Fig. 9(a) and 9(b). The device characteristics with double HTLs were also found better than those with the single HTL spin-coated structure (NPB:α-NPD)+THF only (Device G4). Because spin-coating the film in the amorphous mode on the ITO surface does not easily generate pin holes, and the organic layer contact with the ITO becomes tighter, the HTL surface roughness can be greatly improved [18-21]. Therefore, the holes are injected more easily from the ITO electrode into the NPB layer. Furthermore, the optoelectronic characteristics of Device G1 were better than Device G2 because spin-coating (NPB:α-NPD)+THF has similar p-type doping properties that enable holes to be injected efficiently [22]. NPB and α-NPD co-doping will produce band bending and enable the holes to tunnel from the ITO electrode into the HTL, forming an Ohmic contact interface [23]. Consequently the Device G1 exhibited the best performances than the other devices. Fourier Transform Infrared Spectroscopy (FTIR) was also performed on the HTL films to study the variation in molecular structure. Moreover, from the lifetime test results shown in Fig. 10, the lifetime of Device G1 was the longest, increased by about 41 % longer than Device G3 (fabricated without spin-coating (NPB:α-NPD)+THF layer).

Devices	HTL thickness	
	(NPB:α-NPD) dissolved in THF then spin-coating	evaporate
G1	spin (NPB:α-NPD)+THF	NPB
	58 nm	41 nm
G2	spin NPB +THF	NPB
	37 nm	41 nm
G3	spin (NPB:α-NPD)+THF	NPB
	0 nm	41 nm
G4	spin (NPB:α-NPD)+THF	NPB
	58 nm	**0 nm**

Table 3. Different parameters of hole transport layer structures (unit: nm)

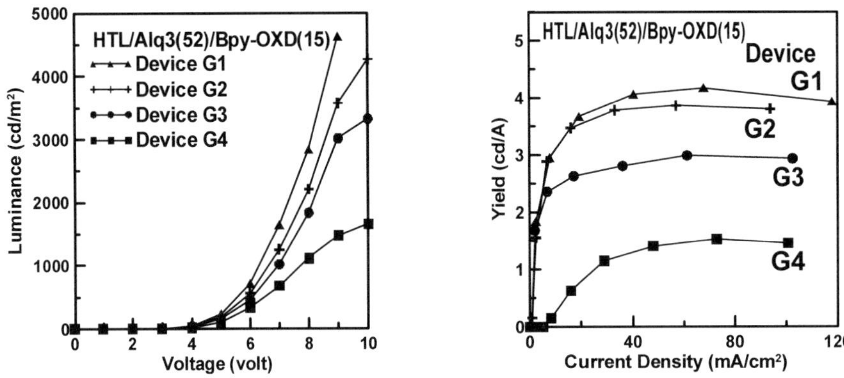

(a) (b)

Fig. 9. (a) L–V and (b) Y–J characteristics for different HTL structures

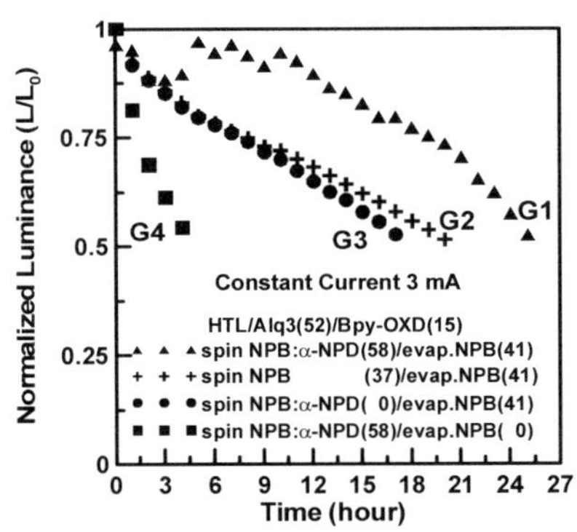

Fig. 10. The lifetime of flexible OLEDs with different HTL structures, and the devices were packaged by evaporating m-MTDATA 500 nm on the device surface

4. Conclusion

This research successfully improved the luminance efficiency and lifetime of flexible fluorescent and phosphorescent organic light emitting diodes by optimizing organic layer thicknesses or inserting a spin-coated buffer layer. From the results, the best phosphorescent device structure (ITO/ NPB (50nm)/ $Ir(ppy)_3$:CBP (40nm)/ TPBi (10nm)/ Alq3 (50nm)/ LiF

(0.5nm)/ Al (65nm)) has a maximum luminance efficiency of 34.2 cd/A by optimizing organic layer thicknesses. This device performance improvement is attributed to increased hole injection and improved hole-electron balance. In addition, extending the charge carrier recombining area and using more suitable HBL material and suitable thickness of its thin film will gradually increase and improve the luminance efficiency of the device. Comparing devices with TPBi and BCP as the HBL, it was found that devices fabricated with TPBi as the HBL exhibited better luminance efficiency than BCP.

We also discussed the effect with/without a spin-coated buffer layer on device performance. It was found that devices inserted with a spin-coated buffer layer had better brightness, luminance efficiency and lifetime than conventional devices (without spin-coating buffer layer). Furthermore, the research demonstrated the advantage of using spin-coating co-doping (NPB:α-NPD)+THF (exhibit the similar p-type doping properties and metal-like Ohmic contact interface) plus thermal-evaporation in fabricating flexible fluorescent organic light emitting diodes with double-hole transport layer structures. The performance of the best Device G1 exhibited a luminance of 4634 cd/m^2 at 9 V and a luminance efficiency 4.18 cd/A at 68 mA/cm^2. Compared with the single thermal-evaporated NPB only HTL structure, the luminance efficiency of device with spin-coated (NPB:α-NPD)+THF as buffer layer was increased by about 1.32 cd/A and its half-lifetime was increased by about 41 % longer than the device without spin-coated buffer layer.

5. References

[1] C. W. Tang, and S. A. VanSlyke, (1987). Organic electroluminescent diodes. Appl. Phys. Lett., Vol. 51, pp. 913-915.

[2] Z. Liu, J. Pinto, J. Soares, E. Pereira, (2001). Efficiency multilayer organic light emitting diode. Synth. Met., Vol. 122, pp. 177-179.

[3] Y. Shirota, Y. Kuwabara, H. Inada, T. Wakimoto, H. Nakada, Y. Yonemoto, S. Kawami, K. Imai, (1994). Multilayered organic electroluminescent device using a novel starburst molecule, 4,4 ,4 -tris(3-methylphenylphenylamino)triphenylamine, as a hole transport material. Appl. Phys. Lett., Vol. 65, No. 7, pp. 807-809.

[4] J. Shi, C.W. Tang, (1997). Doped organic electroluminescent devices with improved stability. Appl. Phys. Lett., Vol. 70, No. 13, pp. 1665-1667.

[5] C. Giebeler, H. Antoniadis, D. D. C. Bradley, Y. Shirota, (1999). Influence of the hole transport layer on the performance of organic light-emitting diodes. J. Appl. Phys., Vol. 85, No. 1, pp. 608-615.

[6] Z. Liu, Z. Weiming, J. Rongbin, Z. Zhilin, J. Xuejing and X. Minzhao, (1996). Organic thin film electroluminescent devices with ZnO:Al as the anode. J. Phys.: Condens. Matter., Vol. 8, pp. 3221-3228.

[7] F. Zhu, B. Low, K. Zhang, and S. Chua, (2001). Lithium–fluoride-modified indium tin oxide anode for enhanced carrier injection in phenyl-substituted polymer electroluminescent devices. Appl. Phys. Lett., Vol. 79, pp. 1205-1207.

[8] W. L. Yu, J. Pei, Y. Cao, and W. Huang, (2001). Hole-injection enhancement by copper phthalocyanine (CuPc) in blue polymer light-emitting diodes. J. Appl. Phys., Vol. 89, pp. 2343-2350.

[9] S. A. Carter, M. Angelopoulous, S. Karg, P. J. Brock, and J. C. Scott, (1997). Polymeric anodes for improved polymer light-emitting diode performance. Appl. Phys. Lett., Vol. 70, pp. 2067-2069.

[10] J. C. Scott, S. A. Carter, S. Karg, and M. Angelopoulous, (1997). Polymeric anodes for organic light-emitting diodes. Synth. Met., Vol. 85, pp. 1197-1200.

[11] T. M. Brown, J. S. Kim, R. H. Friend, F. Cacialli, R. Daik, and W. J. Feast, (1999). Built-in field electroabsorption spectroscopy of polymer light-emitting diodes incorporating a doped poly(3,4-ethylene dioxythiophene) hole injection layer. Appl. Phys. Lett., Vol. 75, pp. 1679-1681.

[12] M. A. Baldo, S. Lamansky, P. E. Burrows, M. E. Thompson, and S. R. Forrest, (1999). Very high-efficiency green organic light-emitting devices based on electrophosphorescence. Appl. Phys. Lett., Vol. 75, pp. 4-6.

[13] C. Adachi, M. A. Baldo, M. E. Thompson, and S. R. Forrest, (2000). High-efficiency organic electrophosphorescent devices with tris(2-phenylpyridine)iridium doped into electron-transporting materials. Appl. Phys. Lett., Vol. 77, pp. 904-906.

[14] M. Ikai, S. Tokito, Y. Sakamoto, T. Suzuki and Y. Taga, (2001). Highly efficient phosphorescence from organic light-emitting devices with an exciton-block layer. Appl. Phys. Lett., Vol. 79, pp. 156-158.

[15] T. D. Anthopoulos, Jonathan P. J. Markham, E. B. Namdas, Ifor D. W. Samuel, S. C. Lo and P. L. Burn, (2003). Highly efficient single-layer dendrimer light-emitting diodes with balanced charge transport. Appl. Phys. Lett., Vol. 82, pp. 4824-4826.

[16] B. D. Chin, M. C. Suh, M. H. Kim, S. T. Lee, H. D. Kim, and H. K. Chung, (2005). Carrier trapping and efficient recombination of electrophosphorescent device with stepwise doping profile. Appl. Phys. Lett., Vol. 86, pp. 133505-1~4.

[17] M. A. Lambert and P. Mark, Current Injection in Solids (Academic Press, New York, 1970).

[18] Y. S. Tsai, S. H. Wang and S. L. Chen, (2009). Performance Improvement of Flexible Organic Light-Emitting Diodes with Double Hole Transport Layers by Spin-Coating and Evaporation. Jpn. J. Appl. Phys., Vol. 48, pp. 052103-1~3.

[19] Y. S. Tsai, S. H. Wang, S. Y. Chen, S. Y. Su, F. S. Juang, (2009). Efficiency improvement of flexible fluorescent and phosphorescent organic light-emitting diodes by inserting a spin-coating buffer layer. Thin Solid Films, Vol. 517, pp. 5334~5342.

[20] S. Y. Su, F. S. Juang and Y. S. Tsai, (2006). High Efficiency Flexible Organic Light-Emitting Diodes by Phosphorescent Dopant. Master thesis of Graduate Institute of Electro-Optical and Materials Science, National Formosa University.

[21] S. L. Chen, F. S. Juang and Y. S. Tsai, (2007) Double Hole Transport Layers Deposited by Spin-Coating and Thermal-Evaporating for Flexible Organic Light Emitting Diodes. Master thesis of Graduate Institute of Electro-Optical and Materials Science, National Formosa University.

[22] A. Yamamori, C. Adachi, T. Koyama and Y. Taniguchi, (1998). Doped organic light emitting diodes having a 650-nm-thick hole transport layer. Appl. Phys. Lett., Vol. 72, pp. 2147~2149.

[23] J. Blochwitz, M. Pfeiffer, T. Fritz, and K. Leo, (1998). Low voltage organic light emitting diodes featuring doped phthalocyanine as hole transport material. Appl. Phys. Lett., Vol. 73, pp. 729~731.

9

a-Si:H TFT and Pixel Structure for AMOLED on a Flexible Metal Substrate

Chang-Wook Han, Chang-Dong Kim and In-Jae Chung
LG Display Co. Ltd.
Korea

1. Introduction

Flexible displays have attracted enormous interest from research institutes. The images of the flexible displays include bendable display, and electronic paper. The technology of flexible displays includes many domains: the display modes, flexible substrates, and backplanes.
Organic light emitting diodes (OLEDs) are considered the natural choice for flexible displays because of wide viewing angle, no cell gap problem, and the potential for using solution process. In addition, OLEDs can also be fabricated at relatively low temperature, which enable the use of plastic substrate. However, a key challenge for flexible OLEDs display is oxygen and moisture sensitivity. A device lifetime of 10,000 hrs requires maximum moisture permeability of $5x10^{-6}$ $g/m^2/day$. Plastic has 10~1000 $g/m^2/day$ of moisture permeability. Other challenge for OLEDs is high power consumption, which is also one of key issues for portable flexible displays along with weight and thickness. In the long run, OLEDs are excellent candidate for flexible application.
Thin metal foils have an alternative flexible substrate for the fabrication of OLED displays. They have potential advantages over plastic such as lower moisture permeability and higher process temperature. They have much better dimensional stability, and present a perfect diffusion barrier to water and oxygen. Because the metal foils have higher elastic modulus (~200 GPa), and lower coefficient of thermal expansion (18 ppm/°C), they have excellent dimensional stability, which allows multi-photo process comparable to those of glass. Conductive substrate can provide for common power signal and shielding as well as heat dissipation. However, the metal surfaces are usually rougher than most plastics, and the foils can wrinkle if flexed to very small bending radius.
In this paper, hydrogenated amorphous silicon thin film transistors (a-Si:H TFTs) fabricated on a metal foil substrate is reported for active matrix OLEDs (AMOLEDs) displays. The electrical properties of a-Si:H TFTs fabricated on a metal foil substrate are introduced. To increase the stability of a-Si:H TFTs fabricated at low temperature on a metal foil substrate, negative bias applied to metal foil substrate can also recover the shifted threshold voltage during idle time. A new cathode-contact structure employing a normal top-emitting OLED, which is suited to n-type a-Si:H TFT backplane is proposed.

2. Electrical performances of a-Si:H TFT fabricated on a metal foil

2.1 Fabrication of a-Si:H TFT on a metal foil

Metal foils have potential advantage over plastic. They tolerate high temperature processes in contrast to glass and even more to plastics. They also have much better dimensional stability because the metal foils have higher elastic modulus (~200 GPa), and lower coefficient of thermal expansion (~10 ppm/°C).
However, flexible metal foil substrates do not easily process with traditional TFT equipments such as stepper, PECVD, and etchers that are compatible with rigid glass substrates. To utilize traditional TFT equipments, a flexible metal foil substrate can bond to a rigid glass plate. In this case, the adhesive used for bonding the metal foil to a rigid glass plate reduces the maximum process temperature from the value that the substrate itself would withstand.
The adhesive covered on the rigid glass plate and 76-μm-thick metal foil was stuck on them. A spin-on polymer (SOP) or spin-on glass (SOG) as an initial surface planarization spread on a metal foil substrate in order to reduce the surface roughness of the metal substrate. 200-nm-thick silicon nitride (SiNx) formed by plasma-enhanced chemical vapor deposition (PECVD). This layer insulates electrically TFT layers from metal substrate. The electrical insulation layer also served as the mechanical bond between the TFT layers and the substrate. The root-mean-square (RMS) roughness of the metal foil substrate after planarization was 5 nm. The structure of the a-Si:H TFT was an inverted staggered-type structure, which was fabricated by a conventional five-mask process. The flexible metal substrate carrying TFT array separated from the rigid glass plate. To protect mechanical stress from being able to degrade the electrical performance of a-Si:H TFT, the adhesive between the metal foil and the glass plate was softened by chemical before separation. The performance of TFT did not change before and after separation the flexible metal substrate carrying TFT array from the rigid glass plate. Fig. 1 shows a cross section of a-Si:H TFT, which was fabricated on flexible metal foil.

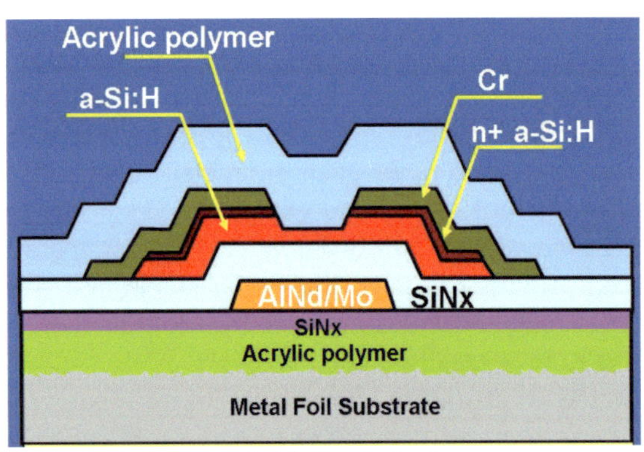

Fig. 1. Structure of TFT fabricated on metal foil substrate

2.2 Electrical performances of a-Si:H TFT dependent on planarization layer

In flexible displays employing metal foil substrate, planarization is required to reduce the surface roughness of metal foil and electrically insulate TFT array from conductive metal foil. Spin-on glass (SOG) and spin-on polymer (SOP) were selected as planarization dielectric. 2

㎛-thick siloxane as SOG or 2 ㎛-thick acryl as SOP were coated and 0.2 ㎛-thick SiNx insulation was deposited by PECVD.
Table 1 shows the result for capacitance and leakage current for SOG/SiNx, SOP/SiNx, and single SiNx layers. Both SOG/SiNx and SOP/SiNx are acceptable as planarization and insulation layer in the viewpoint of the insulation and capacitive coupling characteristics. For large display, the unit capacitance of less than 2 nF/cm^2 was suggested.
Hence, the capacitance of 1.2 and 2.1 nF/cm^2 for SOG/SiNx and SOP/SiNx, respectively is allowable. Surface root-mean-square (RMS) roughness of bare metal foil was 110 nm. It improved to 12 and 10 nm after the planarization layer such as SOG/SiNx and SOP/SiNx was coated on the metal foil substrate, respectively.

Planarization	Curing Condition(°C)	Insulation Layer	Capacitance (nF/cm^2)	Leakage Current (nA/cm^2)
SOG 2 ㎛ (Siloxane)	230	SiNx 200 nm	1.2	24
SOP 2 ㎛ (Acryl)	150	SiNx 200 nm	2.1	23
-	-	SiNx 200 nm	27.1	720

Table 1. Capacitance and leakage current for a package of planarization and insulation layers

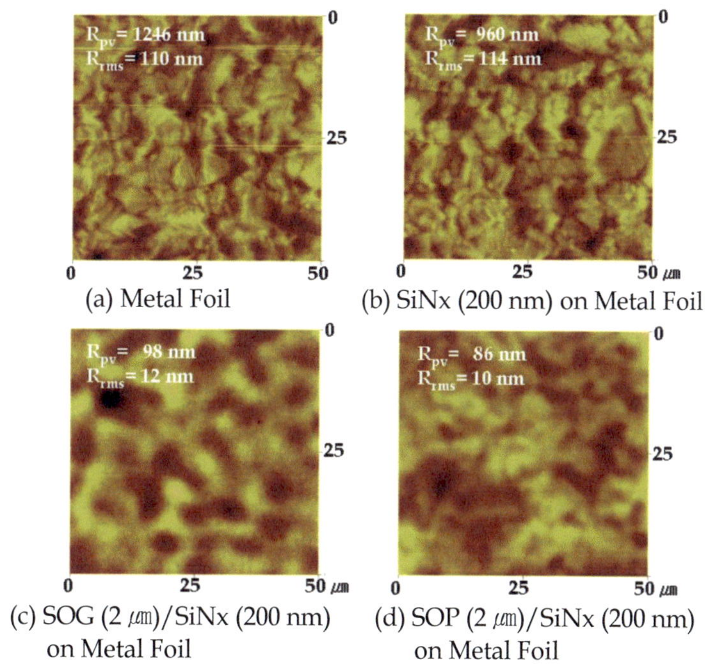

Fig. 2. AFM images and surface roughness of (a) metal foil and planarized with (b) SiNx 200 nm, (c) SOG 2 ㎛ / SiNx 200 nm, and (d) SOP 2 ㎛ / SiNx 200 nm

The surface images measured by atomic force microscope (AFM) are shown in Fig. 2.
Inverted staggered a-Si:H TFTs were fabricated by the process flow as mentioned in the previous section. Fig. 3 shows the transfer characteristics of a-Si:H TFTs as planarization layers. Electrical properties measured by semiconductor characterization system (Keithley 4200). The electrical performances of a-Si:H TFTs as planarization layers is compared as shown in Table 2.
There were not remarkable differences of the electrical performances of a-Si:H TFTs as planarization layers of SOG/SiNx and SOP/SiNx. For both TFTs, the off current is about 10^{-13} A and the on current at gate voltage of 20 V is about 10^{-6} A at a drain voltage of 10 V, resulting in an on-off current ratio of 10^{7}. We obtain a subthreshold slope 0.62 ~ 0.64 V/decade, demonstrating a sharp device turn-on. We also obtain a threshold voltage and mobility of 0.3 ~ 0.6 V and 0.27 ~ 0.45 cm²/Vs, respectively, in the saturated regime.

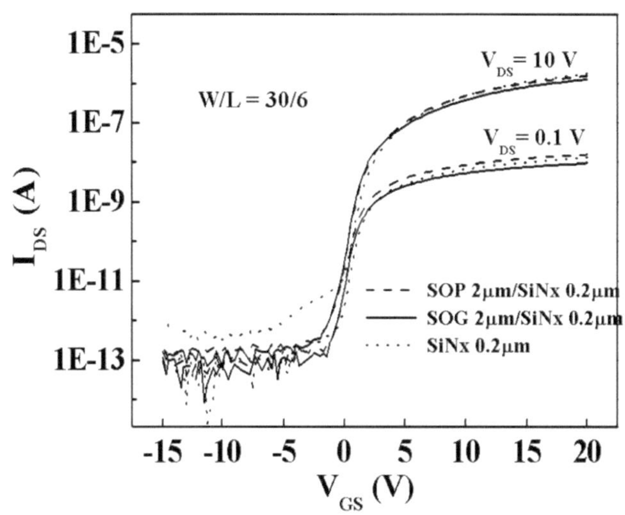

Fig. 3. Transfer characteristics of a-Si:H TFTs as planarization layers: SiNx 200 nm, SOG 2 μm / SiNx 200 nm, and SOP 2 μm / SiNx 200 nm. The size of TFT is W = 30 μm and L= 6 μm.

No	Planarization	Insulation Layer	TFT Characteristics				Remark
			Mobility (cm²/Vs)	Threshold Voltage(V)	Subthreshold Slope(V/dec.)	Off Current(A)	
1	SOG 2 μm (Siloxane)	SiNx 200 nm	0.30~0.44	0.33	0.639	2.1×10^{-13}	From 5 TFTs
2	SOP 2 μm (Acryl)	SiNx 200 nm	0.27~0.45	0.61	0.618	2.0×10^{-13}	From 5 TFTs
3	-	SiNx 200 nm	0.40	1.08	0.647	8.0×10^{-13}	From 1 TFT

Table 2. Electrical performances of a-Si:H TFTs as planarization layers

However, SOG/SiNx wrinkled after TFT fabrication as shown in Fig. 4 (a). Many TFTs on the wrinkled SOG failed to measure. All TFTs employing SOP/SiNx secured electrical performances. The SOG/SiNx wrinkled after passivation process. We used acrylic polymer for passivation, which is the same material as a planarization. Dimension mismatch may occur between SOG and acrylic polymer during thermal treatment of passsivation material. The length of SOG is the same as that of acrylic polymer. As the sample cools to room temperature, the SOG and acrylic polymer want to contract by different amounts due to the difference of the coefficient of thermal expansion (CTE). The CTE of the SOG is usually smaller than that of the acrylic polymer ($\alpha \approx 1 \times 10^{-6}/°C$ for the SOG, $\alpha \approx 50\sim70 \times 10^{-6}/°C$ for the acrylic polymer). The strain of the SOG is negative as compared with that of acrylic polymer, this means that the SOG is under compressive stress. This compressive stress can make the SOG film wrinkle. When a single SiNx was used as a planarization, the off current is high as 8×10^{-13} A, and the threshold voltage is also high as 1.08 V because of the large surface roughness of single SiNx on the rough metal foil. The rms roughness of 200 nm-thick SiNx on rough metal foil is 114 nm. After fabrication of TFT on metal foil planarized by SiNx, most of TFTs were delaminated from substrate as shown Fig. 4 (c).

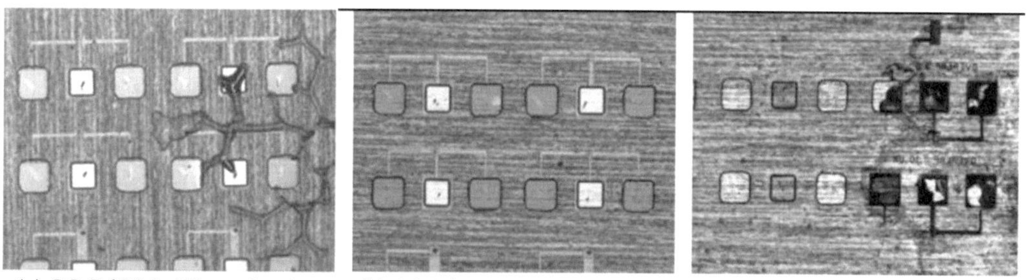

(a) SOG/SiNx Planarization (b) SOP/SiNx Planarization (c) SiNx Planarization

Fig. 4. Optical images of a-Si:H TFT as planarization materials

3. Effect of passivation materials on electrical performances of a-Si:H TFTs fabricated on a metal foil

3.1 Electrical performances of a-Si:H TFTs passivated with SiNx or acrylic polymer

a-Si:H TFTs were fabricated on flexible metal foil as mentioned in previous section. 200 nm-thick SiNx and acrylic polymer were chosen as passivation layer on TFT channel. Fig. 5 shows the transfer characteristics of a-Si:H TFTs dependent on passivation materials.

In the case of SiNx-passivation, the off current is about 10^{-13} A and the on current at gate voltage of 20 V is about 10^{-6} A at a drain voltage of 10 V, resulting in an on-off current ratio of 10^{7}. We obtain a subthreshold slope ~ 0.69 V/decade, demonstrating a sharp device turn-on. We also obtain a threshold voltage and mobility of 2.1 V and 0.61 cm^2/Vs, respectively, in the saturated regime. In the case of acryl polymer employed as a passivation layer, most electrical characteristics of the acryl-passivated TFT except threshold voltage were comparable to those of the SiNx-passivated TFT. The off current is about 10^{-13} A and the on current at gate voltage of 20 V is about 10^{-6} A at a drain voltage of 10 V, resulting in an on-off current ratio of 10^{7}. We obtain a subthreshold slope ~ 0.61 V/decade, demonstrating a sharp device turn-on. We also obtain a threshold voltage and mobility of 1.0 V and 0.54 cm^2/Vs, respectively, in the saturated regime.

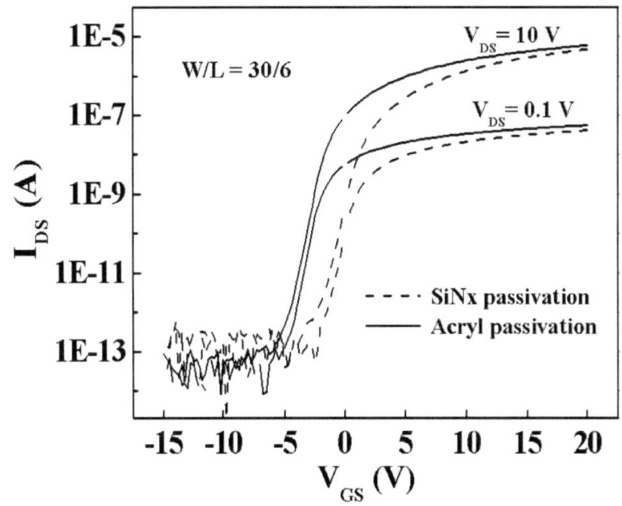

Fig. 5. Transfer characteristics of a-Si:H TFTs as passivation materials: SiNx and acrylic polymer. The size of TFT is W = 30 ㎛ and L = 6 ㎛.

The lower threshold voltage of the acryl-passivated TFT was explained by more positive fixed charge at the back interface. Less hydrogen and nitrogen in the acryl-passivation could not eliminate the fixed charges. Then some free electrons which were accumulated in the back interface were controlled by the gate bias.

3.2 Effect of passivation materials on electrical performances of a-Si:H TFTs under mechanical stress

To investigate the effect of passivation layer on the performances of a-Si:H TFTs under mechanical stress, we stressed both TFTs by outward cylindrical bending at the radius of curvature R = 5 mm as shown in Fig. 6. Less than R = 4 mm, most of TFTs employing acrylic polymer or SiNx as passivation layer were failed. When the radius of curvature was 5 mm, the strain on the surface of acryl and SiNx was 0.0082 (0.82 %) and 0.0077 (0.72 %), respectively. Eq. 1 considering single acryl (3 ㎛) or SiNx (substrate-insulation, gate-insulation and backchannel passivation 0.9 ㎛) layer calculated the strain.

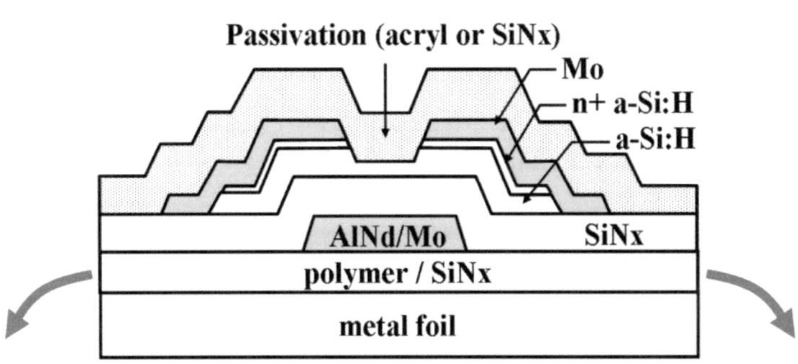

Fig. 6. Cross section of an inverted staggered type TFT fabricated on a flexible metal substrate. The arrows indicate the bending direction.

$$\varepsilon_{top} = \left[\frac{\left(d_f = d_s\right)}{2R} \right] \frac{\left(1+2\eta+\chi\eta^2\right)}{(1+\eta)(1+\chi\eta)} \qquad (1)$$

where $\eta = d_f/d_s$, $\chi = E_f/E_s$, here d_f (acryl) = 3, d_f(SiNx) = 0.9 μm and d_s = 76 μm are film and substrate thickness, respectively. E_f (acryl) = 3.2, E_f (SiNx) = 183 and E_s = 200 GPa are Young's modulus of film and substrate, respectively.

Before bending the TFTs, electrical performances were measured. The TFTs were stressed for 24 hours. The transfer characteristics of TFTs in the bended condition (R = 5 mm) were measured at arbitrary intervals as shown in Fig. 7. A zero hour duration time meant that TFTs measured as flattened before bending them.

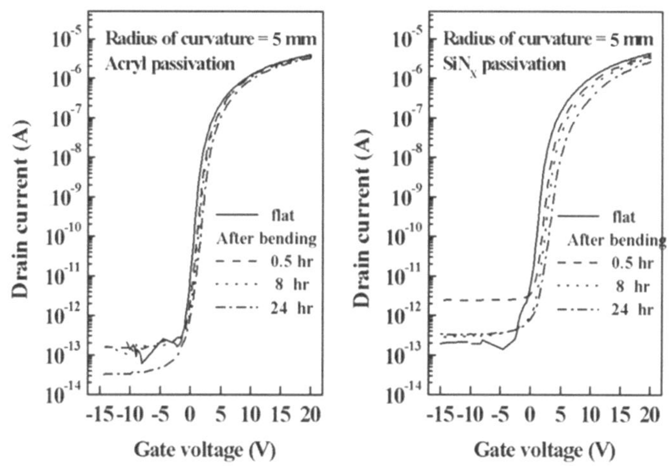

Fig. 7. Transfer characteristics for the acryl and SiNx-passivated TFT as bending duration time. The radius of curvature R is 5 mm.

As the bending duration time increased, the transfer curves shifted to the right direction and the electrical performances deteriorated as shown in Fig. 8. The field effect mobility decreased and the sub-threshold slope increased. The field effect mobility and the subthreshold slope are much dependent on the width of the tail state. The field effect mobility is smaller and the subthreshold slope is larger as the a-Si:H tail state increases. This increase of the width of the tail states might be due to the relative increase of the deep states in tail states. When the flexible TFTs are bended outward, Si-Si bonds in a-Si:H, close to the gate dielectric interface, are shrunk. The increase in the number of the strained Si-Si bonds implies a wider conduction-band-tail distribution. This may be the reason why the field effect mobility decreases and the subthreshold slope increases as the TFTs are bended. The threshold voltage also increased as the bended-duration time increased. Stutzmann proposed that the mechanical stress could influence the metastable defect creation in an a-Si:H film and that defect formation was enhanced in strained regions. The increased threshold voltage decreased again as time passed after the bended-TFTs flattened and almost recovered after thermal annealing. This meant that the degraded electric characteristics were curable. Hence, we may also conclude that the increased threshold voltage by an increase in strained regions is attributed to increasing the number of the meta-stable dangling bonds.

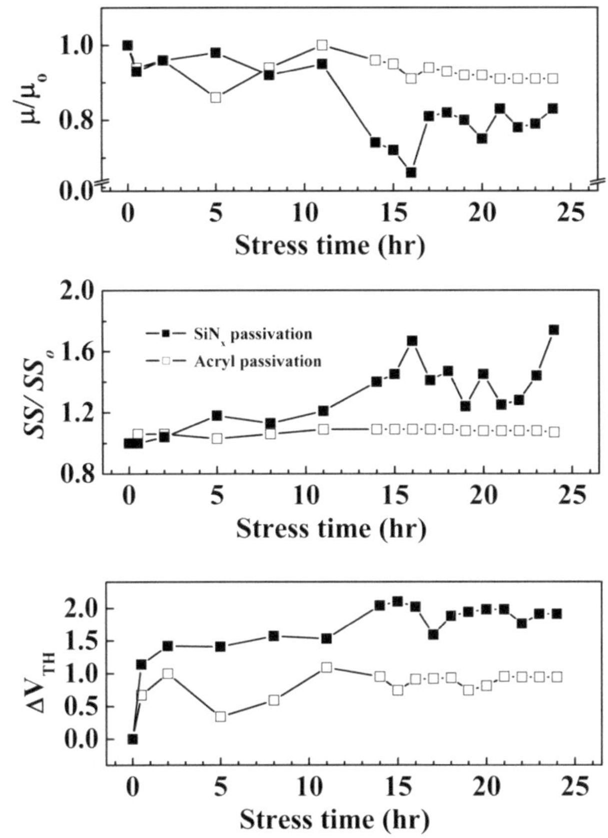

Fig. 8. Relative field-effect mobility, subthreshold slope and threshold voltage plotted as functions of stress time. TFTs were measured as bended except 0-hour stress time. The data at 0-hour stress time were measured from the flattened-TFTs before bending.

The deterioration of the electrical performances in the SiNx-passivated TFT under mechanical stress was more severe than that in the acryl-passivated TFT. Under an outward bending, the bending momentum elongates the TFT-films in the upper part of TFT on a metal substrate and compresses the flexible metal substrate in the lower part. Neutral plane is free from any stress between the elongated and the compressed part. The Young's modulus of acrylic polymer (3.2 GPa) is lower than that of SiNx film (183 GPa). As the 3 μm-thick acrylic polymer was employed as the passivation layer in place of 0.3 μm-thick SiNx film, a total thickness of TFT-films was increased. Therefore, the position of the neutral plane may shift from mid-surface toward the TFT-films. It accompanies with decrease of stress applied on the TFT-films. Hence, the acryl-passivated TFT could endure mechanical stress better than the SiNx-passivated TFT.

3.3 Effect of passivation materials on electrical performances of a-Si:H TFTs by humidity permeation

To investigate moisture permeation phenomenon dependent on the passivation materials, a humidity-temperature test was carried out at 65 °C and 90 % relative humidity (RH). We compared the electrical performances of the SiNx-passivated and the acryl-passivated TFTs after 3-hour humidity-temperature test. Fig. 9 shows transfer curves before and after the humidity-temperature test. We could not find any significant shift of the threshold voltage in the SiNx-passivated TFT. The SiNx passivation did not allow moisture to penetrate into TFT. On the other hand, threshold voltage shifted as large as 2.8 V for the acryl-passivated TFT. The acrylic polymer could not keep the moisture from permeating. Other researchers have reported the effect of moisture on a-Si:H TFT.

Moisture increases the conductance of a-Si:H films. When an a-Si:H TFT without passivation was exposed to the air, the off current increased due to the absorbed air humidity. However, we could not find any increase of off current after the humidity-temperature test. The effect of the moisture on the acryl-passivated TFT might be different from the unpassivated-TFT. The moisture can not affect a front interface at the bottom of a-Si:H layer because it cannot penetrate a-Si:H layer. The negative fixed charges may increase at the back interface after the moisture permeates the acryl passivation. Then the negative charges may perhaps make the threshold voltage shift to the positive direction. More investigations are required to analyze what affects the threshold voltage increment after the humidity-temperature test.

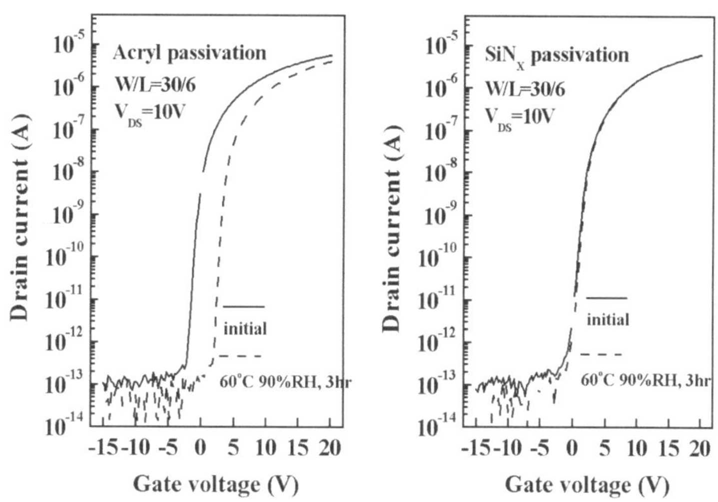

Fig. 9. Transfer characteristics for the acryl and SiNx-passivated TFT were measured before and after they were stored at 60°C and 90 % RH for 3 hours.

3.4 Effect of hybrid passivation layer on a-Si:H TFTs electrical characteristics

A hybrid structure which was composed of 50 nm-thick SiNx and 3 μm-thick acrylic polymer was chosen as passivation layer as shown in the Fig. 10. TFTs were stressed by outward cylindrical bending at the radius of curvature R = 5 mm. Before bending the TFTs, electrical performances measured. The TFTs were stressed for 24 hours. The transfer characteristics of

TFTs in the bended condition (R = 5 mm) were measured at arbitrary intervals as shown in Fig. 11 (left) and 12. A zero hour duration time meant that TFTs were measured as flattened before bending them. The mobility change (μ_{fe}/μ_{feo}) of 0.92, the subthreshold slope change (SS/SS_o) of 1.04, and the threshold voltage shift (ΔV_{TH}) of 1.03 are almost same as those of TFT employing the single acryl passivation. As the 50 nm-thick SiNx and 3 ㎛-thick acrylic polymer were employed as the passivation, the position of the neutral plane may shift from mid-surface toward the TFT-films. These results are similar to that of a single acrylic polymer passivation.

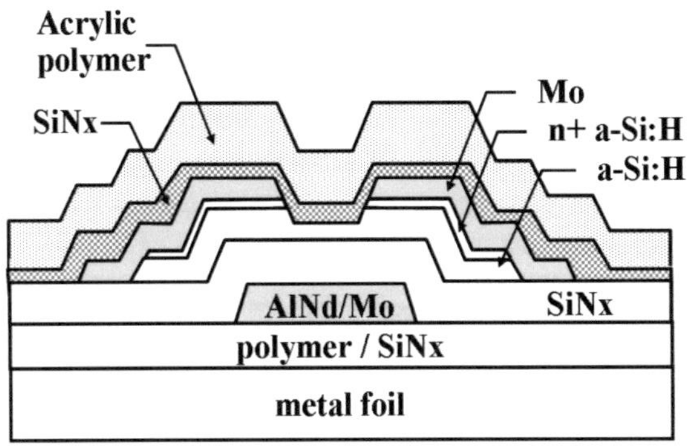

Fig. 10. Cross section of an inverted staggered type TFT employing 50 nm-thick SiNx and 2 ㎛-thick acrylic polymer passivation

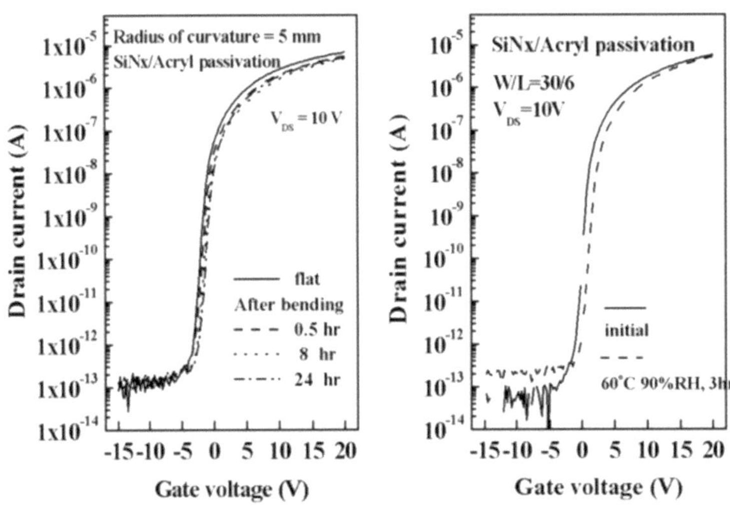

Fig. 11. Transfer characteristics for the hybrid passivation TFT as bending duration time. The radius of curvature R is 5 mm (left). Transfer characteristics for the hybrid passivation TFT stored at 60°C and 90 % RH for 3 hour (right)

To investigate moisture a permeation phenomenon dependent on the passivation materials, TFTs were kept at 65 °C and 90 % relative humidity (RH). The electrical performances of the

SiNx/acryl-passivated TFTs were measured after 3-hour humidity-temperature test. Fig. 11 (right) shows transfer curves before and after the humidity-temperature test. The threshold voltage shift was 1 V after 3-hour humidity-temperature test. It was reduced compared to the threshold voltage shift of the single acryl passivated TFT (2.8 V). 50 nm-thick SiNx could somewhat prevent moisture penetrating into TFT.

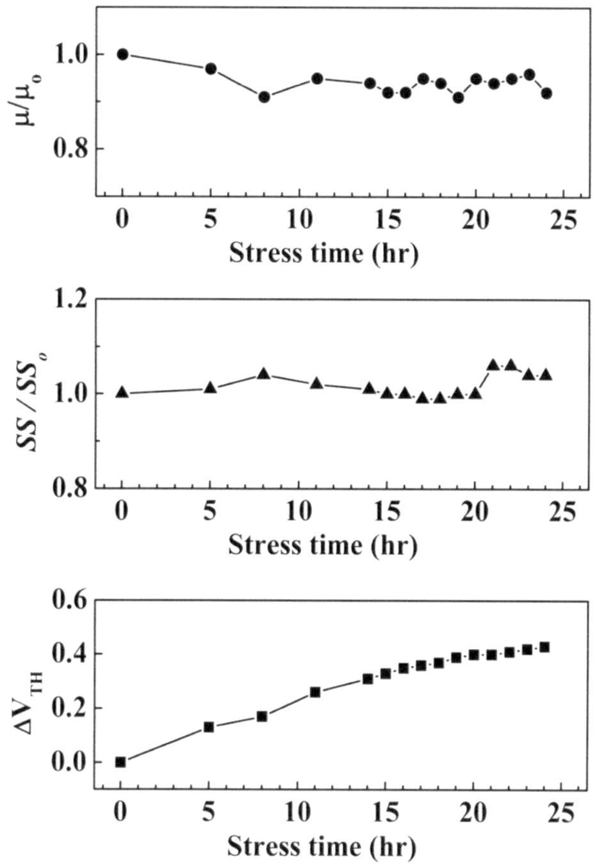

Fig. 12. Electrical performances for the hybrid passivation TFT are shown as bending duration time. TFTs were measured as bended except 0-hour stress time. The data at 0-hour stress time were measured from the flattened-TFTs before bending.

4. Substrate bias effect on the recovery of shifted threshold voltage in a-Si:H TFTs

4.1 Negative gate bias effects

TFT stability is a key issue for AMOLED displays. The threshold voltage of the a-Si:H TFTs changes with the application of the gate bias voltage. In a conventional 2-TFT pixel circuit, the output current is sensitive to the ΔV_{TH} of the driving TFT. This leads to a degradation of the luminance of the OLED pixel over time. The ΔV_{TH} is because of carrier trapping and the creation of dangling bonds at the interface of the a-Si:H and SiNx layers. Electrical performances measured under a prolonged gate bias of +15 V and a drain bias of +15 V at 65 °C applied for 3,500 s. Fig. 13 shows a drain current drop, a mobility change, and a ΔV_{TH} under BTS measurements. The BTS decreased the drain current by 32 % in comparison with the initial drain current. The threshold voltage shift was about 2.3 V whereas the mobility almost maintained under BTS measurements. The drain-current reduction was due to the threshold voltage shift not to the mobility change.

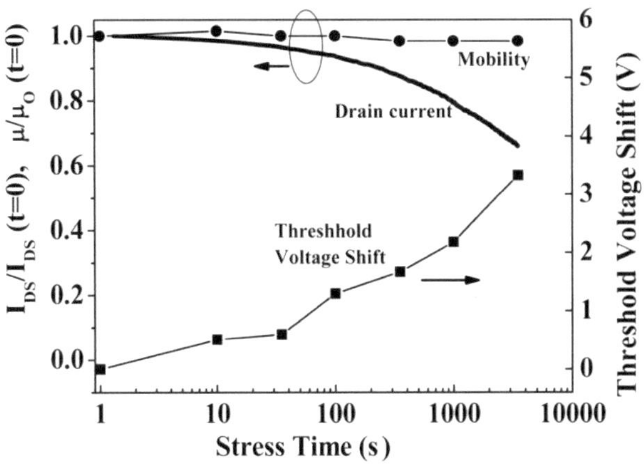

Fig. 13. Drain current, mobility, and threshold voltage transition of a-Si:H TFT under a bias-temperature-stress measurement (V_{GS}=V_{DS}=15 V, 65 °C). The current drop is attributed to the threshold voltage shift in a-Si:H TFT.

Researchers have introduced several methods to reduce the sensitivity of the current to ΔV_{TH}. They are the current-programmed circuit, voltage-programmed circuit, and reverse bias annealing methods. In particular, the reverse bias annealing method used the effect of negative gate bias applied to a-Si:H TFT on the negatively ΔV_{TH}. The electrons which were trapped at the interface of a-Si:H and SiNx by the positive gate bias were injected from the interface so that the gate potential for a desired electron current is reduced by de-trapped electrons. In reverse bias annealing method, ΔV_{TH} reduced by a fraction time annealing. During the fraction time, we applied a reverse bias to the gate electrode of a driving transistor. Then ΔV_{TH} was suppressed and the reduced drain-current can recover as the negative gate voltage. However, all of these methods need more than three TFTs and many bus lines for each pixel. These methods are too complicated in a driving scheme to achieve a highly reliable display.

When we used metal foil as a substrate, the surface of the metal foil should be planarized and insulated by dielectric materials, which is called as the planarization layer. We have to consider the captive coupling between the display circuits and the conductive metal foil substrate for the design of displays. The captive coupling from the metal foil substrate caused more data and scan line delay than in the glass substrate case. Therefore, the planarization layers preferred a larger thickness and a low dielectric constant to reduce the captive coupling. On the other hand, we can use the metal foil substrate as a common power. If a negative bias is applied to the metal foil substrate the captive coupled gate electrode is induced as a negative bias, which recovers the ΔV_{TH} of the a-Si:H TFT without additional complicated circuits.

4.2 Substrate bias effect on the recovery of shifted threshold voltage in a TFT array

We planarized the metal foil substrate by dielectric materials in order to reduce the surface roughness of the substrate. This planarization layer acted as a capacitor between the gate electrode and the metal foil substrate. When we grounded V_{DD} and biased the substrate as a negative voltage during idle time, the floating gate electrode of the driving transistor was charged with a negative voltage by the dielectric capacitor. If we use the substrate bias effect in AMOLED displays, we should consider the pixel circuits with two transistors and one capacitor (2T1C) as shown in Fig. 14. Just before idle time, a constant V_{DATA}, for example 0 V, is required to set the initial voltage of the floating gate electrode of the driver TFT. Next, a negative select voltage is required to maintain the off state of the switch TFT during the idle time. During the idle time, we grounded V_{DD} and V_{SUB} supplied a negative voltage to enable recovery of the degraded driver TFT. In the 2T1C circuit, the voltage of the floating gate (V_{FG}) is also influenced by the storage capacitance (C_{ST}). After the driving transistors were degraded, which meant a drain-current drop and a threshold voltage increase, the negative voltage at the substrate and grounded-V_{DD} lines during idle time enabled recovery of the degraded transistors.

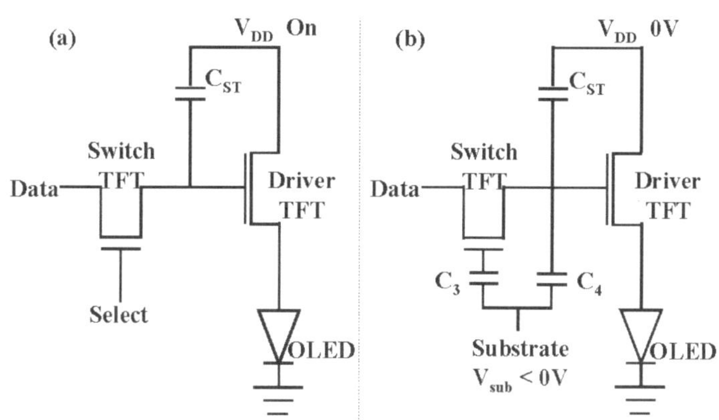

Fig. 14. Two-transistor and one-capacitor pixel circuits for AMOLED: (a) the operating circuit and (b) the recovery circuit during idle time (V_{SUB}<0 V, V_{DD}=0 V)

4-inch diagonal TFT array was designed for AMOLED display of which pixel circuit was 2T1C configuration. Fig. 15 shows pixel circuit considering parasitic and substrate capacitances. Node A indicates the floating gate electrode of the driving TFT, which will be induced negative bias by the negative substrate bias. The capacitance per unit area of planarization, SiNx/a-Si:H, and SiNx passivation is 2.4×10^{-17}, 1.27×10^{-16}, and 1.51×10^{-16}, respectively. The voltage at node A, which was charged by substrate bias, was verified by SPICE simulation. One pixel design and the simulation parameters are shown in Fig. 16.

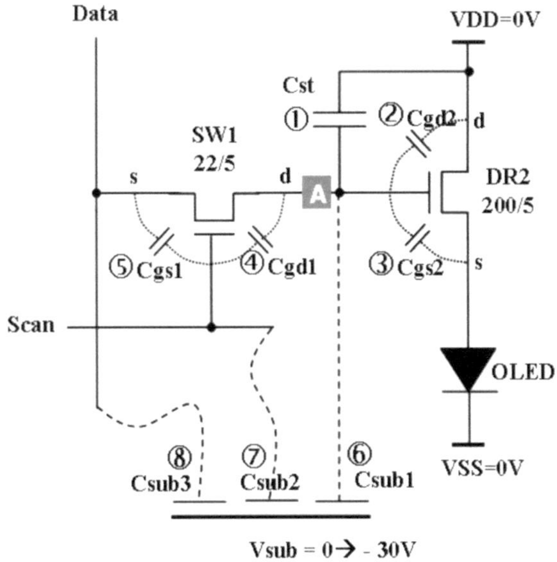

Fig. 15. Pixel circuit considering parasitic capacitances and substrate capacitances. The size of switching and driving TFT is 22/5 ㎛ and 200/5 ㎛, respectively. Node A means the floating gate electrode of the driving TFT.

Before supplying a negative bias at the metal foil substrate, V_{DD} was set as 0 V. Data and scan line were applied 0 and -5 V, respectively, then both lines were cut off from electrical sources. The gate electrode of the driving TFT floated. V_{SUB} swung from 0 to -30 V then the floating gate electrode of the driving TFT was negatively biased. The level of voltage is defined as capacitance sharing among capacitances connected to the node A. The voltage of the node A can be calculated by a following relationship.

$$V_A = \frac{C_{sub1}}{C_{st} + C_{gd2} + C_{gs2} + C_{gd1} + C_{sub1}} \bullet V_{sub} \quad (2)$$

Fig. 17 shows the simulation results of the voltage applied to node A as V_{SUB} swung from 0 to -30 V. The voltage of Node A (V_A) was compared from data simulated by SPICE and data calculated from Eq. (2).

4-inch diagonal TFT array which had 320 x 3 x 240 pixels was fabricated for AMOLED display of which pixel circuit was 2T1C configuration. Fig. 18 shows a-Si:H TFT backplane fabricated on a metal foil substrate. The a-Si:H TFT array and magnification of one pixel including 2-transistor and 1-capacitor is shown in Fig. 19. The recovery of the drain current and the negative shift of threshold voltage, as a function of the substrate bias, are shown in Fig. 20. The size of driving TFT is W = 200 and L = 5 ㎛. The V_{TH} of the TFT increased as 0.95 V and the drain-current decreased as 83 % of the initial value under a gate bias of +15 V and a drain bias of +15 V at 65 ℃ applied for 3,500 s. V_{TH} decreased as -1.19 and 1.74 V and the drain-current recovered 111 and 154 % of its initial value when the substrate voltage was -25 and -30 V at 65 ℃ applied for 3,500 s, respectively.

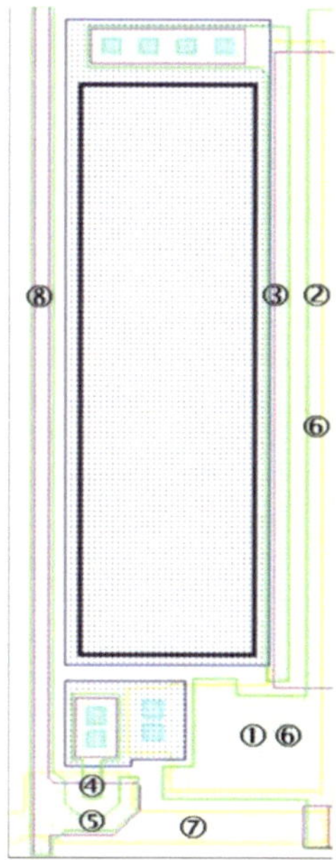

①	Cst	1.67E-13 F
②	Cgd2	1.00E-13 F
③	Cgs2	1.24E-13 F
④	Cgd1	5.31E-15 F
⑤	Cgs1	3.64E-14 F
⑥	Csub1	1.03E-13 F
⑦	Csub2	2.83E-14 F
⑧	Csub3	3.57E-14 F

Fig. 16. Lay-out of one pixel and simulation parameters to get the voltage at the floating gate electrode by the substrate bias

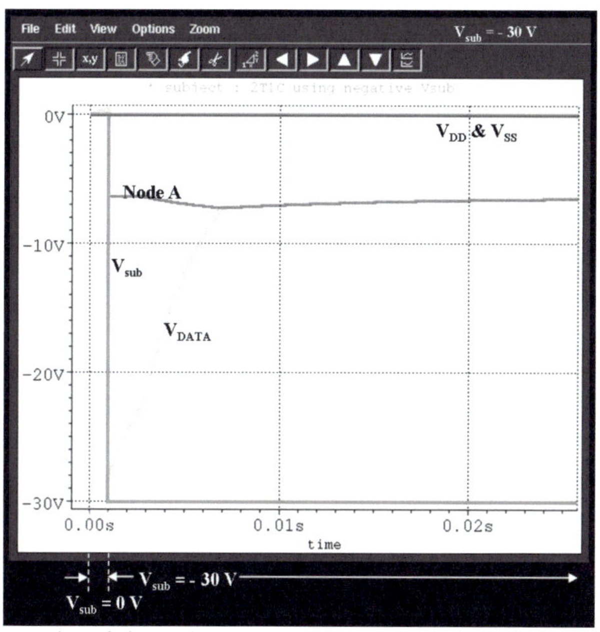

Fig. 17. Simulation results of the voltage applied to node A as V_{SUB} swung from 0 to -30 V. About -7.2 V is induced at the Node A.

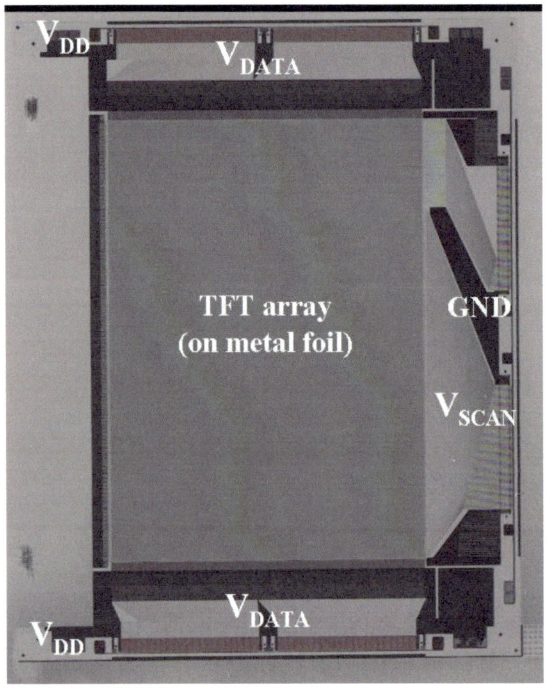

Fig. 18. 4 inch diagonal a-Si:H TFT backplane fabricated on a metal foil substrate. This is designed for AMOLED display.

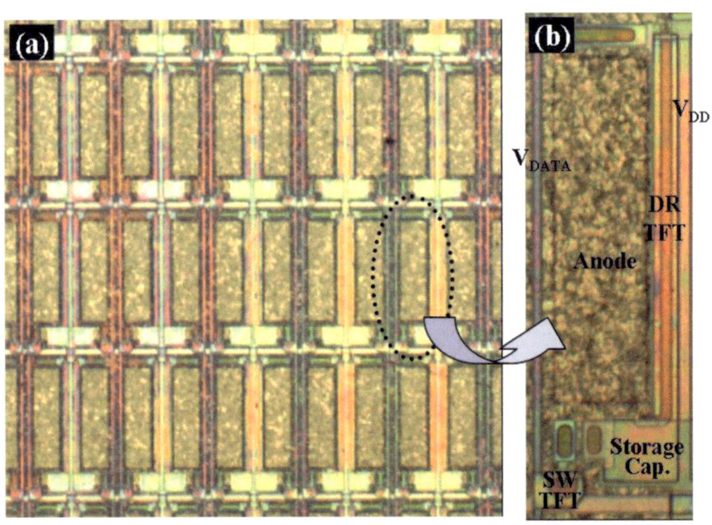

Fig. 19. (a) a-Si:H TFT array fabricated on a metal foil substrate (b) magnification of one pixel including 2-transistor and 1-capacitor.

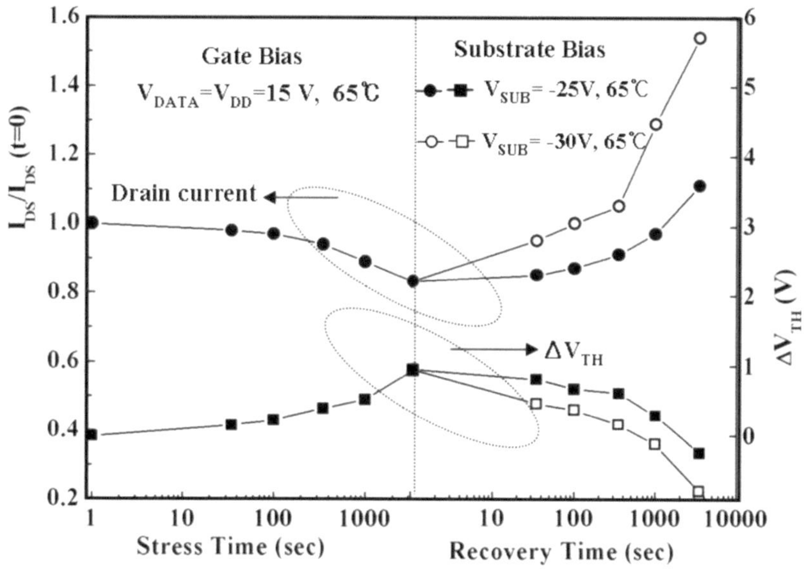

Fig. 20. Drain current and threshold voltage transition of a driving TFT as a function of stress and recovery time. $V_{DATA} = V_{DD} = 15V$ is applied to the driving TFT at 65 °C for stress time and V_{SUB} = -25 and -30 V is applied to metal foil substrate at 65 °C for recovery time.

5. Cathode contact top emitting OLED

5.1 Comparison between anode-contact and anode-contact configuration

To control the current of the OLED through the a-Si:H TFT, optimum pixel circuit designs should be considered. Since the conventional top emitting OLED (TOLED) has an anode at bottom and a cathode on top, it is electrically connected to the driving TFT through the anode. We call this structure an anode-contact structure with the TOLED (ACTOLED) as shown in Fig. 21 (a). In this structure, the current of the driving TFT is dependent of the operating voltage of the OLED (V_{OLED}). The situation can get even worse during the degradation of the display since the V_{OLED} increases affecting the gate-source voltage of the driving TFT. Some groups have proposed a cathode-contact structure to solve the problems generated from the ACTOLED. They employed an inverted-TOLED, i.e. making TOLED that has a cathode at bottom and an anode on top. However, one major challenge in the inverted-TOLED has been to prepare a reflective cathode providing an effective electron injection. Based on such considerations, low work-function metals, such as Mg and Li were introduced to reduce the turn-on voltage. However, there remained issues in employing highly reactive metals particularly their handling and operational stability. Another group has demonstrated a more complicated structure using an ultra thin Alq_3-LiF-Al trilayer as the electron-injection layer.

In this section, we propose a new cathode-contact structure employing the normal TOLED (CCTOLED) that has an anode at bottom and a cathode on top (Fig. 21 (b)). We also compare the electrical characteristics between the CCTOLED and ACTOLED pixel structure.

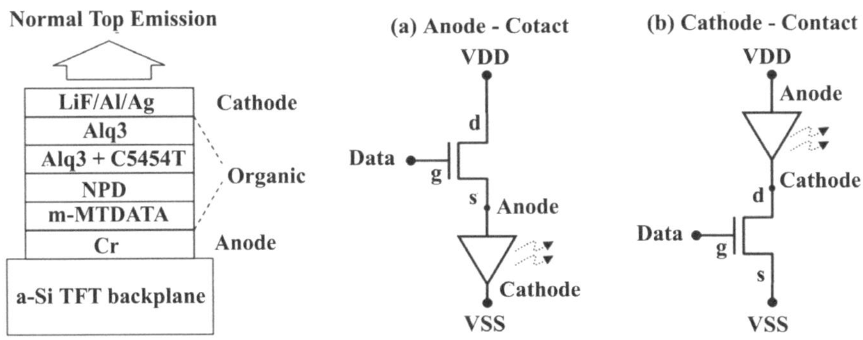

Fig. 21. Structure and circuit implementation of normal top-emission AMOLED (TOLED) pixel: (a) anode-contact with a-Si:H TFT (ACTOLED) and (b) cathode-contact with a-Si:H TFT (CCTOLED)

5.2 Process flow to make cathode-contact pixel structure

The schematic of the fabrication process is illustrated in Fig. 22. The a-Si:H TFT was fabricated on the glass substrate (Fig. 22 (a)). The structure of a-Si:H TFT was an inverted staggered type, which was made by a conventional 5-photomask process. We deposited a reflective anode by a sputter process and patterned by photolithography. It covered all the pixel-area as a common electrode keeping away from the contact area on the drain electrode of the TFT (Fig. 22 (b)). A step-covering layer was located over the step area of the anode to minimize the probability of the breakdown of the emission layer at the step area of the anode. It was made by 1 ㎛-thick polyimide which was spin-coated and photo-patterned opening the drain electrode of TFT. A separator layer which separates cathode layer as sub-pixels was made by 2 ㎛-thick negative photo-resist from spin coating and photolithography (Fig. 22 (c)). All organic layers including common layers for each color, such as hole-injection, hole-transport, and electron-transport layer were thermally evaporated through the shadow mask on the anode, not evaporated on the drain electrode of TFT (Fig. 22 (d)). Finally, electron-injection layer, cathode aluminum (Al) and silver (Ag) were thermally evaporated and then were made to contact the drain electrode of the TFT (Fig. 22 (e)). Each of the cathode layers of sub-pixel is automatically patterned during evaporation by separator. Then, the cathode-contact structure, employing a normal TOLED, was completed. The organic layers of the TOLED were prepared with the following structures: Cr (100 nm) /m-MTDATA (30 nm)/α-NPD (30 nm)/Alq_3+C545T (25 nm)/Alq_3 (35 nm)/LiF (0.5 nm)/Al (1 nm)/Ag (15 nm). The organic multilayer structure sequentially consisted of 4,4′,4″-tris(3-methylphenylphenylamino) triphenylamine (m-MTDATA, 30 nm) as the hole-injection layer, α-naphthylphenylbiphennyl (α-NPD 30 nm) as the hole-transport layer, tris-(8-hrydroxyquinoline) aluminum doped with 1 wt% 10-(2-Benzothiazolyl)-2,3,6,7-tetrahydro-1,1,7,7-tetramethyl-1H,5H,11H-(1)-benzopyropyrano (6,7-8-i,j)quinolizin-11-one (Alq_3+C545T, 25 nm) as the emitting layer, and tris-(8-hrydroxyquinoline) aluminum (Alq_3, 35 nm) as the electron-transport layer.

Fig. 23 shows a SEM image of the fabricated pixels. The cathode layer of sub-pixel is successfully isolated by separator (Fig. 23 (a)). And it is connected with the drain of a-Si:H TFT through the via hole which is formed by step-covering layer (Fig. 23 (b)).

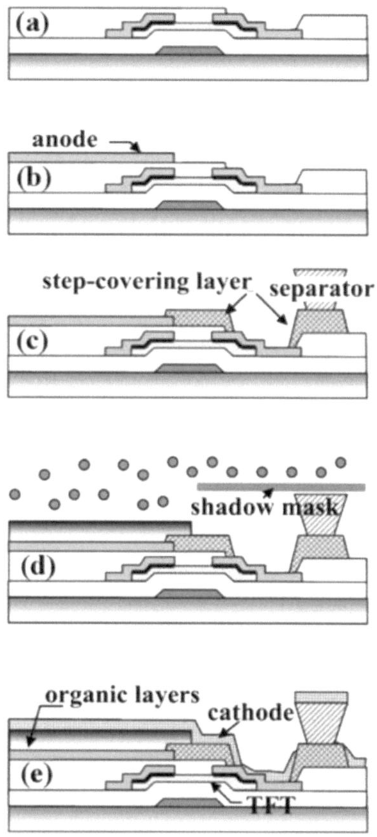

Fig. 22. Fabrication process flow of a newly proposed normal top-emission OLED pixel employing cathode-contact structure (a) a-Si:H TFT, (b) reflective anode, (c) step-covering layer and separator, (d) organic layer evaporation through the shadow mask on the anode, (e) cathode evaporation.

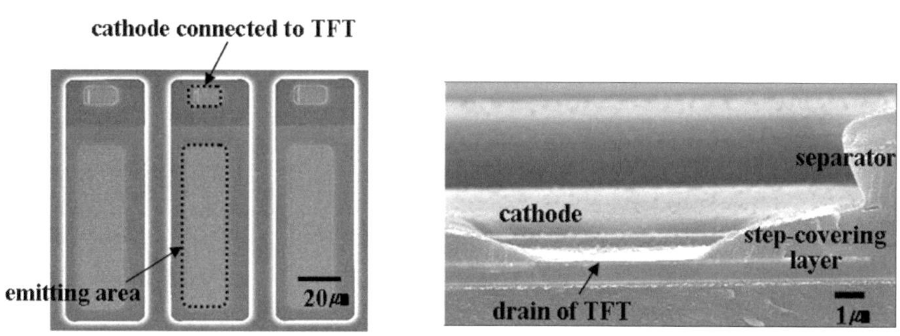

(a) Top view of CCOLED pixels (b) Cross section of contact area

Fig. 23. SEM image of fabricated cathode-contact type OLED pixel

5.3 Electro-optic characteristics

To investigate the pixel performances of the CCTOLED and ACTOLED cells employing the same TFT and TOLED, we designed and fabricated a unit cell having an emitting area of 1x1 mm^2. The off current of TFT was about 10^{-9} A. The on current, at a gate voltage of 20 V, was about 10^{-3} A at a drain voltage of 10 V resulting in an on-off current ratio of 10^6. We obtained a subthreshold slope of approximately 0.74 V/decade demonstrating a sharp device turn-on. The threshold voltage and the saturation mobility were 1.8 V and 0.34 cm^2/Vs, respectively. Fig. 24 shows the current of the OLED (I_{OLED}) as a function of the V_{DATA}. When the V_{SS} was grounded, the ACTOLED showed lower I_{OLED} as compared with the CCTOLED. The I_{OLED} of the ACTOLED and the CCTOLED at V_{DATA} = 14 V and V_{DD} = 27 V were 1.2 x 10^{-4} A and 9.5 x 10^{-4} A, respectively.

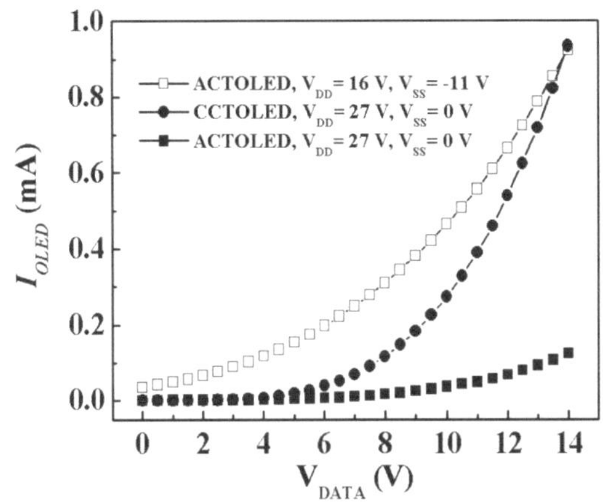

Fig. 24. Current of the OLED as a function of V_{DATA} compared between the ACTOLED and the CCTOLED

In the case of the ACTOLED, the effective gate voltage (V_{GE}) of the driving TFT decreased, which was defined as the difference of the V_{DATA} and the source voltage of the driving TFT (V_S) as shown in Fig. 21. The lower current of the ACTOLED was attributed to this lowered-V_{GE}. As a result, the ACTOLED was inappropriate for a high luminance display when the V_{SS} was grounded. When a negative voltage was supplied at the V_{SS} in order to increase the current value in the ACTOLED as shown in Fig. 21, the I_{OLED} of the ACTOLED could reach the same amount as that of the CCTOLED at V_{DATA} = 14 V. However the I_{OLED} of the ACOLED at V_{DATA} = 0 V, V_{DD} = 16 V, and V_{SS} = -11 V and the CCOLED at V_{DATA} = 0 V, V_{DD} = 27 V, and V_{SS} = 0 V were 3.4 x 10^{-5} A and 3.6 x 10^{-8} A, respectively. In the case of the ACTOLED even though the V_{DATA} was set as 0 V, the V_{GE} was not zero because the V_S of the driving transistor was induced as a negative voltage when the V_{SS} was set as a negative value. The contrast ratio, which means the ratio of the white and black level, is low because of a leakage light at the black level. On the other hand, the I_{OLED} of the CCTOLED independent of the V_{OLED}, this meant that the V_{GE} was always equal to the V_{DATA}. Therefore, the CCTOLED was suitable for better image performances having high luminance and contrast ratio at the same driving conditions. Fig. 25 shows the current density characteristics of the CCTOLED as a function of the V_{DATA} and the V_{DD}. Well-saturated characteristics were shown over V_{DD} = 15 V and less than V_{DATA} = 10 V which were the driving condition for real displays.

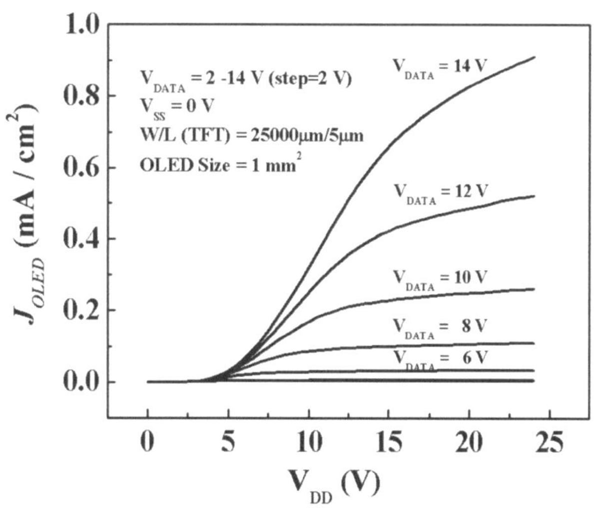

Fig. 25. Current density of the OLED as a function of V_{DD} and V_{DATA}

6. Conclusion

In this paper, electrical performances and new approaches to increase the stability of a-Si:H TFT fabricated on a metal foil substrate were reported. A new cathode-contact structure employing a normal top emitting OLED also was proposed and compared with an anode-contact structure by experimental data.

76-μm-thick metal foil laminated on the rigid glass plate. On top of this foil, the rough surface was planarized and the inverted staggered a-Si:H TFT was fabricated at 150°C. The acrylic polymer as a planarization layer was well matched to a-Si:H TFT fabricated at 150°C. The a-Si:H TFT of which size was W=30 μm and L=6 μm showed the good electrical performances. The off current was about 10^{-13} A and the on current at gate voltage of 20 V is about 10^{-6} A at a drain voltage of 10 V, resulting in an on-off current ratio of 10^{7}. We obtained a threshold voltage and mobility of 1.0 V and 0.54 cm^2/Vs, respectively, in the saturated regime.

The effect of passivation layer on the performances of a-Si:H TFT under mechanical stress was investigated. The acryl-passivated TFT could endure mechanical stress better than the SiNx-passivated TFT. However, a larger threshold voltage shift was observed for the acryl-passivated TFT when a humidity-temperature test was carried out. The hybrid passivation, which was composed of SiNx and acrylic polymer was proposed. It secured the degradation of electrical performances under the mechanical stress and somewhat prevented moisture penetrating into TFT.

We have studied a negative bias effect using the substrate bias without additional circuits to enable recovery of the degraded drain-current of a driving TFT in 2T1C pixel circuit, which was fabricated on a metal foil substrate. When V_{DD} was grounded and the substrate was biased as a negative voltage during idle time, the floating gate electrode of the driving transistor was induced as a negative voltage by the dielectric capacitor. The degraded drain current of the driving transistor can be recovered during the idle time by simply applying a negative substrate bias. The power consumption can be neglected during the idle time because no current flows.

Cathode-contact structure pixel structure employing normal TOLED was proposed for a-Si:H TFT backplane. The new top-emission AMOLED pixel structure employing the TOLED as well as the cathode-drain contact structure was proposed and fabricated. The structure of TOLED had a cathode at bottom and an anode on top. The negative photo-resist separator wall successfully patterned the pixel cathode layers. As the electrical performances of CCTOLED and ACTOLED were compared, the CCTOLED was verified more suitable for better display performance having a high luminance and a high contrast ratio.

7. References

Ashtiani, S.,J.; Servati, P.; Striakhilev, D. & Nathan, A. (2005). A 3-TFT Current-Programmed Pixel Circuit for AMOLEDs. *IEEE Trans. Electron Devices*, Vol. 52, (July, 2005) 1514-1518, ISSN 0018-9383

Burrows, E.; Graff, G. L.; Gross, M. E.; Martin, P.M.; Hall, M.; Mast, E.; Bonham, C.; Bennet, W.; Michalski, M.; Weaver, M. S.; Brown, J. J.; Fogarty, D.& Sapochak, L. S. (2001). Gas permeation and lifetime tests on polymer-based barrier coatings, *Proceedings of SPIE*, pp. 75-83, ISBN 9780819437501, Feb. 2001, Society of photo-optical Instrumentation Engineers, Bellingham

Chandler, H. H.; Bowen, R. L. & Paffenbarger, G. C. (1971). Physical properties of a radiopaque denture base material. *J. Biomed. Mater. Res.*, Vol. 5, (July, 1971) 335-357, ISSN 1549-3296

Chen, C. W.; Lin, C. L. & Wu, C. C. (2004). An effective cathode structure for inverted top-emitting organic light-emitting devices, *Appl. Phys. Lett.*, Vol. 85, 2469-2471, ISSN 0003-6951

Dobbertin, T.; Werner, O.; Meyer, J.; Kammoun, A.; Schneider, D.; Riedl, T.; Becker, E.; Johannes, H. H. & Kowalsky, W. (2003). Inverted hybrid organic light-emitting device with polyethylene dioxythiophene-polystyrene sulfonate as an anode buffer layer, *Appl. Phys. Lett.*, Vol. 83, 5071-5073, ISSN 0003-6951

Fu, L.; Lever, P.; Tan, H. H.; Jagadish, C.; Reece, P. & Gal, M. (2002). Suppression of interdiffusion in GaAs/AlGaAs quantum-well structure capped with dielectric films by deposition of gallium oxide, *Appl. Phys. Lett.*, Vol. 82, 3579-3583, ISSN 0003-6951

Goh, J. C.; Jang, J.; Cho, K. S. & Kim, C. K. (2003). A New a-Si:H Thin-Film Transistor Pixel Circuit for Active-Matrix Organic Light Emitting Diodes, *IEEE Electron Device Lett.*, Vol. 24, 583-585, ISSN 0741-3106

Hicknell, T.,S.; Fliegel, F. M. & Hicknell, F. S. (1990). The Elastic Properties of Thin-Film Silicon Nitride, *Proceedings of IEEE Ultrasonic Symposium*, pp. 445-448, Institute of Electrical & Electronics Enginee

Hiranaka , K.; Yoshimura, T. & Yamaguchi, T. (1989). Effects of the Deposition Sequence on Amorphous Silicon Thin-Film Transistors, *Jpn. J. Appl. Phys.*, Vol. 28, 2197-2200, ISSN 0021-4922

Hong, M. P.; Seo, J. H.; Lee, W. J.; Rho, S. G.; Hong, W. S.; Choi, T. Y.; Jeon, H. I.; Kim, S. I.; Kim, B. S.; Lee, Y. U.; Oh, J. H.; Cho, J. H. & Chung, K. H. (2005) Large Area Full Color Transmissive a-Si TFT-LCD Using Low Temperature Processes on Plastic Substrate, *Proceedings of SID Symposium*, Vol. 36, pp.14-17, Boston, MA, May 2005, SID, San Jose, CA, ISSN 005-966x

Hong, Y. T.; Heiler, G.; Kerr, R.; Kattamis, A. Z.; Cheng, I. C. & Wagner, S. (2006) Amorphous Silicon Thin-Film Transistor Backplane on Stainless Steel Foil Substrate for AMOLEDs, *Proceedings of SID Symposium*, Vol. 37, pp.1862-1865, San Francisco, CA, June 2006, SID, San Jose, CA, ISSN 006-966x

Jones, B. L. (1985). The Effect of Mechanical Stress on Amorphous Silicon Transistors, *J. Non-Cryst. Solids*, Vol. 77&78, 1405-1408, ISSN 0022-3093

Lee, J. H.; You, B. H.; Han, C. W.; Shin, K. S.& Han, M. K. (2005) A New a-Si:H TFT Pixel Circuit Suppressing OLED Current Error Caused by the Hysteresis and Threshold Voltage Shift for Active Matrix Organic Light Emitting Diode, *Proceedings of SID Symposium*, Vol. 36, pp. 228-231, Boston, MA, May 2005, SID, San Jose, CA, ISSN 005-966x

Liao, W. S. & Lee, S. C. (1997). Novel Low-Temperature Double Passivation Layer in Hydrogenated Amorphous Silicon Thin Film Transistors, *Jpn. J. Appl. Phys.*, Vol. 36, 2073-2076, ISSN 0021-4922

Lim, B. C.; Choi, Y. J.; Choi, J. H. & Jang, J. (2000). Hydrogenated Amorphous Silicon Thin Film Transistor Fabricated on Plasma Treated Silicon Nitride, *IEEE Trans. Electron Device*, Vol. 47, 367-371, ISSN 0018-9383

Lin, Y. C.; Shieh, H. P. D. & Kanicki, J. (2005). A Novel Current-Scaling a-Si:H TFTs Pixel Electrode Circuit for AM-OLEDs, *IEEE Trans. Electron Devices*, Vol. 52, 1123-1131, 0018-9383

Lustig, N.& Kanicki J. (1989). Gate dielectric and contact effects in hydrogenated amorphous silicon-silicon nitride thin-film transistors, J. Appl. Phys., Vol. 65, 3951-3957, ISSN 0003-6951

Park, S. K.; Han, J. I. & Kim, W. K. (2001). Mechanics of indium-tin-oxide films on polymer substrate with organic buffer layer, *Proceedings of Mater. Res. Soc. Symp.*, Vol. 695, pp. 223-230, ISBN 1-55899-631-1, Boston, MA, Nov. 2001, MRS, Warrendale, PA.

Stutzmann, M. (1985). Role of mechanical stress in the light-induced degradation of hydrogenated amorphous silicon, *Appl. Phys. Lett.*, Vol. 47, 21- 23, ISSN 0003-8979

Suo, Z.; Ma, E. Y.; Gleskova, H. & Wagner, S. (1999). Mechanics of rollable and foldable film-on-foil electronics, *Appl. Phys. Lett.*, Vol. 74, 1177- 1179, ISSN 0003-6951

Tanielian, M.; Fritzsche, H.; Tsai, C. C.& Symbalisty, E. (1978). Effect of adsorbed gases on the conductance of amorphous films of semiconductor silicon-hydrogen alloys, *Appl. Phys. Lett.*, Vol. 33, 353 -356, ISSN 0003-6951

Tsujimura, T. (2004). Amorphous/Microcrystalline Silicon Thin Film Transistor Characteristics for Large Size OLED Television Driving, *Jpn. J. Appl. Phys.*, Vol. 43, 5122-5128, ISSN 0021-4922

Wagner, S.; Cheng, I. C.; Kattamis, A. Z.; Cannella, V. & Hong, Y. T. (2006). Flexible Stainless Steel Substrates for a-Si Display Backplanes, *Proceedings of IDRC Symposium*, pp. 13-15, Kent, Ohio, Sep. 2006, SID, San Jose, CA, ISSN 1083-1312

Wehrspohn, R. B.; Deane, S. C.; French, I. D.; Gale, I.; Hewett, J.; Powell, M. J. & Robertson, J. (2000). Relative importance of the Si–Si bond and Si–H bond for the stability of amorphous silicon thin film transistors, *J. Appl. Phys.*, Vol. 87, 144-154, ISSN 0021-8979

Yoon, J. K. & Kim, J. H. (1998). Device Analysis for a-Si:H Thin-Film Transistors with Organic Passivation Layer, *IEEE Electron Device Lett.*, Vol. 19, 335-337, ISSN 0741-3106

10

Organic Light Emitting Diode for White Light Emission

M.N. Kamalasanan, Ritu Srivastava, Gayatri Chauhan, Arunandan Kumar, Priyanka Tayagi and Amit Kumar
Center for Organic Electronics, Polymeric and Soft Materials Section, National Physical Laboratory (Council of Scientific and Industrial Research), Dr. K.S. Krishnan Road, New Delhi 110012, India

1. Introduction

During the last few years, research based on energy saving technologies is being given high priority all over the world. General lighting is one area in which large quantity of electrical energy is being spend and substantial energy saving is possible by using energy saving technologies. Conventional light sources like incandescent filament lamps in which a major part of the energy is wasted as heat and is a less energy efficient technology is being phased out. Other technologies like gas filled electrical discharge lamps are more efficient but are polluting. Therefore there is a need for energy efficient and clean light source and solid state lighting is one of the ways to address the problem

Organic light emitting diodes (OLED) is a new technology which has the potential to replace the existing lighting technologies. The attraction to organic semiconductors for lighting and display application has started during 1950-1960 because of the high fluorescence quantum efficiency exhibited by some organic molecules and their ability to generate a wide variety of colors. Study of electroluminescence (EL) in organic semiconductors have started in 1950s by Bernanose et.al (1953) using dispersed polymer films This was followed by the study of electroluminescence in anthracene single crystals by Pope et al (1963) and W.Helfrich et.al. (1965) who has studied the fundamental aspects of light generation in OLEDs. Since the single crystal based anthracence OLEDs fabricated by Pope et al (1963) were very thick and worked at very high voltages, the devices were not commercialized. In 1987, Tang and VanSlyke (1987) of Eastman Kodak has demonstrated a highly efficient multi layer OLED device based on vacuum evaporated aluminum tris 8-hydroxy quonoline (Alq_3)as the emitter material. The device had different layers for hole transporting, electron transporting and light emission. Transparent Indium Tin Oxide (ITO) and aluminum metal were the anode and cathode respectively. Quantum efficiency and luminescence efficiency of 1% and 1lm/W respectively were considered enough for commercial application. This work has stimulated a very intense activity in the field of Organic electroluminescence. Numerous improvements in device structure and addition of more layers having different functionalities were incorporated and are now on the verge of commercialization. Further, the developments in π- conjugated polymers by Heeger, MacDiarmid, and Shirakawa in

1977 for which they shared the 2000 Noble Prize in Chemistry as well as the report by Burroughes et al. (1990)of the first polymer (long chain molecules) light-emitting diode has also given a boost to the already expanding field of OLEDs. The new discovery of polymer light emitting diodes(PLEDs) have shown that even solution grown thin layers of a conjugated polymer can be used as an emitter material which has given new device concepts like ink jet printing and roll to roll processing of OLEDs. In 1998, Baldo et al (1998) showed that the efficiency of OLEDs can be improved by the incorporation of phosphorescent dyes. In this way, the triplets generated in the electron-hole recombination process (~75%) which are otherwise not used in light generation can be harvested to get light emission. This new development has enhanced the internal quantum efficiency of organic LEDs to nearly 100%. Sun et al (2006) introduced a different device concept that exploits a blue fluorescent in combination with green and red phosphor dopants, to yield high power efficiency and stable colour balance, while maintaining the potential for unity internal quantum efficiency. Two distinct modes of energy transfer within this device serve to channel nearly all of the triplet energy to the phosphorescent dopants i.e, retaining the singlet energy exclusively on the blue fluorescent dopant and eliminating the exchange energy loss to the blue fluorophore by direct excitation which allows for roughly 20 per cent increased power efficiency compared to a fully phosphorescent device. The device challenges incandescent sources by exhibiting total external quantum and power efficiencies that peak at 18.7 +/- 0.5 per cent and 37.6 +/- 0.6 lm/W, respectively, decreasing to 18.4 +/- 0.5 per cent and 23.8 +/- 0.5 lm/W at a high luminance of 500 cd/m^2.

Further, introduction of new technological concepts like electrical doping of transport layers has enhanced the OLED efficiency to more than 100 lm/W and enhanced life time of the devices to more than 100,000 hours which is better than the gas filled discharge lamps (Murano et al 2005). However, efficiency and lifetime are still considered widely as the big obstacles on the road of OLED development. A further improvement in the OLED performance relies on the more detailed understanding of the EL physics and the new development in the OLED materials, structure and fabrication.

Even though OLEDs of different colours have been developed with enough efficiency for commercialization, white light emitting organic LEDs have a special significance. It can be used for general lighting, back light for LED displays and for display applications. Since Organic materials are band emitters, OLEDs using these materials are mono chromatic and have low half width. Single broad band emitters developed so far has low efficiencies. To get white light emission from organic materials special efforts have to be made. Many methods like optical doping using fluorescent and phosphorescent materials as well as down conversion using inorganic phosphors have been used to get white light emission. Compared to other sources, OLEDs are thin, flat, lightweight, flexible and emitts cold light. WOLED having high energy efficiency of 62 lm/W have been demonstrated on R&D level by OSRAM Opto Semiconductor GmbH (Nov. 2009) and >100 lm/W reachable in future. They can produce high quality white light (CRI ~ 80), which are diffuse and non glaring large area light source. Further, they can be instantly on/off and are driven at low voltages. They have various colors and different color temperatures functionality.

Numerous white OLEDs have been fabricated (Kido et al 1994, 1996, Dodabalapur et al 1994, Yang et al 1997). In the fabrication of full colour display all three primary colours have equal importance but white light emission has drawn particular attention because any desired colour range can be achieved by filtering of white light (Strukeji et al 1996, Zhang et al 2001). To obtain high quality (high CRI) white light, all the three primary colors red, green, and blue have to be produced simultaneously. Since it is difficult to obtain all primary emissions from a single molecule, excitation of more than one organic species is often necessary, thus introducing color stability problems. Due to the different degradation rate of the employed organic compounds, the emission color of the device can, in fact, change with time.

The first white OLED was produced by Kido and his colleagues in 1994. This device contained red, green and blue light emitting compounds that together produce white light. But there were some problems with these devices such as their efficiency was less than 1 lm/W, required large driving voltage and burned out quickly. But now the efficiency of these devices has increased very fast. White emission from OLEDs can now be achieved in both small molecule and polymer systems (Strukeji et al 1996, Granstom et al 1996, Jordan et al 1996). The yearly progress in the efficiencies of conventional LEDs, nitride LEDs and white OLEDs is shown in Fig.1.

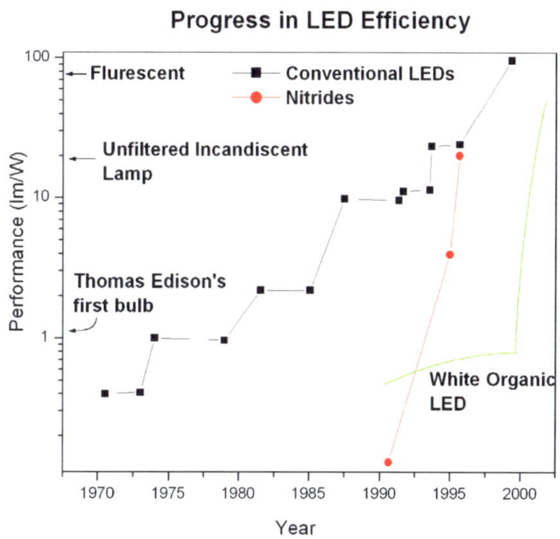

Fig. 1. The yearly progress in the efficiencies of conventional LEDs, nitride LEDs and white OLEDs

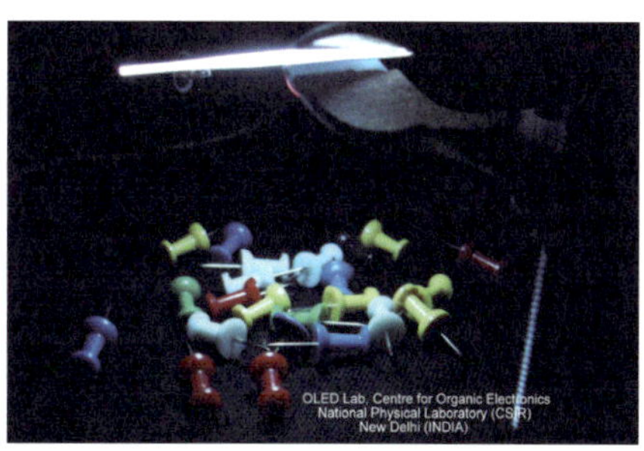

Fig. 2. 1"x1" proto type of a multilayer phosphorescent efficient WOLED developed at National Physical Laboratory, New Delhi, India

National Physical Laboratory New Delhi has taken up a program for developing WOLEDs for general lighting applications. In this effort a 1″x1″ proto type of a multilayer phosphorescent efficient WOLED has been demonstrated (Fig.2). In this review, we like to highlight on the development of white organic LEDs for general lighting.

2. Basic OLED Structure and Operation principles

White organic light emitting diodes are thin-film multilayer devices in which active charge transport and light emitting materials are sandwiched between two thin film electrodes, and at least one of the two electrodes must be transparent to light. Generally high work function (~4.8 eV), low sheet resistant (20 Ω/□) and optically transparent indium tin oxide (ITO) is used as an anode, while the cathode is a low work function metal such as Ca, Mg, Al or their alloys Mg:Ag, Li:Al. An organic layer with good electron transport and hole blocking properties is typically used between the cathode and the emissive layer. The device structure of an OLED is given in Fig. 3. When an electric field is applied across the electrodes, electrons and holes are injected into states of the lowest unoccupied molecular orbital (LUMO) and the highest occupied molecular orbital (HOMO), respectively and transported through the organic layer. Inside the semiconductor electrons and holes recombine to form excited state of the molecule. Light emission from the organic material occurs when the molecule relaxes from the excited state to the ground state. Highly efficient OLEDs which are being developed at present, contains many layers with different functionality like hole injection layer(HIL), hole transport layer (HTL),electron blocking layer(EBL), emissive layer(EML), hole blocking layer(HBL), electron transport layer(ETL) and electron injection layer(EIL) etc apart from electrodes. A schematic diagram of multilayer structure is shown in Fig. 4.

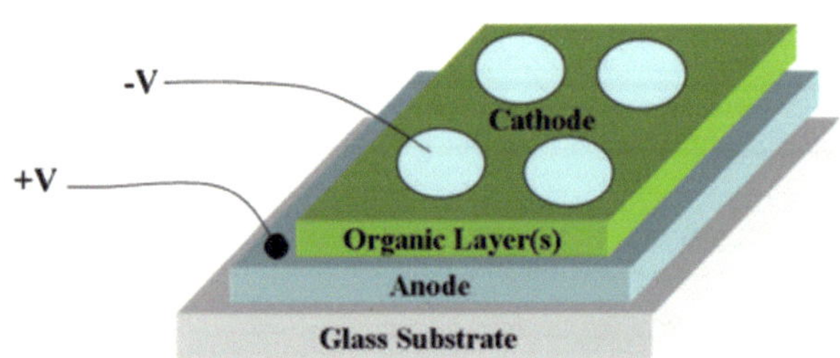

Fig. 3. The device structure of an OLED

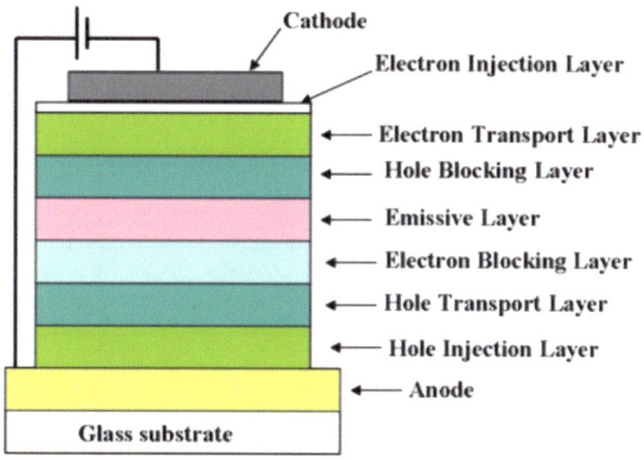

Fig. 4. A schematic diagram of multilayer structure of OLED

3. Characterization of White OLEDs

3.1 Colour quality

In order for a light-emitting device to be acceptable as a general illumination source, it clearly must provide high-illumination-quality light source. White light has three characteristics (i) the Commission International d'Eclairage (CIE) coordinates (ii) the co related colour temperature (CCT) and (iii) the colour rendering index (CRI)

3.1.1 Commission International d'Eclairage (C-I-E) co ordinates

The color of a light source is typically characterized in terms of CIE colorimetry system. Any colour can be expressed by the chromaticity coordinates x and y on the CIE chromaticity diagram (Fig. 5). The boundaries of this horseshoe-shaped diagram are the plots of monochromatic light, called spectrum loci, and all the colours in the visible spectrum fall within or on the boundary of this diagram. The arc near the centre of the diagram is called the Planckian locus, which is the plot of the coordinates of black body radiation at the temperatures from 1000 K to 20 000 K, described as CCT. The colours of most of the traditional light sources fall in the region between 2850 and 6500 K of black body. For general illumination a light source should have high-energy efficiency and CIE-1931 chromaticity coordinates (x, y) close to the equal energy white (EEW) (0.33, 0.33).

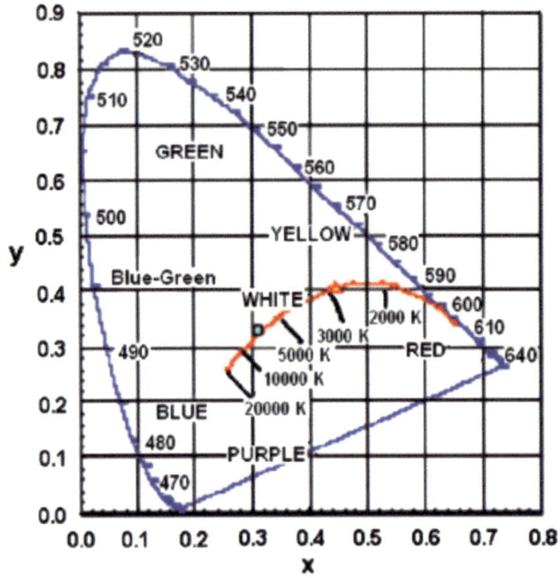

Fig. 5. CIE (x, y) chromaticity diagram.

3.1.2 Colour rendering index (CRI)

For a given light source, the CRI attempts to quantify how different a set of test colors appears when illuminated by the source compared to when the same test colors are illuminated by the standard illuminant with the same correlated color temperature. It is measured in 0-100 scales and the highest possible CRI value is 100, and this occurs when there is no difference in color rendering between the light source and the standard illuminant. An example of such a light source is the incandescent lamp. When a color rendering difference exists, the CRI is less than 100. Achieving illumination-quality white light generally requires a CRI value of 80 or greater.

3.1.3 Correlated colour temperature (CCT)

The color of a light source is typically characterized in terms of its color temperature. If the x,y coordinates of an illumination source do not exactly sit on the blackbody locus, the color of a light source is characterized in terms of its CCT. The CCT is the temperature of a blackbody radiator that has a colour that most closely matches the emission from a non-blackbody radiator. For high quality white light illumination the CCT should between 2500K and 6500 K. There is an accepted method (Wyszelki et al 1982) to determine lines of constant correlated color temperature in x, y space. CIE, CCT and CRI for common white light sources are given in Table 1 for comparison purpose (Misra et al 2006).

Light source	Chromaticity coordinate x	y	CCT (K)	CRI
High pressure sodium lamp	0.519	0.417	2100	24
Xenon lamp	0.324	0.324	5920	94
Tungsten Halogen lamp	0.448	0.407	2856	100
Daylight	0.313	0.329	6500	90
Fluorescent cool white lamp	0.375	0.367	4080	89
Fluorescent warm white lamp	0.440	0.403	2940	72
Incandescent bulb	0.448	0.408	2854	100

Table 1. Chromaticity coordinates (CIE), correlated colour temperature (CCT) and colour rendering indices (CRI) for common white light sources.

3.2 Device Efficiency

The efficiency of OLEDs is characterized by quantum efficiency, power efficiency and luminous efficiency. Over the past several years, the power (η_P) and external quantum (η_{ext}) efficiencies of white OLEDs have been steadily improving.

3.2.1 Quantum efficiency.

The quantum efficiency of a device can be differentiated into two categories i.e internal and external quantum efficiencies.

Internal quantum efficiency (IQE)- This is the total number of photons generated inside the device per electron- hole pair injected into the device. It is represented by η_{int}.
For OLEDs the internal quantum efficiency in the case of fluorescent materials is given by (OIDA 2002)

$$\eta_{int} = \gamma\, \eta_s \Phi_f\,, \tag{1}$$

where γ is the fraction of injected charges that produce excitons and is called the charge balance factor, η_s is the fraction of singlet excitons called singlet exciton efficiency and Φ_f is the fraction of energy released from material as light and called the quantum efficiency of fluorescence.
Generally based on spin statistics fluorescent organic materials exhibit 25% singlet and 75% triplet states in EL and 100% singlet states in PL (Baldo et al 1998, Friend et al 1999). In fluorescent materials triplet energy states have a low emission quantum yield and thus do not contribute to electroluminescence. This means the quantum efficiency for EL can only be about 25% of the PL efficiency. But some organometallic complexes (phosphors) have a strong triplet emission quantum yield and provide the possibility of a high efficiency EL device by using these materials. A research group from Princeton University demonstrated the efficiency limitation breakthrough in OLEDs by energy transfer from fluorescent host to a phosphorescent guest material (Baldo et al 1998). The phosphorescent dopants are doped in host materials with a wide energy gap. In electrophosphorescence the energy from both the singlet and triplet states of the fluorescent host is transferred to the triplet state of the phosphorescent guest molecule or the charges are directly trapped to the phosphor triplet. This harvesting of both singlet and triplet states has been resulted result in 100% internal

quantum efficiency (Adachi et al 2001). But exciton–exciton quenching (Baldo et al 2000), polaron– exciton quenching (Young et al 2002) and exciton dissociation (Szmytkowski et al 2002) may reduce the internal quantum efficiency to much lower values.
One of the important developments of WOLEDs is the demonstration of phosphorescence sensitization of EL (Kanno et al 2006). It has been observed that addition of a small quantity of a phosphorescent dopant in a guest host system enhances the fluorescence efficiency of a co-dopant. Cheng et al (2003) reported that the internal efficiency of fluorescence can be enhanced to 100% by using a phosphorescent sensitizer to excite the fluorescent dye through resonant energy transfer between the triplet excitons in the phosphor and singlets in the fluorescent dye. Using the blue emission from a spatially separated hole transport layer NPB and $Ir(ppy)_3$ sensitized DCJTB in a CBP host the authors obtained the high efficiency white OLEDs. The colour tuning has been achieved by varying the concentration of the sensitizer as well as the thickness of the co-doped emitter layer.

External quantum efficiency (EQE)- This is defined as the total number of photons emitted from the device per electron–hole pair injected into the device. It is represented by η_{ext}.
The external quantum efficiency is related to the internal quantum efficiency and is given by (OIDA 2002)

$$\eta_{ext} = R_e \, \eta_{int}, \quad (2)$$

where R_e is the extraction or outcoupling efficiency representing the number of photons emitted from the device per number of photons generated in the device.

3.2.2 Power Efficiency

The luminous efficacy or power efficiency is the lumen output per input electrical power of the device. It is measured in lumen per watt (lm/ W) or candela per ampere (cd/ A). It is represented by η_p. In order to compete with the fluorescent lighting market, the efficiency of OLED sources should be 120 lm/ W or more. To meet the above requirement the OLED sources must have an electrical to optical power conversion efficiency of 34%. For white light with a CRI of 90 the maximum value is 408 lm/W and for a CRI of 100 it is 240 lm/W (Kamtekar 2010).
The projection for WOLED is that by 2015, efficiency will exceed 100 lm/W with desirable life time and brightness and will start to replace indoor and outdoor light. Murano et al demonstrated white pin-OLEDs based on phosphorescent and fluorescent emitters and stacked OLEDs. This intentional doping of the transport layer led to a very high power efficiency of well above 20 lm/W at 1000 cd/m^2 (Murano et al 2005). The CRI properties of emitted light are very high, between 85 and 95. Universal Display Corporation (UDC) announced the demonstration of a white OLED lighting panel with a high power efficiency of 30 lm/ W using the company's phosphorescent OLED technology. This efficiency was achieved at a colour temperature of 4000 K, which is comparable to the colour temperature and power efficiency of a cool fluorescent lamp. The colour-rendering index was >80 across the measured colour temperatures because of the broad spectral output of the combined colours. D'Andrade et al (2004) reported power efficiency of 42 lm/ W for a white OLED that exceeds that of incandescent lamps. Therefore WOLEDs have great potential for energy saving and the replacement of traditional incandescent light sources.

3.2.3 Improvement of Efficiency

One of the measure problems in OLEDs is its low efficiency. Various techniques are used to improve the efficiency of OLED devices.

3.2.3.1 Triplet Harvesting

Due to spin statistics the efficiencies of OLEDs are limited, as only the singlets are responsible for light emission in EL in undoped devices. The recent developments in harvesting of triplet states, using phosphorescent materials, led to an increase in the efficiency and selectivity of colours. Electrophosphorescence achieved by doping an organometallic phosphor into a host has been successfully used for generating the primary colours necessary for display applications (Baldo et al 1998, Holmes et al 2003, Adachi et al 2001). Due to extensive work, the power efficiency of white organic light emitting devices (WOLEDs) has continuously increased over the past decade and it has attained the level required for WOLEDs acceptance into the lighting market. Universal Display Corporation is a world leader in developing and commercializing innovative OLED technologies and materials for use in the electronic flat panel display and other markets. Universal Display is working with a network of world-class organizations including Princeton University, the University of Southern California, DuPont Displays, Samsung SDI Co., Seiko Epson Corporation, Sony Corporation, Tohoku Pioneer Corporation and Toyota Industries Corporation. NOVALED GmbH, Dresden Germany, is another emerging company in the field of organic displays. NOVALED works in close cooperation with Technical University Dresden and Fraunhofer Dresden Institute IPMS. According to a press release in 2005 from Dresden, Germany, NOVALED has developed a green emitting OLED with efficiency of 110 lm / W.

3.2.3.2 Optical doping

The doping of the emissive layer in an OLED has been used extensively as a way of improving efficiency and lifetime, in addition to being used to modify the emission color (Optical doping has been explained in guest host system). Tang et al. (1989) first introduced fluorescent dyes, 3-(2-benzothiazolyl)-7-diethylaminocoumarin (coumarin 540 or coumarin 6 and DCMs, as dopants in Alq_3 to improve the efficiency and color purity of devices. Since then, a wide range of fluorescent dopants have been used in OLEDs (Sano et al 1997, Hamada et al 1995). The ground state of most materials has a single spin state. Emission of a photon in fluorescent materials conserves spin, therefore only singlet excited states typically emit light. Decay from the triplet excited states is typically a nonradiative process for most organic materials and so these triplet excitons are lost from the perspective of light emission. The maximum possible internal quantum efficiency that can be obtained in an OLED using fluorescent material is limited by the ratio of these excited states or the so-called exciton singlet-to-triplet ratio, which is approximately 1:3 (Baldo et al 1999, 1999a). This limits fluorescent OLEDs to a maximum internal quantum efficiency of approximately 25%.

3.2.3.3 Electrical doping

In typical OLEDs, the applied voltage (V) is usually 5-8 V, when illuminated at 500-1000 cd/ m^2, i.e., greater than twice the voltage of the emitted photon V_g. The voltage drop across the emission layer itself is usually 2 to 3 V, depending upon the emission wavelength.

The remaining voltage is dropped predominantly across the ETL, across the HTL, and at the heterojunction interfaces. Charge transport in low-mobility organic films is space-charge limited (Marks et al 1993) and high electric fields are required to inject the necessary charge to generate the desired photon flux. Band misalignments at the heterojunction interfaces also result in voltage loss. However, the drive voltage can be significantly reduced by electrical doping of the transport layers (Blochwitz 2002). Electrical doping was demonstrated using green $Ir(ppy)_3$-doped PHOLEDs by Pfeiffer et al (2002). It was observed that the drive voltage necessary to produce 100 cd/ m^2 was 2.65 V, i.e., only slightly higher than Vg. This device used p-type (tetra fluoro tetra cyano quino dimethane (F_4-TCNQ)) and n-type (Li) doping of the HTL and ETL, respectively. In this effort Murano et al from NOVALED demonstrated efficient white OLEDs based on an intentional doping of the charge carrier transport layers and the usage of different state of the art emission principles (Murano et al 2005).

Large quantity of (~50%) low work function metal like Li is usually co-evaporated with conventional electron transport materials like Bphen to achieve n-type doping. In corporation of low work function metals induces stability problems in OLEDs. Recently Tyagi et al (2010) have demonstrated efficient n-type doping by doping Liq in electron transport material Alq_3. An increase in current density by two orders of magnitude has been achieved with 33 wt% of Liq doped in Alq_3. Organic light emitting diode with p–i–n structure was fabricated using F_4-TCNQ doped α-NPD as hole transport layer, $Ir(ppy)_3$ doped CBP as emitting layer and 33 wt% Liq doped Alq_3 as electron transport layer. Comparison of OLEDs fabricated using undoped Alq_3 and 33 wt% Liq doped Alq_3 as electron transport layer shows reduction in turn on voltage from 5 to 2.5V and enhancement of power efficiency from 5.8 to 10.6 lm/W at 5V.

3.2.3.4 Improving out coupling efficiency

It is well understood that the generated light from the active OLED medium propagates via various modes, that is, external modes (escape from the substrate surface), substrate-, and ITO/organic-waveguided modes due to total internal reflection (TIR) (Gu et al 1997, Madigan 2000, Moller and Forrest 2002). According to the ray optics theory, about 80% of the generated light is lost in waveguided modes due to glass substrate and ITO/organic material which means that the majority of generated light is either trapped inside the glass substrate and device, or emitted out from the edges of an OLED (Gu et al 1997, Madigan 2000, Moller and Forrest 2002). For the purpose of applications in general illumination and flat panel displays, light emitted from the substrate surface (external modes) is most useful which is only about 20% of the total emitted light from the OLED.

Detailed optical modelling (Kim et al 2000) predicted that the fraction of the light emitted in the forward direction is reduced by a factor of $(4/3)n^2$, where n is the index of refraction of the emitter layer. Through a series of experiments using an integrating sphere, Cao et al (Cao et al 1999) demonstrated that the measured reduction factor is approximately 2–2.5, less than the theoretical value $2n^2 \sim 6$. Forrest and coworkers found that the total external efficiency is larger by a factor of 1.7–2.3 than observed in the forward viewing direction (D'Andrade et al 2004). The poor light extraction is the most important factor which limits the external quantum efficiency of devices and hence better outcoupling methods are to be developed to get higher efficiencies. To extract the trapped and waveguided light into external modes, various approaches based on light refraction and scattering to reduce TIR at the interfaces have been reported, such as, the use of a shaped substrate (Gu et al 1997,Madigan et al 2000), use of micro-lenses on the backside of substrate surface (Moller et al 2002, Peng et al 2005, Lim et al 2006), formation of mono-layer of silica micro-spheres as scattering medium (Yamasaki et al 2000, Neyts and Nieto 2006), and use of high refractive index substrate (Lu et al 2000). In another approach, an extremely low refractive index silica-aerogel layer (Tsutsui et al 2001) was inserted between the ITO transparent electrode and glass substrate.

A 50% light extraction efficiency from OLEDs was recently reported by insertion of a two-dimensional photonic crystal structure (Do et al 2004, Kitamura et al 2005, Liu et al 2005, Lee et al 2005), and using nano-porous and nano-patterned films (Lee et al 2003, Kim et al 2005). More recently, use of diffusive layer lamination (Nakamura et al 2004), holographic technique (Liu et al 2005), and shaped substrate OLED luminair (Andrade et al 2006) has also been investigated for the improvement of out-coupling efficiency in conventional OLEDs. An index-matching layer has also been used for top emitting OLED (Hung et al 2001).

To extract the trapped light, Saxena et al (2008) used simple AR coating technique and demonstrated pronounced enhancement in light out-coupling of conventional OLED. Single-layer MgF_2 was coated on backside of glass substrate of conventional OLED with thickness of $\lambda/4$. About two-fold enhancement in luminance with anti-reflection coating of MgF_2 has been observed. Fig. 6 shows the schematic diagram of the phenomenon of anti-reflection (AR) coating using single-layer MgF2 for the extraction of substrate-waveguided modes.

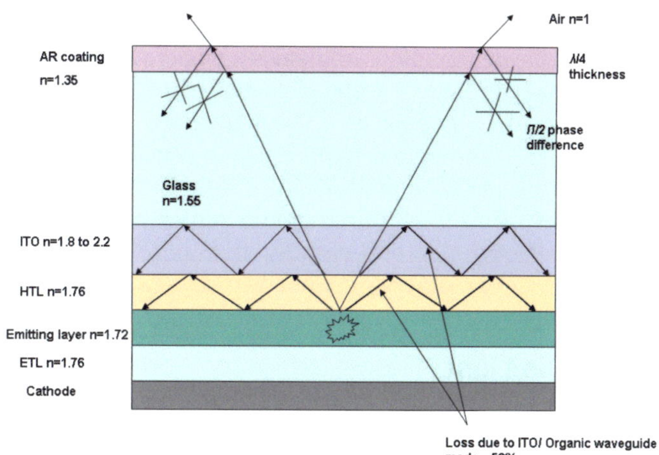

Fig. 6 Schematic diagram of the phenomenon of anti-reflection (AR) coating using single-layer MgF_2 for the extraction of substrate-waveguided modes.

3.3 Stability

One issue that limited the early adoption of OLEDs in commercial products was device stability both during storage and in operation. Suggested causes of degradation include indium migration from the ITO anode (Lee et al 1999), morphological instability of the organic materials (Higginson et al 1998), fixed charge accumulation within the device (Kondakov et al 2003), damage to the electrodes, and the formation of non emissive dark spots (Burrows et al 1994, Aziz et al 1998, Cumpston et al 1996). Water and oxygen are known to cause problems in OLEDs. Therefore, a great deal of effort has been directed toward the encapsulation of devices. Encapsulation is typically carried out under a nitrogen atmosphere inside a glove box.

In addition to extrinsic environmental causes of degradation in OLEDs, some groups have explored the stability problem related to the individual device materials to transport charge and emit light. For example, Aziz et al 1999 have proposed that in simple Alq_3 devices hole transport through the Alq_3 layer is the dominant cause of device degradation due to the instability of the Alq_3^+ cationic species. A useful overview of the factors affecting device reliability is given by Forrest et al. (1997) and Popovic and Aziz (2002).

3.4 Encapsulation

In the OLED fabrication process encapsulation is the final step to ensure a long device lifetime. OLEDs (Tang and Vanslyake 1987, Adachi et al 2000, Burroughs et al 1990) built on glass substrates have been shown to have lifetimes of tens of thousands of hours (Shi and Tang 1997, Burrows et al 2000). There have been many proposed mechanisms for the decay in luminance, but most theories agree that one of the dominant degradation mechanisms in unencapsulated OLEDs, which have far shorter lifetimes than encapsulated devices, is the exposure of the organic–cathode interface to atmospheric oxygen and water. This leads to oxidation and delamination of the metal cathode (Liew et al 2000, Kolosov et al 2001) as well as potential chemical reactions within the organic layers. The device acts like electrochemical cell producing H_2 and O_2 at the electrodes there by, degrading the device.

As most of the OLED work, to date, has been focused on the development and manufacture of glass-based displays, the degradation problem is a meliorated by sealing the display in an inert atmosphere, e.g., in a nitrogen or argon glove box (< 1 ppm water and oxygen), using a glass or metal lid attached by a bead of UV cured epoxy resin (Burrows et al 1994). A desiccant such as CaO or BaO is often located in the package to react with any residual water incorporated in the package or diffusing through the epoxy seal. In addition to encapsulation techniques using a lid, thin-film encapsulation techniques are also possible. Wong et al have done effective thin film encapsulation of OLED by altering and repeating deposition of multilayers of CF_x and Si_3N_4 films (Wong et al 2008).

4. Generation of White Light

As discussed earlier for generation of white light, all the three primary colors have to be produced simultaneously and for illumination purpose should have good colour rendering index (>75) and good position close to (0.33, 0.33) on the CIE-1931 diagram. Since it is difficult to obtain all primary emissions from a single molecule, excitation of more than one organic species are often necessary. Generally two methods are used to generate white light from OLEDs i.e (a) Colour mixing and (b) Wavelength conversion.

4.1 Colour mixing

In the colour mixing technique, no phosphors are used, and therefore the losses associated with the wavelength conversion do not occur and this approach has the potential for the highest efficiency. This method uses multiple emitters in a single device and mixing of different lights from different emitters produces white light. White light can be obtained by mixing two complementary colours (blue and orange) or three primary colours (red, green and blue). The typical techniques used for the production of white light by colour mixing are (a) Multilayer structure consisting of red, green and blue emissive layers, (b) Single emissive layer structure (c) exciplex/excimer structure and (d) microcavity structure.

4.1.1 Multilayer device structure

Most widely used approach to achieve white light is a multilayer structure where simultaneous emission of light from two or more separate emitting layers with different emission colours results in white light. This technique is based on the consecutive deposition or co-evaporation of different emitting materials and control of the exciton recombination zone. This structure consists of many organic–organic interfaces leading to interface barriers, which may result in the inhibition of carrier injection and joule heating. Therefore to minimize the charge injection barriers and joule heating at the organic/organic interfaces the emissive materials are chosen in such a way that the highest occupied molecular orbital (HOMO) and the lowest unoccupied molecular orbital (LUMO) of different adjacent emissive materials closely match with each other. The emission from the device depends on the thickness and composition of each layer, and is to be precisely controlled to achieve color balance. The exciton recombination zone is controlled by inserting blocking layers that block only one type of carrier between the hole transporting layers (HTL) and electron transporting layers (ETL), so that the recombination takes place in two or three different layers (Deshpande et al 1999, Li and Shinar 2003, Ko and Tao 2001, Tokito et al 2003, Yang et al 2002, Kim et al 2004, Zugang and Nazar´e 2000, Lee et al 2002, Xie et al 2003, Cheng et al 2004, Zhang et al 2005, Guo et al 2005). This results in emission from different layers (Fig. 7). By controlling the recombination current within individual organic layers, emission from red, green and blue light emitting layers is balanced to obtain white light of the desired colour purity.

Deshpande et al (1999) achieved white light emission by the sequential energy transfer between different layers. The device was fabricated in the configuration ITO/α-NPD/α-NPD:DCM2(0.6–8 wt%)/BCP/Alq3/Mg:Ag (20:1)/Ag. Here 4,4′ bis (N-(1-napthyl-N-phenyl-amino)) biphenyl (α-NPD) was used as a hole injection layer, α-NPD: DCM2 (2, 4-(dicyanomethylene)-2-methyl-6-(2- (2, 3, 6, 7-tetrahydro-1H, 5H benzo(I, j)quinolizin-8-yl)vinyl)- 4H-pyran) layer was used as a hole transport layer (HTL) as well as an emitting layer, 2,9-dimethyl-4,7-diphenyl-1,10- phenanthroline (BCP) layer was deposited for the purpose of hole blocking, Alq_3 was used as a green emitting electron transporting layer (ETL) and Mg:Ag alloy followed by a thick layer of Ag was deposited as the cathode.

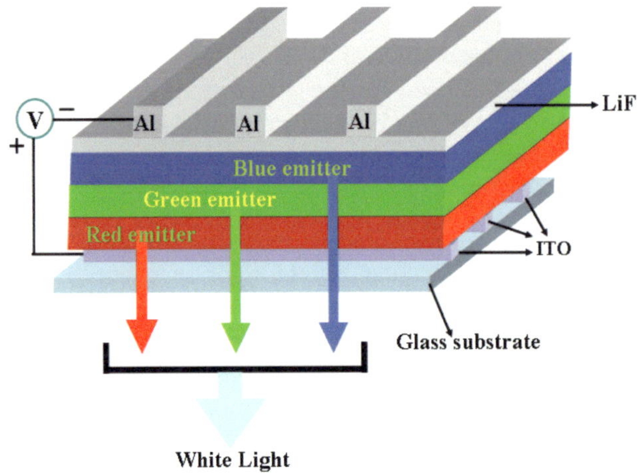

Fig. 7. Schematic diagram of multilayer white OLED

A maximum luminance of 13 500 cd/m^2, a maximum external quantum efficiency>0.5% and an average power efficiency of 0.3 lm/W were reported for the above configuration. Recently Wu et al (2005) reported white light emission from a dual emitting layer OLED with and without blocking layers. The device with a blocking layer exhibited better performance with an external quantum efficiency of 3.86%. The emission colour of these devices strongly depends upon the thickness of the emissive layer and the applied voltage. The drawback of this technique is that it requires complex processing and a large amount of wasted organic materials resulting in relatively high fabrication cost. The CIE coordinates are often dependent upon the driving current due to shift of the exciton recombination zone. Brian et al (2002) have demonstrated that multi-emissive layer fully electrophosphorescent WOLEDs can take advantage of the diffusion of triplets to produce bright white devices with high power and quantum efficiencies. The device color can be tuned by varying the thickness and the dopant concentrations in each layer, and by introducing exciton blocking layers between emissive layers.

Gong et al (2005) have reported that high performance multilayer white light emitting PLEDs can be fabricated by using a blend of luminescent semiconducting polymers and organometallic complexes as the emission layer and water soluble (or ethanol soluable)PVK-SO_3Li as the hole injection/transport layer (HIL/HTL) and t-Bu-PBD-SO_3Na as the electron injection/electron transport layer (EIL/ETL). Each layer is spin-cost sequentially from solution. Illumination quality white light is emitted with stable CIE coordinates, stable colour temperature and stable clour rendering indices.

Tayagi et al (2010) have demonstrated a WOLED by double layers of blue Zn(hpb)$_2$ and yellow Zn(hpb)mq emitting materials. Broad electroluminescence spectrum has been observed and as the thickness of Zn(hpb)mq layer increases the dominant wavelength shifts from bluish region to yellowish region. Three peaks have been observed in the EL spectrum at wavelengths 450 nm, 485 nm and 550 nm. The peak at 450 nm and 485 nm are due to the recombination of electrons and holes in Zn(hpb)$_2$ layer and the peak at 550 nm is due to the recombination in Zn(hpb)mq layer. The peak at 485 nm has been attributed to the excimer formation in Zn(hpb)$_2$. The EL spectrum of duoble layer was found to be an overlap of the EL spectrum of Zn(hpb)$_2$ and Zn(hpb)mq layers. CIE coordinates (0.29, 0.38) were well within the white region and have low turn on voltage (5V).The highest brightness obtained was 8390 Cd/m^2 at a current density of 518 mA/cm^2.

White OLEDs which comprised of separate emitters having independent electrodes stacked one over the other in which separate voltage source control the emission from each device is known as stacked OLED. Stacking is advantageous due to better luminous efficiency, better color contrast and good color rendering over a wide range. Furthermore, this tuning strategy can delay the onset of differential aging of the several emitting layer. It has been shown that by layering several devices in this manner, a high total brightness OLED can be achieved without driving any particular element in the stack at such a high intensity that its operational life time is reduced (Lu and Sturn 2002, Brian et al 2002).

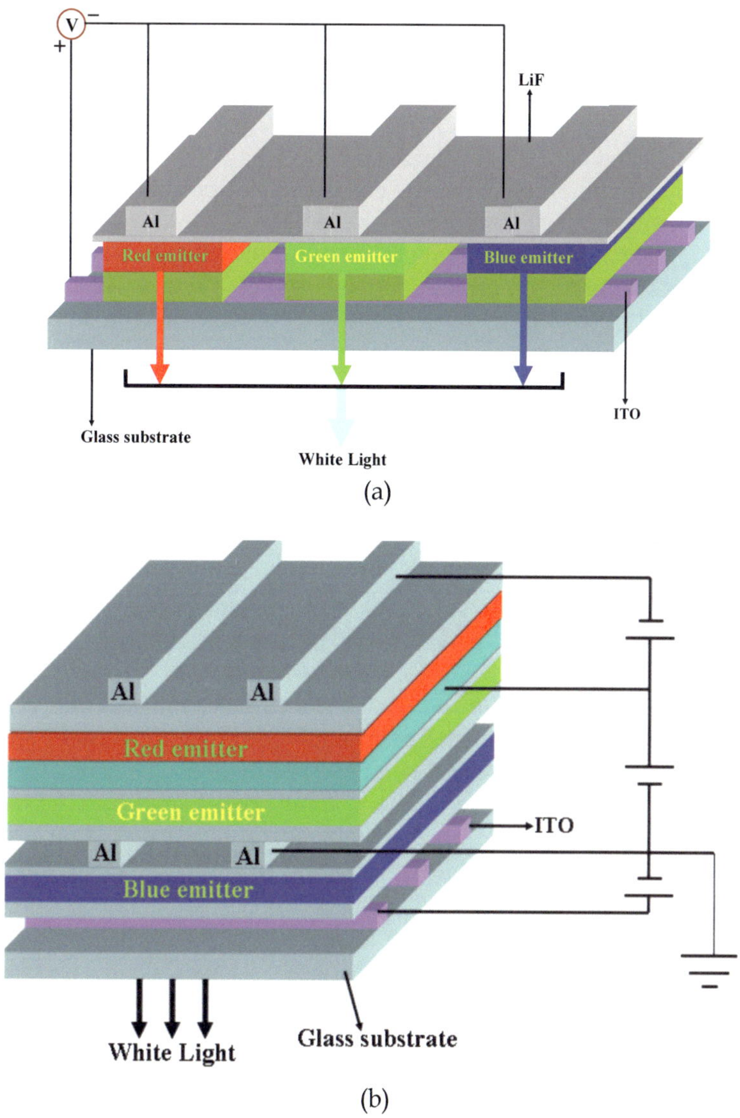

Fig. 8. Schematic diagram of (*a*) horizontally and (*b*) vertically stacked OLED.

In a similar concept to the stacked OLED, tunable emitters of different colours (red-, green-, and blue) are placed side by side in strips. If spaced sufficiently very closely the colors will merge, as in full color display, producing bright and efficient white light similar to SOLED emitter with less complexity (Brian et al 2002). This technology is similar to liquid crystal flat panel displays. Here the pixels of the three principal colours are patterned separately either horizontally or vertically and addressing them independently (Burrows et al 1997, Forrest et al 1997, Burrows et al 1998) (see Fig. 8). In the horizontally stacked pattern the individual colour emitting pixels are deposited either in the form of dots, squares, circles, thin lines or very thin strips. As a result of mixing of these colours any desired range of colours can be produced in the same pane. As each colour component is addressed individually, the differential colour ageing can be mitigated by changing the current through the components. Each pixel can be optimized to operate at a minimum operating voltage and for highest efficiency. Also by reducing the size of the pixels the lifetime of the device can be controlled to the maximum.

Stacked white OLEDs usually produce higher brightness and efficiency than those of conventional WOLED and can be a good candidate as a light source because double or even triple current efficiency can be obtained in such devices as compared to the single emitter device. Recently Sun et al (2005) reported an efficient stacked WOLED using a novel anode cathode layer (ACL) for connecting a blue phosphorescent and red phosphorescent emissive unit. This ACL layer was used as a middle electrode and EL characteristics of two individual emissive units were also studied. By biasing the two emissive units in a proper ratio white emission was obtained. They reported a maximum luminescence of 40000 cd/ m^2 at 26 V with CIE coordinates of (0.32, 0.38). The luminescence efficiency was 11.6 cd /A at 28 mA/ cm^2.

Liao et al(2004) and Kido et al (2003) have demonstrated a variant of the SOLED that allows the contacts between intermediate OLED in the stack to electrically "float" and performs as a series of independent OLEDs, with a single electron exciting the multiple OLEDs as it passes through the circuit.

Chang et al (2005) fabricated two types of stacked/tandem WOLEDs containing an interconnecting layer of Mg:Alq_3/WO and one control white emitting device for comparison. In these devices white emission was obtained by mixing complementary blue and yellow colours. Device 1 was obtained by connecting blue and yellow devices in series, while device 2 stacked two white emitting devices with the same blue and yellow dopants as used in device 1. Device 2 shows better performance compared to device1 and the control device. An interesting amplification effect was observed in device 2 such that it exhibited the highest efficiency of 22 cd /A, which was almost three times that of the control device. This was due to the microcavity effect, which enhances the amount of light emitted in the forward direction. This shows that by just connecting two devices higher efficiency can be achieved. It was found that the driving voltage increases with increasing number of active units. Device 2 was the least stable, while the control device showed the longest half-life. This was due to the fact that device 2 suffered more driving power than the control and device 1. The thermal breakdown process may be present in these stacked devices due to non-ohmic contact of the interconnecting layers. However the half-life of device 2 at 100 cd/ m^2 was projected to be greater than 80000 h. In these stacked devices the emissive intensity and colour were dependent on the viewing angle. This viewing angle dependence of emissive intensity and colour was attributed to the microcavity effect. Therefore it is important to have a good optical design for the stacked devices. Such device structures had disadvantages of having complex layer structure and lack of known methods for damage free post deposition patterning of organic layers at resolution required for color displays. Another approach for white light emission from multilayer OLEDs is the multiple quantum well structure (Liu et al 2000) (Fig. 9), which includes two or more emissive layers separated by blocking layers. Electrons and holes tunnel through the potential barriers of the blocking layers and distribute uniformly in different wells and emit light. Matching of the energy levels of different organic materials is not so critical in this system. Excitons are formed in different wells and decay to produce different coloured lights in their own wells. The confinement of charge carriers inside the quantum well improves the probability of exciton formation and they do not move to other zones or transfer their energy to the next zone. But this approach is very complicated and requires the optimization of thicknesses of various light emitting and blocking layers. This multilayer architecture has relatively high operating voltage due to the combined thickness of many layers used.

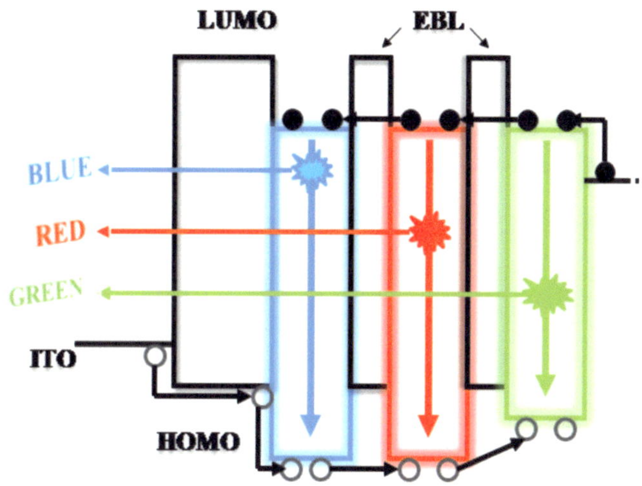

Fig. 9. Schematic diagram of a multiple quantum well white OLED

4.1.2 Single emissive layer structure

The fabrication process and device operation of white OLEDs through multilayer structure is very complex and several parameters need to be optimized for good colour rendering and to have luminescence efficiency. Also, these devices have high operating voltage because of the thick profile due to the several stacked organic layers used to perform different functions for efficient WOLEDs. The device profile must be as thin as possible to ensure low voltage operation. Single layer white light emitting devices consist of only one active organic layer can emit in the entire visible range and can overcome all such complexities. In comparison to other structures single layer structure can achieve higher emission colour stability. White emission from a single layer consisting of a blue emitter doped with different dyes or blending two or more polymers has been reported by many authors (Mazzeo et al 2003, Lee et al 2002, Al Attar et al 2005, Tasch et al 1997, Ko et al 2003, Chuen and Tao 2002, Shao and Yang 2005, Yang et al 2000, Chang et al 2005, Tsai et al 2003).

4.1.2.1 Host Guest structure

One of the most widely used methods to generate white light is host- guest structure. In this structure often a higher energy-emitting host (donor) material is doped with lower energy emitting guest (dye, dopant or acceptor) materials to cause energy transfer from the host to the guests. The dopant site can be excited directly by capturing the charge carriers or by energy transfer from the host to guest, as a result light emission can come from both the host and guests, the combined effect of which produces white light and is called emission due to the incomplete energy transfer. There are many examples where blue and red/orange color emitting dyes are co-deposited to form the emission layer (Chuen and Tao 2002, Koo et al 2003, Zheng et al 2003, Jiang et al 2002).

An important aspect of host-guest systems is the choice of host and guest materials for both single and multidoped systems. The energy transfer from host to guest can be either Förster (Lakowicz 1999) type energy transfer or Dexter type (Turro 1991) charge transfer or due to the formation of excimer or exciplexes (the principles are discussed in section 5). The primary conditions for such energy transfers are overlap of the emission spectrum of the host and absorption spectrum of the guest (Fig. 10). Therefore, the host material is always one with emission at higher energies, generally a blue-emitting material.

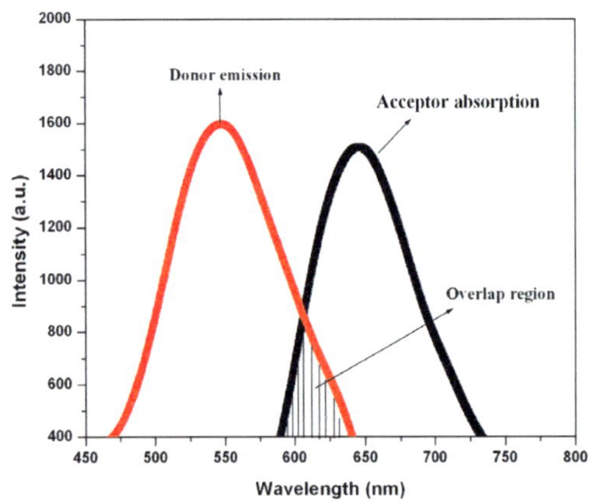

Fig. 10. Spectral overlapping between emission of donor and absorption of acceptor.

The host-guest system for white light generation can be either a single-doped or a multi-doped system in a single layer (D'Andrade et al 2004) or a multilayer structure (Lim et al 2002). The simplest device structure with a single emitting layer is obtained by doping primary (Kido et al 1994, Hu and Karasz 2003,) or complementary (Kawamura 2002, Zhang et al 2003, 2003a) color emitting dyes in a conductive polymer/small molecule host. In these devices, the concentration of the dopants was so maintained that emission from the host was small or negligible.

It is not necessary to use only dyes to take advantage of the energy transfer; blends of two polymers can also be used as host-guest systems (Lee et al 2002). The guest molecules can be florescent or phosphorescent in nature. However, phosphorescent dyes based on Ir and Pt complexes have provided significantly higher efficiency of OLEDs because of their ability to emit from both singlet and triplet excitons of the host molecule (Kamata et al 2002), whereas a florescent dye can only utilize the singlet exciton. The devices based on phosphorescent dyes are named as electrophosphorescent devices. Representative examples of various host materials, florescent and phosphorescent dyes are listed in Table 2.

Host materials		1. Poly(N-vinylcarbazole) (PVK)
		2. 1,1,4,4-Tetraphenyl-1,3-butadiene (TPD)
		3. 4,4′,N,N′-Dicarbazole-biphenyl (CBP)
		4. 9,10-Bis(3′5′-diaryl)phenyl anthracene (JBEM)
		5. 9,10-Bis(2′-naphthyl)anthracene (BNA)
		6. Bis(2-methyl-8-quinolato) (triphenylsiloxy) aluminum (III) (SAlq)
		7. 4-{4-(N-(1-Naphthyl)-N-phenylaminophenyl)}-1,7-diphenyl-3,5-dimethyl-1,7-dihydro-dipyrazolo(3,4-b;4′3′-e)pyridine (PAP-NPA)
		8. Bis (2-(2-hydroxyphenyl)benzothiazolate)zinc (Zn(BTZ)$_2$)
		9. 4,4′Bis(N-(1-naphthyl)-N-phenyl-amino)-biphenyl (α-NPD)
Florescent dyes	Red	1. 4-(Dicyanomethylene)-2-methyl-6-(p-dimethyl-aminostyryl)-4H-pyran (DCM1)
		2. 4-(Dicyanomethylene)-2-methyl-6-(2-(2,3,6,7-tetrahydro-1H, 5H-benzo(I,j)quinolizin-8-yl)vinyl)-4H-pyran (DCM2) (-)
		3.4-(Dicyanomethylene)-2,6-di-(4-dimethylaminobenzaldehyde)-γ-pyran (DCDM)
		4.4-(Dicyanomethylene)-2-tert-butyl-6(1,1,7,tetramethyljulolidyl-9-enyl)-4H-pyran (DCJTB)
		5. 5,6,11,12-Tetraphenyl-naphthacene (Rubrene) (orange)
		6. Zinc tetraphenylporphyrin (ZnTPP)
	Green	1. Coumarin6
		2. 9-Cyanoanthracene (CNA)
		3. Tris(8-quinolato)aluminum (III) (AlQ$_3$)
	Blue	1. (perylene)
		2. 4,4′-Bis(2,2′-diphenylvinyl)-1,1′-biphenyl(DPVBi)
		3. 9,10-Bis(3′5′-diaryl)phenyl anthracene(JBEM)
Phosphorescent dyes	Red	1. Fac-tris(2-phenyl)-bis(2-(2′-benzothienyl)-pyridinato-N,C′)(acetylacetonate)Ir(III) (Bt$_2$Ir (acac))
		2.Bis(2-(2′-benzothienyl)-pyridinato-N,C$^{3'}$)(acetylacetonate)Ir(III)(Btp$_2$Ir (acac))
		3.Bis(2-phenylbenzothiozolato-N,C$^{2'}$)(acetylacetonate)Ir(III)(Bt$_2$Ir (acac))
	Green	Fac-tris(2-phenylpyridyl)iridium(III)(Ir(ppy)$_3$)
	Blue	1.Bis((4,6-difluorophenyl)-pyridinato-N,C)(picolinato)Ir(III)(FIrpic)
		2.Bis{2-(3,5-bis(trifluoromethyl)phenyl)-pyridinato-N,C$^{3'}$}iridium(III)picolinate ((CF$_3$ppy)$_2$Ir(pic)) (greenish-blue)

Table 2. List of various host materials and fluorescent and phosphorescent dyes used for fabrication of WOLED

In most of the electrophosphorescence based OLEDs the device quantum efficiencies drop rapidly with increasing current density and consequently with the brightness due to triplet-triplet annihilation at high current densities. WOLED based on phosphorescent material had a maximum forward viewing power efficiency of 26 ± 3 lm W^{-1} at low luminosity, decreasing to 11 ± 1 lm W−1 at 1000 cd m^{-2} (Kamata et al 2002, D'Andrade et al 2004).

The color tenability and spectral characteristics in host–guest systems is achieved by changing the concentration of the dopants and the energy transfer rate to each dopant and energy transfer between the dopants in multi-doped systems respectively (Kido et al 1994, Kamata et al 2002, Kawamura et al 2002). The range in which the dopant concentration can be varied is limited, usually less than 1 wt.% and 10 wt.% for florescent and phosphorescent dyes, respectively and the upper limit for dopant concentration is due to aggregate formation at higher concentration or quenching of luminescence due to non-radiative processes. For example, in a single dopant system, energy transfer from host to guest can be fast enough to saturate all the guest sites leading to change in spectral characteristics for higher current densities in a device or higher excitation intensity in PL measurements (Cheun and Tao 2002, Zheng et al 2003). Similarly, in case of multi-doped systems the emission from the higher energy dopant increases due to the filled lower energy states (Kamata et al 2002). Therefore, the concentration ratio of the dopants has to be carefully balanced in order to have stable white emission over the entire operating conditions of the device.

Theoretically, for single layer white OLEDs, the organic material should have chromophores that emit in different visible regions but most of the single molecule used as emitting material show the photoluminescence (PL) peak in the high-energy blue region (Tsai et al 2003, Paik et al 2002). It is their electroluminescence (EL) that is white or near white, which implies that some other emitting species like aggregates (Tsai et al 2003) or intramolecular charge transfer complex (Paik et al 2002) form in the solid state of the film during operation of the device, which is responsible for the additional peaks in the longer wavelength regions. Also, the formation of red-shifted peaks and their relative intensity is highly dependent on the applied bias and thus the emission spectrum is again voltage dependent (Tsai et al 2003, Paik et al 2002). In the case of emission through aggregates, the relative intensity of the peaks becomes further dependent on the solvent used for spin coating and the morphology of the film (Tsai et al 2003). Various molecules that are reported to give white or near-white emission are listed in Table 3.

Materials	Reference
Anthracene fused norbornadiene derivatives	(Tsai et al 2003)
Silicon-based alternating copolymers containing carbazole and oxadiazole moieties	(Paik et al 2002)
1,4-Bis-(9-anthrylvinyl)-benzene polymer	(Romdhane et al 2003)

Table 3. List of organic molecules that are reported to give white or near-white electroluminescence

Rai et al (2009) reported the fabrication of a WOLED by using $Zn(hpb)_2$ doped with an orange fluorescent dye DCM in the configuration ITO/α-NPD/ $Zn(hpb)_2$:DCM/BCP/Alq_3/LiF/Al and obtained white light emission with broad spectrum for very low concentration of the dye (0.01%). Since Förster type energy transfer (Rai et al 2008a, Shoustikov et al 1998) was improbable at such low dye concentration, the reason for emission from such low concentration was ascribed as due to trapping of carrier on to dye molecule followed by recombination. The white EL spectrum (Fig11) of device with suitable color coordinates was independent of the applied voltage.

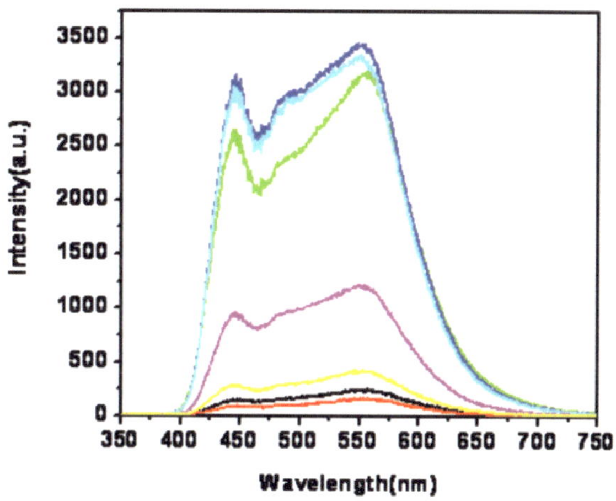

Fig. 11. Electroluminescence spectrum of WOLED at 6–10 V.

The most important benefit of OLEDs with only one emission zone over the others is the fact that high emission colour stability can be achieved. But the approach of white emission by two or three different light emitting dopants in a single layer has its own problem that different rates of energy transfer between dopants may lead to colour imbalance. Some fraction of the highest energy (blue) will readily transfer energy to the green and red emitters and the green emitter can transfer energy to the red emitter. If the three emitters are at equal concentrations the red emitter will dominate the spectrum. So the doping ratio must be blue > green > red at a very carefully balanced ratio.

Shao et al (2005) demonstrated the achievement of highly colour stable WOLED using a single emissive layer containing a uniformly doped host. To avoid the difficulties in the precise control of dopants by thermal co-evaporation, the host *α*-naphthylphenylbiphenyl diamine (*α*-NPD) was uniformly doped by the fused organic solid solution method prior to the deposition with 4,4′-bis(2,2-diphenylethen-1-yl) biphenyl (DPVBi) for the blue emission, and 10-(2- benzothiazolyl)-2,3,6,7-tetrahydro-1,1,7,7,-tetra methyl-1H, 5H,11H benzopyrano(6,7,8-ij) quinolizin-11-one (C545T) for the green emission, 5,6,11,12 tetraphenylnaphthacene (rubrene) for the yellow emission and 4-(dicyanomethylene)- 2-tertbutyl-6-(1,1,7,7 -teramethyljulolidyl -9 –enyl)-4H-pyan (DCJTB) for the red emission. The correct weight ratio of α-NPD, DPVBi, rubrene, DCJTB and C545T for stable white light emission was 100:5.81:0.342:0.304:0.394. The excitons generated from the blue dopant easily transfered their energy to other dopants. But the energy transfer from host to guest exhibits energy losses which has been avoided by the process of direct triplet exciton formation in the phosphorescent dyes. This leads to reduction in the operating voltage and hence increases the power efficiency.

D'Andrade et al (2004) reported white light emission from a single emissive layer WOLED. The emissive layer contained three organometallic phosphorescent dopants: tris(2-phenylpyridine) iridium(III) ($Ir(ppy)_3$) for green light emission, iridium (III)bis(2-phenylquinolyl-N, C2') (acetylacetonate) (PQIr) for red light emission and iridium(III)bis(4', 6'-difluorophenylpyridinato) tetrakis(1-pyrazolyl) borate (FIr6) providing blue light emission. The materials were simultaneously codoped into wide energy gap p-bis(triphenylsilyly)benzene (UGH2) host. The triplet doped WOLED exhibited a peak power efficiency of 42 lm / W with a colour rendering index 80 and a maximum external quantum efficiency of 12%.

Srivastava et al (2009) used single emission layer device structure in which two phosphorescent materials were co-doped in suitable ratio and fabricated organic LEDs to get the white light emission from the devices. The greenish blue and red emission came from the single emitting layer by an incomplete energy transfer process in which a mixture of highly efficient phosphorescent materials (FIrPic) (Bis(2-(4,6-difluorophenyl)pyridinato-N,$C^{2'}$) iridium(III)) (greenish blue) and (Ir-BTPA) (bis(2-(2'-benzothienyl) pyridinato-N,$C^{3'}$) (acetyl-acetonate) iridium(III)) (red) were used as guest molecules and 4,4' bis 9 carbozyl (biphenyl) (CBP) as host. BCP (2, 9 dimethyl 4, 7 diphenyl 1, 1' phenanthrolene) was used as hole blocking material. A suitable combination of charge carrier transport material and electrode materials were used to fabricate white light emitting diodes. Varying dopant concentrations controls the color of the device (Fig. 12). The maximum luminance of the device is 4450 cd/m^2. The CIE coordinates of the device are (0.27, 0.32) which is well within the white region.

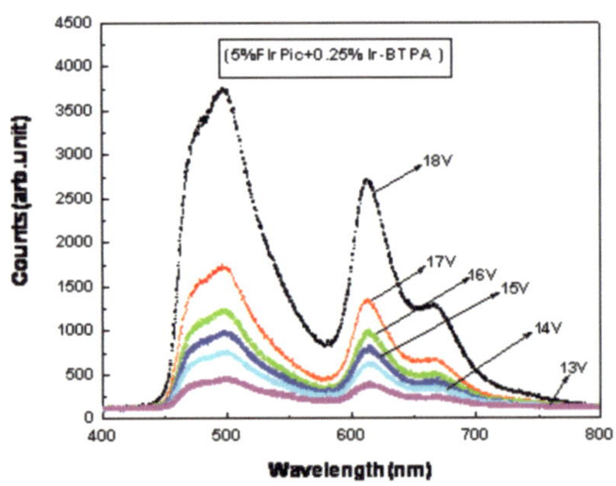

Fig. 12. Electroluminescence spectrum of WOLED at different applied voltages

Further, Rai et al (2010) fabricated an efficient WOLED using a blue light emitting material namely $Zn(hpb)_2$ and tuning its spectral response for white light emission by optimally doping it with bis(2-(2'-benzothienyl) pyridinato-N,C30) iridium(acetylacetonate) ($(Ir(btp)_2acac)$) that results in emission from both the host and the guest. The blue component for the white emission has been obtained from the singlet state of the host material $Zn(hpb)_2$

and red component from the triplet energy transfer from the triplet state of the host to the triplet state of the guest as shown in Fig. 13. The color coordinates of the white emission spectrum was controlled by optimizing the concentration of red dopant in the blue fluorescent emissive layer. Organic light-emitting diodes were fabricated in the configuration ITO/α-NPD/$Zn(hpb)_2$:0.01 wt%$Ir(btp)_2acac$/BCP/Alq3/LiF/Al. The J-V-L characteristic of the device shows a turn on voltage of 5 V. The electroluminescence (EL) spectra of the device cover a wide range of visible region of the electromagnetic spectrum with three peaks around 450, 485 and 610 nm. A maximum white luminance of 3500 cd/m^2 with CIE coordinates of (x, y=0.34, 0.27) at 15 V has been achieved. The maximum current efficiency and power efficiency of the device was 5.2 cd/A and 1.43 lm/W respectively at 11.5 V.
EL spectrum of the white emitting device (0.01wt% $Ir(btp)_2acac$) at various voltages i.e. 6 to 12V is shown in Fig.14 which consist of emission in red, green and blue of the electromagnetic spectra.

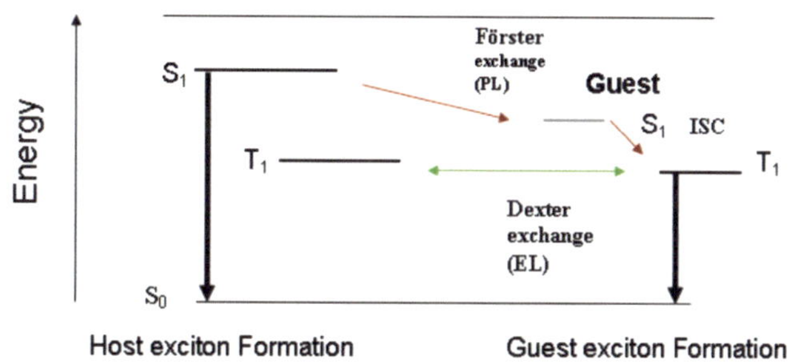

Fig. 13. Energy transfer mechanism for $Zn(hpb)_2$ doped with phosphorescent dopant $Ir(btp)_2acac$ in electroluminescence process.

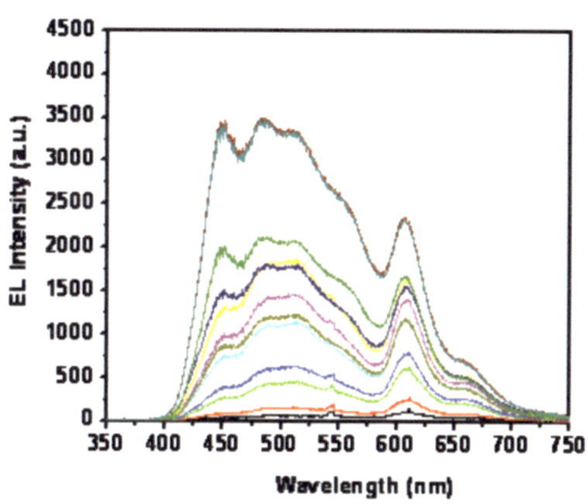

Fig. 14 EL spectrum of WOLED at different bias voltage (6 to 12 V).

4.1.2.2 Solution processed WOLED

One of the ways to get white light emission from conjugated polymers is by using blends of two polymers to extend their emission spectrum (Lee et al 2002, Gong et al 2005, Granstrom and Inganas 1996). Gong et al (2005) achieved WOLED by using a blend of conjugated polymers (PFO-ETM and PFO-F (1%)) and organometallic complex ($Ir(HFP)_3$) as an emissive layer. The device exhibited a maximum brightness of 10 000 cd/m^2 at 25 V. The emission of white light can be understood as the electrons and holes are recombined by two processes: direct recombination on the main chain (PFO-ETM) to produce blue and green emission in parallel with electron and hole trapping on the fluorenone units and on the $Ir(HFP)_3$ followed by radiative recombination with green light from PFO-F (1%) and red light from the triplet excited states of $Ir(HFP)_3$. As a result the mixture of these primary colours gives white light. The devices had a CCT value of ~4500 K, which is very close to that of sunlight (~4700 K) at a solar altitude of 22° and a CRI value of 86. Both CCT and CRI values were insensitive to applied voltage and current density. It has been seen that the quality of emission colour in doped/blend devices is very sensitive to doping/blending concentration and a minor shift in the dopant or polymer ratio will significantly affect the quality of colour. This problem can be solved if a single material is used as an emissive layer and the material has chromophores emitting in the different visible regions. Research is in progress on the development of white OLEDs based on a single molecule as emissive material (Tsai et al 2003, Bai et al 2004, Tu et al 2004). Mazzeo et al (2005) reported a bright single layer white OLED by spin coating a single emitting molecule 3,5 dimethyl 2,6-bis (dimesitylboryl)-dithieno(3,2′ b:2′,3′-d)thiophene. White emission was achieved by the superposition of intrinsic blue-green light emission of the single molecule with red shifted emission from cross-linked dimers. Bright white electroluminescence was obtained with a maximum luminance of 3800 cd/ m^2 at 18 V and an external quantum efficiency of 0.35%. Tu et al (2006) reported a successful development of a WOLED by using a single polymer: polyfluorene derivatives with 1,8-naphthalimide chromophores chemically attached on to the polyfluorene backbones. Optimization of the relative content of 1,8-naphthalimide derivatives in the polymer resulted in pure white-light electroluminescence from a single polymer. The external quantum efficiency of the single emissive WOLEDs is significantly affected by the thickness of emissive and transport layers. Better device efficiency requires the optimization of these layers for balanced charge recombination within the emissive layer.

4.1.3 Exciplex –Excimer structure

OLED characteristics are largely affected by the chemical and physical interaction at organic/organic interfaces. An interaction of organic materials at interface forms a charge-transfer excited-state complex which is known as exciplex/excimer (Li et al 2006, Su et al 2007). An exciplex/excimer is a transient charge transfer complex formed due to the interaction between the excited states of one molecule with the ground state of neighbouring molecule. The resulting electron-hole pair complex decays radiatively, the emission of which is considerably red shifted and broadened as compared to the individual molecules. When the two molecules are same, the transient complex is known as excimer on the other hand if they are different, they are termed as exciplex. The schematic diagram of the emission from the exciplex/excimer is shown below (Fig. 15).

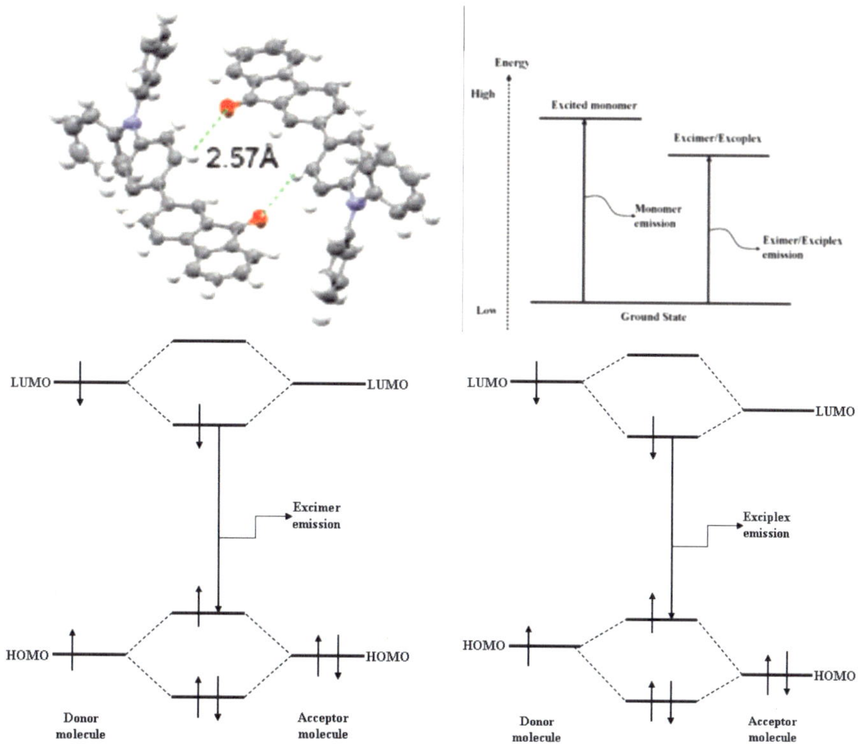

Fig. 15. Schematic diagram showing the formation of excimer/exciplex in organic molecule and light emission from excimer/exciplex molecule is red shifted from the excited monomer emission.

Depending upon the spin multiplicity, excimer and exciplexes can be fluorescencent or phosphorescencent. When singlet excited state of the donor molecule interact with the singlet ground state of acceptor molecule, fluorescence excimer/exciplex are formed where as interaction of triplet excited state of donor and triplet state of acceptor gives phosphorescence excimer/exciplex (Fig. 16).

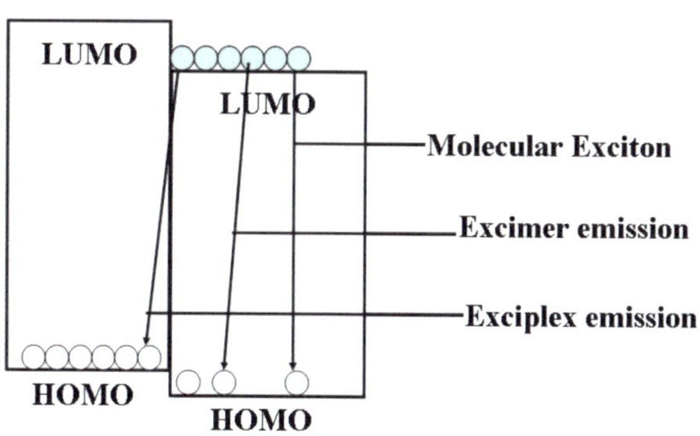

Donor and acceptors, from same molecule excimer are formed
Donor and acceptor, from different molecules exciplex are formed

Fig. 16. Formation of excimer and exciplexes

In the OLEDs there is a high possibility that exciplex formation occurs at the ETL/EML or HTL/EML interfaces because HTL and electron transport layer (ETL) usually have an electron-donating and an electron-accepting nature, respectively. There have been some researches on the application of exciplexes for the tuning of emission colors (Li et al 2006, Liang and Choy 2006) and white emitting OLEDs (Tong et al 2007). Extensive studies on exited bi-molecular complexes and their application in electrophosphorescent devices have been done by Kalinowski et al (2007) and Cocchi et al (2006).

Mazzeo et al. (2002, 2003, 2003a, Blyth et al 2003) obtained white-light emission by spin coating the blend of two different blue-emitting molecules having significant spatial overlap between their LUMOs. The green–red emission from the exciplex combined with the blue emission from the individual molecules gives white- or near-white-light emission. A host material can also be doped with two blue emitting and a red emitting material resulting in green emission from the exciplex, blue and red emission of guests through the energy transfer from host (Kim et al 2003). The concept of exciplex formation between two blue-emitting molecules can be extended to multi-layer device design in which these layers are placed adjacent to each other and the exciplex formation occurs at their interface (Cha and Jin 2003, Cocchi et al 2002 Liu et al 2002, Fang et al 2004). The exciplex emission is more favored if the difference between the HOMOs and LUMOs of the two molecules is large. This will tend to accumulate the charge carriers at the interface, causing increase in the probability of recombination near the interface. The emission color of these devices is highly dependent upon the thickness of the layers (Feng et al 2003) and the applied electric field (Cha and Jin 2003). The layer emitting in the red (Feng et al 2003, 2003a) or green/blue (Liang et al 2003) region can also be placed adjacent to the cathode if the white light is weak in intensity for red or green/blue emission.

WOLEDs can also be fabricated based on phosphorescence excimers. A high energy host organic material is doped with two blue-emitting phosphorescent dyes, namely iridum-bis(4,6,-difluorophenyl-pyridinato- N, C^2)-picolinate (FIrpic) and platinum(II)(2-(4′,6′-difluorophenyl)pyridinato-N, $C^{2'}$)(2,4-pentanedionato) (FPt1) (D′ Andrade et al 2002). The white emission is obtained by combining the blue monomer emission from FIrpic through energy transfer from the host and excimer emission from the square planar complex of FPt1. Another simplification in the device structure is made using a single dopant only and coupling the monomer and excimer emission of the same molecule (D′ Andrade et al 2002, Adamovich et al 2002). The monomer to excimer ratio in these devices is very important to achieve balanced white-light emission and, therefore, the concentration of the dopants becomes crucial (Adamovich et al 2002, D′ Andrade and Forrest 2003).

Kumar et al (2010) has demonstrated the formation of exciplexes at the α-NPD/2-methyl 8-hydroxy quinolinalo lithium (LiMeq) interface using an ultra thin layer of DCM dye as a probe. They have placed the thin layer of DCM dye layer (1nm) at different distance from the interface. The exciplexes formed at the interface transferred their energy to the DCM dye by a Förster type energy transfer. Excitons were formed at the dye molecule and was detected which gave its characteristics emission. As the dye layer is moved away from the interface the intensity of emission from DCM decreased and emission from the exciplex increased indicating that the exciplexes are generated at the interface only. The maximum energy transfer was observed when the dye layer was placed at the interface which was limited by the number of dye molecules. At an optimum distance from the interface the emission from the exciplexes together with that from the dye gave white light emission. The origin of exciplex formation was explained as due to a mismatch of the HOMO and LUMO (Fig17) energies and accumulation of charges at the α-NPD/LiMeq interface.

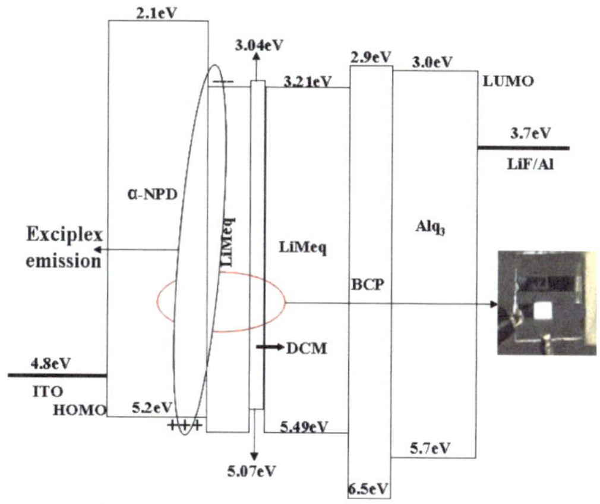

Fig. 17. Energy level diagram showing the formation of exciplexes at the interface

There are several issues involved in formation of exciplexes and excimers that limit the device performance. In the blended emissive layers, the formation of exciplex strongly depends on the concentration, structure and morphology of the film (Mazzeo et al 2002). In such cases fine-tuning of the emission spectrum can be done by concentration variation. However, there is a limit for concentration variation since phase separation or aggregation occurs above an optimum concentration, which in turn, will not allow spatial overlap of the LUMOs of the donor and the acceptor (Mazzeo et al 2002). Another difficulty in exciplex formation is the temperature dependence since at low temperatures the emission by exciton recombination will be predominant while at high temperatures, the exciplex emission will be predominant (Chao and Chen 1998).

The external quantum efficiencies of most of the exiplex emitting devices are usually low (~1%)(Noda et al 1999, Wang et al 1998, Kawabe and Abe 2002, Cocchi et al 2002). Indeed the appearance of exiplex is often considered as a major reason for the poor device performance but the dissociative property of the exiplex of the ground state imparts in to broad featureless emission band which is red shifted from the parent molecule emission spectrum. This is the reason that exiplex light emitting devices can be considered as promising technologies for manufacturing WOLEDs (Feng et al 2001).

4.1.4 Microcavity structure

WOLED fabricated by optical optimization through multimode resonant cavities have also getting some attention (Singha et al 2003, Dodabalapur 1994, 1994a). The microcavity is a system consisting of a pair of highly reflecting mirrors having separation of the order of a micron and utilizes the concept of Fabry-Perot resonant cavity. A resonant microcavity is formed when the emissive material is sandwiched between two metallic mirrors or a metallic mirror and semitransparent distributed Bragg reflector (DBR) (Peng et al 2003). During the operation of device, standing waves are generated, the wavelength of which depends upon the length and refractive index of the cavity. In conventional structures, light is wasted since it leaks in all directions. But in a microcavity light emerges only from one end of the cavity and the structure is more efficient. By varying the thickness of the layer, undesirable light can be filtered out and the emission of light can be obtained at any desired wavelength. Since a microcavity LED is more efficient and uses less current, it lasts longer. A microcavity resonator is one of the most effective ways of enhancing the luminance and brightness of monochromatic OLEDs (Hsu et al 2004, Neyts 2005). Spectral narrowing and intensity enhancement of spontaneous emission in OLEDs by microcavity has been reported (Juang et al 2005). Dodabalapur et al (1994) demonstrated the control of the emission of OLEDs by multimode resonant cavities such that the thickness of the cavity is greater than the single mode cavity devices so that it has several resonant modes within the emission spectrum of the material. White emission from the microcavity structures is obtained by a simple modification of the Fabry-Perot resonator (Singha et al 2003) in which combined emission from the two cavities of different lengths produces white light. Shiga et al (2003) employed a modified Fabry-Perot resonator cavity with different cavities of different lengths (see Fig.18, the terms MM, DM, EML and FL represent metallic mirror, dielectric mirror, emission layer and filter layer, respectively), and produced white light. One disadvantage of this approach is that the colour coordinates change with viewing angle and this angular dependence of the emission limits the microcavity applications for white OLEDs.

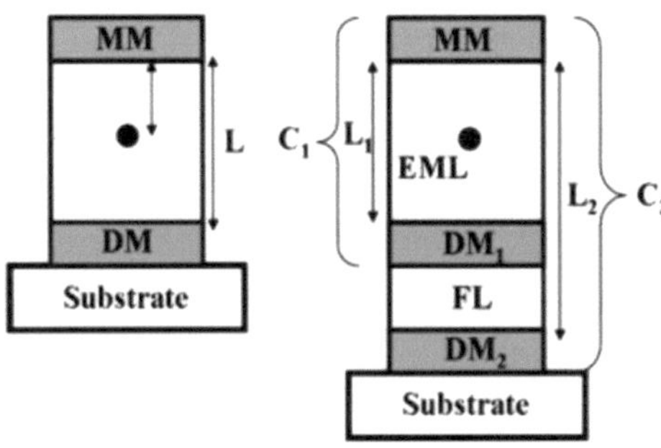

Fig. 18. Concept of (*a*) usual cavity and (*b*) multiwavelength resonant cavity (MWRC).

4.2 Wavelength conversion

Some blue or ultraviolet light from one OLED is used to excite several phosphors, each of which emits a different colour and these different colours are mixed to make a white light with the broadest and richest wavelength spectrum and is called down conversion by phosphors.

4.2.1 Down conversion by phosphor

A difficulty of colour stability due to differential ageing of various species occurs in several methods used for white emission from organic LEDs White emission by down conversion phosphors may be an alternative method in which a blue emitting OLED is coupled with one or more down conversion phosphor layers. During the operation of the device a small portion of the blue light is scattered and goes through the phosphors without down conversion and the phosphor layers absorb emission from the blue OLED and emit according to their intrinsic property. The mixing of unabsorbed emission from the blue

OLED and the emission from the phosphors produces white light. A schematic diagram of white light emission by down conversion methode is shown in Fig19. Here only the blue emitter conducts the charge and is the only site which is directly excited. Once the excitons are generated they excite other phosphors to produce balanced white emission. As the blue light emission decreases with age, light from other coupled phosphors should decrease proportionately because their relative intensities are directly related to those of the blue light emitter there for there is no differential colour ageing in the down conversion technique. The emission spectrum can be adjusted by varying the concentration and thickness of the phosphor layers (Misra et al 2006, Gupta et al 2006).These down conversion WOLEDs demonstrated the highest CRI value, 93 and efficiency 3.8 lm/W

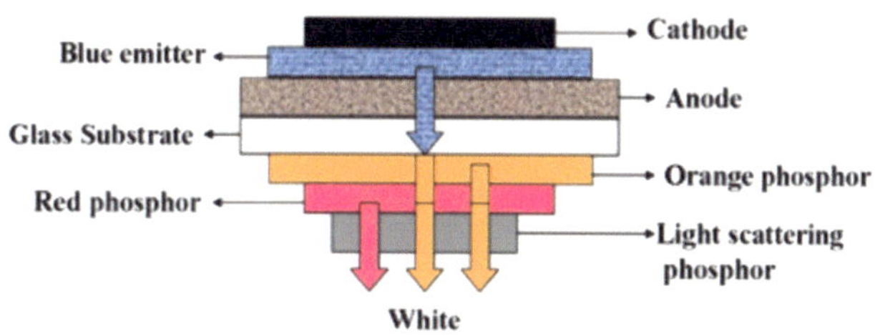

Fig. 19. Schematic diagram of white light emission by down conversion.

Duggal et al (2002) reported white light emission from a blue OLED coupled with down converting orange and the red organic phosphor, namely perylene orange and perylene red dispersed in poly (methacrylate) (PMMA) followed by a layer of inorganic light scattering phosphor, namely Y(Gd)AG:Ce dispersed in polydimethyl siloxane silicon. The quantum efficiency of photoluminescence of dyes in PMMA was found to be >98% and the quantum yield of Y(Gd)AG:Ce was 86%. These down conversion WOLEDs demonstrated the highest CRI value, 93 and efficiency 3.8 lm/W.

In Materials Research Society meeting 2005 (MRS Fall Meeting, 28 November- 2 December 2005, Boston, MA, USA) one of the biggest changes pioneered by Junji Kido and Forrest's groups has been the switching from fluorescent materials to phosphorescent materials. Phosphorescent dyes can convert both singlet and triplet excitons into light, making the devices potentially much more efficient. They reported a new blue OLED based on phosphorescent compound FIrpic, with a record breaking efficiency of 42 lm/ W at 100 cd m^{-2}. They surrounded the FIrpic containing layer with layers of other compounds that allow triplet excitons to reside within the emissive layer. This creates an energetic well in the light emitting material so that excitons cannot get out and they decay in the presence of blue phosphor to give off blue light. To get white light they added a yellow phosphor to their blue light emitting layer, converting some of the emitted light to yellow, which combined with the blue to give off white light. The light photons emitted from OLEDs reflect off the glass–air interface and bounce back inside the device, where many of them are reabsorbed and generate heat instead of light. The efficiency of OLEDs has been boosted (36 lm/W to 57 lm/W) by adding a specialized antireflective coating to the outside of the glass.

White emission from down conversion can also be obtained by coupling UV light with red, green and blue phosphors which excites several phosphors, each of which emits a different colour, as a result of mixing these colours white light emission is obtained. The technique has colour stability but the losses associated with wavelength conversion are the main drawbacks of this technique.

There is a lack of theoretical modelling of electroluminescence in OLEDs. Tyagi et al (2010) has developed a model based on Monte-Carlo simulation technique (Ries et al 1988, Ries and Bässler1987, Movaghar et al 1986, Houili et al 2006) to model the disordered semiconductor (assuming Gaussian density of states) to generate the electroluminescence spectrum of multilayer OLED for white light emission. The electroluminescence (EL) spectrum in an OLED was generated by the recombination of a positive charge carrier with a negative charge carrier in the emitting layer. The emitted photons have energy equal to the difference of energies of negative and positive charge carriers.

5. Photo physics of White OLEDs

Doping of wide band gap materials which emits in the blue region of the spectrum with lower band gap dopants can modify the emission properties of the host molecules. The modification of emission properties upon doping is due to efficient energy transfer process from the host molecules to the guest molecules (dopants) and with careful balancing of the doping it is possible to obtain white light emission. The dopants can be fluorescent or phosphorescent in nature. The dopant site can be excited directly or by energy/charge transfer from the host molecule.

The energy transfer in this matrix occurs in different ways. They are (i) Forster type energy transfer, (ii) Dexter transfer (iii) Exciplex - excimer charge transfer and (iv)Trap assisted recombination. The principles are discussed below.

5.1 Förster Type energy transfer

A molecule that is in an excited singlet or triplet state (Donor) can transfer its energy to a molecule in the ground state (Acceptor) by electronic energy transfer (ET). Energy transfer always involves two molecules that are in close proximity to each other. It is the fundamental process of energy / exciton migration which consists of multiple energy transfer processes. Radiationless energy transfer can occur via a dipole-dipole interaction having a long range separation of about ~30-100A° known as Förster transfer or via exchange of electrons through overlapping orbitals termed as Dexter transfer. The Forster energy transfer requires spectral overlap of the emission spectrum of the donor with the absorption spectrum of the acceptor. The radiation field of the dipole transition of D is coupled with the dipole transition of A through space without the requirement of spatial overlap of wavefunctions and can be explained as

$$D^* + A \rightarrow X \rightarrow D + A + h\nu$$

where D*, A, X, D and hν stand for excited donor, ground state of acceptor, intermediate excited system, ground state of donor and energy of emitted photon respectively. A scheme of Förster transfer is depicted in Fig. 20. The left side of Fig. 20 shows energy transfer between molecules of similar singlet energy. This is possible due to the weak overlap of absorption and emission spectra of identical molecules. The right side shows energy transfer to a molecule which is lower in its singlet energy (trap state). In both cases the ET occurs radiationless.

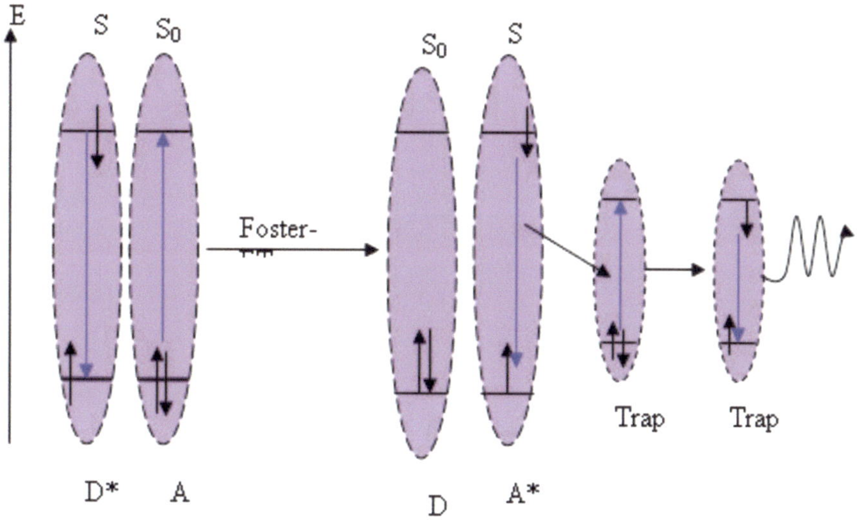

Fig. 20. Simplified scheme of resonant energy Forster energy transfer between a donor (D) and an acceptor (A). Right side shows energy transfer to a trap which is lower in its singlet energy.

Furthermore the fluorescence lifetime of the donor molecules is significantly reduced as a consequence of efficient energy transfer to the lower energy trap. Since Förster energy transfer is mediated by dipole-dipole interaction without the need of direct overlap of orbitals, it can overcome distances up to 10 nm. It allows only singlet-singlet transition at low acceptor concentration and at a much faster rate of $<10^{-9}$ s.

5.2 Dexter transfer

The second possibility of energy transfer is known as exchange type or Dexter energy transfer. Dexter ET is based on quantum mechanical exchange interactions, therefore it needs strong spatial overlap of the involved wavefunctions of D and A. Since the overlap of electronic wavefunctions decays exponentially with distance, it is expected that the rate constant k_{DA} decreases even more rapidly with distance R than observed in the case of singlet transfer. A schematic presentation of Dexter ET is shown in Fig. 21. Dexter ET occurs typically over distances which are similar to the van-der-Waals distance, i.e. R = 0.5 - 1nm. The rate constant drops exponentially with the distance R_{DA} between D and A:

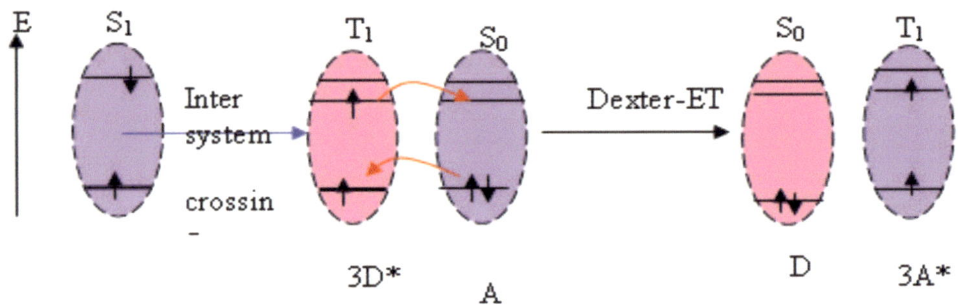

Fig. 21. Schematic presentation of Dexture type ET

Dexter ET is a correlated two electron exchange process. Hence it allows triplet energy transfer without the additional need of intersystem crossing upon energy transfer of a triplet state unlike the Forster energy transfer which requires spin-forbidden ISC for triplet energy transfer. Due to this reason Forster ET is mostly used to describe singlet migration, whereas Dexter ET is used to describe the triplet migration in the solid state.

A lot of effort has been made to achieve white light emission from small molecules (Lim et al 2002, 2004, Kido et al 1995, Mazzeo 2003, Wang 2005, Niu et al 2005) as well as from polymers (Lee et al 2002, Park et al 2005, Al Attar et al 2005, Tasch et al 1997) using the Förster /Dexter energy transfer mechanism. Mazzeo et al (2003) have fabricated OLED from a blend of N, N′-diphenyl-N, N′-bis(3-methylphenyl)-1,1′-biphenyl- 4,4′diamine (TPD) (PL in the blue region) with a thiophenebased pentamer, (3,3′,4‴,3⁗-tetracyclehexyl-3, 4-dimethyl- 2,2′:5′,2:5,2‴;5⁗,2⁗: quinquethiophene-1,1-oxide (T5oCx) (PL in the red region). The incomplete Förster energy transfer occurred from host (TPD) to guest (T5oCx) and as a result, they got emission from both the molecules, which produced white light. This energy transfer was favoured by the overlapping of the strong emission spectra of TPD and absorption spectra of T5oCx. Wang et al (2005) achieved a highly efficient white organic LED using two blue emitters with similar structures 9,10-di-(2-naphthyl)-anthracene(ADN) and 9,10- di-(2-naphthyl)-2-terbutyl-anthracene (TADN) doped with (0.01–0.05%) yellow-orange emitting rubrene. The device had a maximum external quantum efficiency of 2.41% (5.59 cd / A) and a maximum luminance of 20 100 cd/ m^2 at 14.6 V. The advantage of the similar structure of ADN and TADN is that it depresses the molecular aggregation, which leads to better film morphology.

Park et al (2005) have demonstrated white emission from ITO/PVK/(PDHFPPV +MEHPPV)/Li:Al, ternary polymer blended LED. Here poly(N-vinylcarbazole) (PVK) acts as an energy donor as well as electron blocker while poly(9,9-dihexyl-2,7-fluorene phenylenevinylene) (PDHFPPV) + poly(2-methoxy-5-(2-ethylhexyloxyl)-1,4- phenylene vinylene) (MEHPPV) blend acts as an emitting layer. In this bilayer system the spectral overlapping between the emission of PVK and absorption of PDHFPPV and between the emission of PDHFPPV and absorption of MEHPPV, meets the necessary condition for Förster energy transfer. The cascade energy transfer from PVK to PDHFPPV and then to MEHPPV and the emission from PDHFPPV and MEHPPV results in whitish light emission.

Al Atter et al (2005) fabricated an efficient white PLED based on a blue emitting poly(9,9-bis(2-ethylhexyl)fluorine-2,7-diyl) endcapped with bis(4-methylphenyl)phenylamine (PF2/6am4) and doped with yellow-orange phosphor iridium (tri-fluorenyl) pyridine complex ($Ir(Fl3Py)_3$). The white light emission from the system was attributed to a strong Dexter energy transfer from (PF2/6am4) to ($Ir(Fl3Py)_3$). The devices have a with a peak external quantum efficiency of 2.8% and a luminance of 16 000 cd m^{-2} at 5 V.

5.3 Exciplex - excimer charge transfer

The third possibility of energy transfer is known as Exciplex - excimer charge transfer. In the excimer formation the wavefunction of excited states extends over the molecules and the molecules are bound together only in the excited state but not in the ground state. This absence of the bound ground state provides a way for efficient charge transfer from higher energy host to lower energy guest. The charge transfer mechanism can also be explained as

$$D^* + A \rightarrow X \rightarrow D + A + h\nu,$$

where D*, A, X, D and hv stand for excited donor, ground state of acceptor, intermediate excited system, ground state of donor and energy of emitted photon, respectively. Here X is the charge transfer exciplex/excimer complex. The charge transfer takes place at the interface of the charge transport layer and the emitting layer (Chao and Chen1998, Thompson2001, Feng et al 2001, Cocchi et al 2002, Wang et al 2004), because of the mismatched electronic structure of the two molecules (exciplex) and wavefunction overlapping (excimer). The charge transfer excitations occur at energies close to those of excitations localized at the donor and acceptor molecules (Fang et al 2004). The charge transfer occurs due to the interaction between the excited states of one molecule with the ground state of the other molecule (as discuss in section 4.1.3), resulting in a radiative electron- hole recombination pair. The exciplex formation is favoured by a large difference between the HOMOs and LUMOs of the emitter and the charge transport layer. Because of this large difference the injection of the charge carriers from transport layer to the emitter layer and from the emitter layer to the transport layer will be difficult and there will be accumulation of the carriers at the interface. Now the indirect recombination from LUMO of the transport layer to HOMO of the emitter layer is more favoured. The energy of the exciplex is always less than the energy of the excited single molecules and its emission is very broad.

5.4 Trap assisted charge transfer

The Fourth possibility of energy transfer is known as the charge trapping mechanism that requires the energy of the dopant to be in such a way that it is energetically favorable for charge transfer. In the trap assisted charge transfer mechanism the recombination process can be visualized as that the electron and hole gets trapped in the dye molecules which generates excitons which decays for the generation of light.

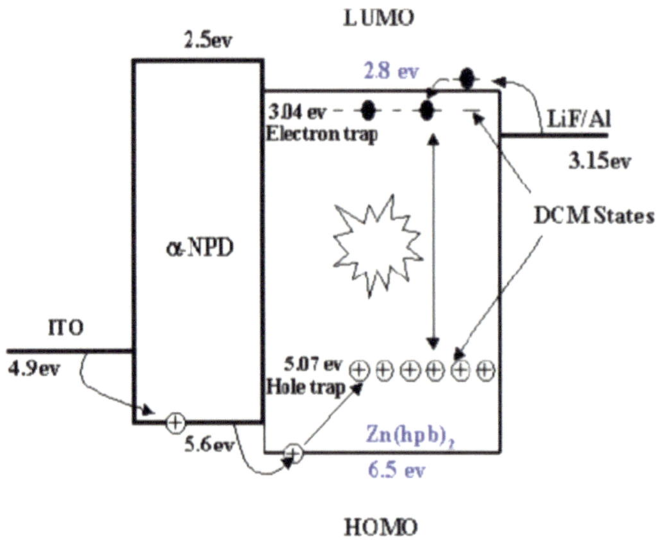

Fig. 22. Energy level diagram for the $Zn(hpb)_2$:DCMsystem.

Fig. 22 shows the energy level diagram of the host and the dye molecules which is used to explain the charge trapping of dye molecules in the $Zn(hpb)_2$ system(Rai et al 2008). The host matrix and the dye have their highest occupied molecular orbital (HOMO) level at~6:5 and ~5:07 eV respectively and their lowest unoccupied molecular orbital (LUMO) at ~2:8 and ~3:04 eV respectively. (Lee et al 2002) According to the energy level diagram, the dye molecules will be forming deep hole traps (1.43 eV) and shallow electron traps (0.24 eV) into the host forbidden energy gap. The hole traps being very deep will be above the Fermi level of the host matrix and will be always remain filled and will not alter hole transport properties. The electron traps being shallow and may lie on the same side of the LUMO compared to the Fermi level should contribute to the carrier trapping and the electrical properties of the guest–host system.

6. Problem to be solved

The main technical challenges that need to be met for OLED technology to displace fluorescent lighting for general illumination have been laid out in detail. The challenges are indeed formidable and will require a long-term investment in technology development. Because OLEDs possess potential features such as conformability to surfaces that are not possible with current lighting technology, it is likely that products will make it into the lighting market before all of the long-term challenges are met. Such shorter-term applications will help to fuel the necessary long-term development for general illumination. There are reasons to be optimistic that an OLED-based solid state light source will become a reality. One reason is simply that while the field has demonstrated incredible progress in the last decade, it has been largely constrained into pursuing certain types of device structures due to the needs of display applications. Once this constraint is lifted, new types of device structures and materials that have so far been ignored can be investigated. These extra parallel approaches can only enhance progress. Another, related, reason for optimism has to do with the fact that OLED technology as a whole is still in a very early stage of development. OLEDs utilize organic molecules that are literally blended together into relatively simple device structures that then yield impressive performance. The number of possible organic molecules, each with tunable functions that can be utilized is virtually unlimited due to the capabilities of modern organic chemistry. In fact, the field is really still in its infancy with regard to understanding what types of molecules should be made. Although the device physics of an OLED is largely understood, the detailed physics of charge transport, exciton spin formation, and energy transfer is not. Similarly, the detailed material science required to understand how molecules interact and produce a characteristic morphology in the solid state is not well understood. These details are necessary to guide the development of new organic molecules/polymers and device structures that optimize performance. Thus, there is a good chance that as basic research in OLED technology continues, and as focused research on solid-state lighting accelerates, the exponential rate of progress seen in the last decade will continue into the next. If so, then by the end of the next decade OLEDs will have a good shot at surpassing fluorescents as the premier lighting technology.

7. Future prospects of WOLED

The prospects of organic LEDs are very good. In the R &D scenario, new efficient emitters are being reported everyday which are far more efficient than those which are in present use. On the technology side, new encapsulation strategies are being introduced particularly those based on of thin film encapsulation which has shown encouraging results. Similarly new ways to reduce the turn on voltage by doping of charge transport layers are also in progress. New organic deposition techniques as well as roll to roll processing of OLEDs are also showing encouraging results. Perhaps the new technologies based on all printed devices may revolutionaries the lighting industry. The efficiency of the best OLED has surpassed that of fluorescent discharge lamps and one can expect that in the coming years we see more efficient devices which replaces the existing lighting concepts.

8. Conclusion

White light sources based on OLEDs are efficient and clean and have the potential to replace the existing lighting system based on incandescent lamp and discharge tubes. Even though the technology has developed to a stage where it can be commercialized, there are many basic issues relating to material science which are not clearly understood and very intense research is required in this direction. Many government funded research agencies and commercial establishment are actively working to improve WOLED efficiency and life time to bring it to acceptable limits. These efforts have started showing results and in the near future we can expect a versatile organic based lighting system replacing the existing light sources.

Acknowledgements

The authors are grateful to Director, National Physical Laboratory, New Delhi, for his keen interest in this investigation. The authors gratefully recognize the financial support from the Department of Science and Technology (DST), Council of Scientific and Industrial Research (CSIR) New Delhi, for providing funds.

9. References

Adachi C, Baldo M A, Forrest S R, and Thompson M E, High-efficiency organic electrophosphorescent devices with tris(2-phenylpyridine)iridium doped into electron-transporting materials, *Appl. Phys. Lett.*, 77 (2000) 904–906

Adachi C, Baldo M A, Forrest S R, Lamansky S, Thompson M E and Kwong R C, High-efficiency red electrophosphorescence devices, *Appl. Phys. Lett.* 78 (2001) 1622

Adachi C, Baldo M A, Thompson M E and Forrest S R Nearly 100% internal phosphorescence efficiency in an organic light-emitting device, 2001,J. Appl. Phys. 90 5048

Adamovich V., Brooks J., Tamayo A., Alexander A. M., Djurovich P.I., Andrade B. W. D., Adachi C., Forrest S. R. and Thompson M. E., "High efficiency single dopant white electrophosphorescent light emitting diodes", *New J. Chem.*, 262002, 1171.

Al Attar H A, Monkman A P, Tavasli M, Bettington S and Bryce M R, White polymeric light-emitting diode based on a fluorene polymer/Ir complex blend system *Appl. Phys. Lett.* 86, (March 2005)121101, doi:10.1063/1.1873046

Aziz H, Popovic Z D, Hu N-X, Hor A-M, and Xu G, Degradation mechanism of small molecule based organic light emitting diodes, *Science*, 283 (1999) 1900–1902

Aziz H, Popovic Z, Xie S, Hor A. M, Hu N-X, Tripp C, and Xu G, Humidity-induced crystallization of tris(8-hydroxyquinoline) aluminum layers in organic light-emitting devices, *Appl. Phys. Lett.*, 72 (1998) 756–758

Bai S J, Wu C C, Dang T D, Arnold F E and Sakaran B, Tunable and white light-emitting diodes of monolayer fluorinated benzoxazole graft copolymers *Appl. Phys. Lett.* 84, (January 2004), 1656, doi:10.1063/1.1667262

Baldo M A, Adachi C and Forrest S. R., Transient analysis of triplet-triplet annihilation Transient analysis of organic electrophosphorescence., *Phys. Rev. B* 62 (2000)10967

Baldo M A, Lamansky S, Burrows P E, Thompson M E, and ForresS R t, Very high-efficiency green organic light-emitting devices based on electro phosphorescence, *Appl. Phys. Lett.*, 75 (1999) 4–6

Baldo M A, O'Brien D F, Thompson M E, and Forrest S F, The excitonic singlet–triplet ratio in a semiconducting organic thin film, *Phys. Rev. B*, 60 (1999)14422–14428

Baldo M. A., O'Brien D. F., You Y., Shoustikov A., Sibley S., Thompson M. E., and Forrest S. R., Highly efficient phosphorescent emission from organic electroluminescent devices, *Nature* 395, (1998) 151

Bernanose A., Comte M. and Vouaux P., *J. Chem. Phys.* 50, (1953) 65 (×)

Blochwitz J, Pfeiffer M, Hofman M, and Leo K, Non-polymeric OLEDs with a doped amorphous hole transporting layer and operating voltages down to 3.2 V to achieve 100 cd/m^2, *Synth. Met.*, 127 (2002)169–173

Blyth R.I.R., Thompson J., Mazzeo M., Gigli G., Cingolani R., "An organic solution to the Kelvin myth: White light with true color temperature via exciplex formation", *Synth. Met.*, 137 2003, 1053.

Brian B, D'Andrade W., Thompson M. E., and Forrest S. R., Controlling Exciton Diffusion in Multilayer White Phosphorescent Organic Light Emitting Devices, *Advanced Materials*, 14, 2, (January 2002), 147-151

Burroughs J H, Bradley D.D C., Brown A R, Marks R N, MacKay K, Friend R H, Burn P L, and Holmes A. B., Light-emitting diodes based on conjugated polymers, *Nature*, 347 (1990) 539–541

Burrows P E, Bulovic V, Forrest S R, Sapochak L S, McCarty D M, and Thompson M E, Reliability and degradation of organic light emitting devices, *Appl. Phys. Lett.*, 65 (1994) 2922–2924.

Burrows P E, Forrest S R, Zhou T X, and Michalski L, Operating lifetime of phosphorescent organic light emitting devices, *Appl. Phys. Lett.*, 76 (2000) 2493–2495

Burrows P E, Gu G and Forrest S R, Stacked organic light-emitting devices for full-color flat panel displays, *Proc. SPIE* 3363 (1998) 269.

Burrows P E, Gu G, Bulovic V, Shen Z, Forrest S R and Thompson M E 1997 IEEE Trans. Electron Devices 44 1188 (×)

Cao Y, Parker I D, Yu G, Zhang C and Heeger A J, Improved quantum efficiency for electroluminescence in semiconducting polymers, *Nature* 397 (1999) 414.

Cha S.W., Jin J.I.,"A field-dependent organic LED consisting of two new high T_g blue light emitting organic layers: a possibility of attainment of a white light source" *J. Mater. Chem.*, 13 2003, 479.

Chang C C, Chen J F, Hwang S W and Chen C H, Highly efficient white organic electroluminescent devices based on tandem architecture, *Appl. Phys. Lett.* 87, (December 2005)253501, doi:10.1063/1.2147730

Chang S M, Tzeng Y J, Wu S Y, Li K Y and Hsueh K L, Emission of white light from 2-(2'-hydroxyphenyl) benzothiazole in polymer electroluminescent devices *Thin Solid Films*, 477, 1-2, (April 2005), 38-41

Chao C I and Chen S A, "White light emission from exciplex in a bilayer device with two blue light-emitting polymers" *Appl. Phys. Lett.* 73 1998 426.

Cheng G, Zhao Y, Li F, Xie W and Liu S, Effect of a thin layer of tris (8-hydroxyquinoline) aluminum doped with 4-(dicyanomethylene)-2-t-butyl-6-(1,1,7,7-tetramethyljulolidyl-9-enyl) on the chromaticity of white organic light-emitting devices *Thin Solid Films*, 467, 1-2, (November 2004), 231-233.

Cheng G., Li F., Duan Y., Feng Y., Liu S., Qiu S., Lin D., Ma Y., Lee S.T., " White organic light emitting devices using a phosphorescent sensitizer" *Appl. Phys. Lett.*, **82**,No. 24, 16 June 2003, 4224- 4226 (DOI: 10.1063/1.1584075)

Cheun C.H., Tao Y.T., Highly-bright white organic light-emitting diodes based on a single emission layer *Appl. Phys. Lett.* 81, (October 2002), 4499, doi:10.1063/ 1.1528736

Cocchi M, Virgili D, Giro G, Fattori V, Marco P D, Kalinowski J and Shirto Y, "Efficient exciplex emitting organic electroluminescent devices" Appl. Phys. Lett. 80 2002 2401

Cocchi M., Virgili D., Sabatini C., Kalinowski J., Organic electroluminescence from singlet and triplet exciplexes: Exciplex electrophosphorescent diode, *Chemical Physics Letters*, 421, 4-6, (April 2006), 351-355.

Cumpston B H and Jensen K F, Electromigration of aluminum cathodes in polymer-based electroluminescent devices, *Appl. Phys. Lett.*, 69 (1996) 3941–3943

D'Andrade B W, Holmes R J and Forrest S R, Efficient Organic Electrophosphorescent White-Light-Emitting Device with a Triple Doped Emissive Layer *Advanced Materials*, 16, 7, (April, 2004), 624-628

D'Androde B.W. and Brown J. J., Organic light-emitting device luminaire for illumination applications, *Appl. Phys. Lett.* 88 (2006) 192908

D; Andrade B.W., Forrest S.R., "Effects of exciton and charge confinement on the performance of white organic p–i–n electrophosphorescent emissive excimer devices", *J. Appl. Phys.*, **94**, 2003, (3101-3108).

D'Andrade B.W., Adachi C., Forrest S. R., Thompson M.E., High efficiency single dopant white electrophosphorescent light emitting diodes *New J. Chem.* 26 (20 02) 1171. DOI: 10.1039/b204301g

D' Andrade B.W., Brooks J., Adamovich V., Thompson M.E., Forrest S.R., "White Light Emission Using Triplet Excimers in Electrophosphorescent Organic Light-Emitting Devices" *Advanced Materials*, 14, 15, August 2002, (1032-1036).

Deshpande R S, Buloviæ V and Forrest S R White-light-emitting organic electroluminescent devices based on interlayer sequential energy transfer. *Appl. Phys. Lett.* **75**, 888 (1999); doi:10.1063/1.124250 (*3 pages*)

Do Y.R., Kim Y. C., Song Y. W. and Lee Y. H., Improvement of coupling-out efficiency in organic electroluminescent devices by addition of a diffusive layer, *J. Appl. Phys.* 96 (2004) 7629

Dodabalapur A, Rothberg L.J., Rothberg L.J., Miller T.M., " Electroluminescence from organic semiconductors in patterned microcavities", *Electron. Lett.*, 30, 1994, 1000.

Dodabalapur A., Rothberg L.J., Miller T.M., Color variation with electroluminescent organic semiconductors in multimode resonant cavities *Appl. Phys. Lett.* 65(1994) 2308

Duggal A R, Shiang J J, Heller C M and Foust D F, "Organic light-emitting devices for illumination quality white light", *Appl. Phys. Lett.*, **80**, 2002, 3470.

Fang Y, Gao S, Yang X, Shuai Z, Beljonne D and Bredas J L," Charge-transfer states and white emission in organic light-emitting diodes: a theoretical investigation" *Synth. Met.* **141** 2004 43

Fang Y., Gao S., Yang X., Shuai Z., Beljonne D., Bre´das J.L., "Polyacetylene film: A new electrode material for photoenergy conversion", *Synth. Met.*, 4, 2004, 43.

Feng J, Li F, GaoWand Liu S, "White light emission from exciplex using tris-(8-hydroxyquinoline) aluminum as chromaticity-tuning layer" *Appl. Phys. Lett.* 78 2001 3947

Feng J., Liu Y., Wang Y., Liu S., "Thickness dependent emission color of organic white light-emitting devices", *Synthetic Metals*, 137, 1-3, 4 April 2003, (1101-1102).

Forrest S R, Burrows P E, and Thompson M E, in Organic Electroluminescent Materials and Devices, Miyata S and Nalwa H S, Eds., (1997) *Gordon & Breach, Amsterdam*, 447–453.

Friend R. H., Gymer R. W., Holmes A. B., Burroughes J. H., Marks R. N., Taliani C., Bradley D. D. C., Dos Santos D. A., Brédas J. L., Lögdlund M., et al., Electro-luminescence in conjugated polymers *Nature* 397 (1999)121

Gong X, Wang S, Moses D, Bazan G C and Heeger A J, Multilayer Polymer Light-Emitting Diodes: White-Light Emission with High Efficiency *Advanced Materials,* ,17, 17,(September, 2005),2053-2058

Granstom M., Inganas O., White light emission from a polymer blend light emitting diode *Appl. Phys. Lett.* 68 (1996) 147

Gu G., Garbuzov D.Z., Burrows P.E., Vankatsh S., Forrest S.R., Thompson M.E., High-external-quantum-efficiency organic light-emitting devices, *Opt. Lett.* 22 (1997) 396

Guo F, Ma D, Wang L, Jing X and Wang F, High efficiency white organic light-emitting devices by effectively controlling exciton recombination region, *Semicond. Sci. Technol.* 20 (2005) 310, doi: 10.1088/0268-1242/20/3/010

Gupta D., Katiyar M., Deepak, Various approaches to white organic light emitting diodes and their recent advancements *Optical Materials*, 28, 4, (March 2006), 295-301

Hamada Y, Sano T, Shibat K a, and Kuroki K, Influence of the emission site on the running durability of organic electroluminescent devices, *Jpn. J. Appl. Phys.*, 34 (1995) L824–L826

Helfrich W. and Schneider W. G., Recombination Radiation in Anthracene Crystals, *Phys. Rev. Lett.* 14, (1965) 229

Higginson K A, Zhang X M, and Papadimitrakopoulos F, Thermal and morphological effects on the hydrolytic stability of aluminum tris(8-hydroxyquinoline) (Alq3), *Chem. Mater.*, 10 (1998) 1017–1020

Holmes R J, Forrest S R, Tung Y J, Kwong R C, Brown J J, Garon S and Thompson M E, Blue organic electrophosphorescence using exothermic host–guest energy transfer, *Appl. Phys. Lett.* 82 (2003) 2422

Houili H., Tutis E., Batistic I., Zuppiroli L., Investigation of the charge transport through disordered organic molecular heterojunctions.*J. Appl. Phys.* 100, (August 2006)033702, doi:10.1063/1.2222041

Hsu S F, Lee C C, Hu A T and Chen C H, "Fabrication of blue top-emitting organic light-emitting devices with highly saturated color", *Current Applied Physics,* 4, 6, November 2004, 663-666.

Hu B., F Karasz.E., Blue, green, red, and white electroluminescence from multichromophore polymer blends, *J. Appl. Phys.* 93, (November 2003)1995, doi:10.1063/ 1.1536018

Hung L.S., Tang C.W., Mason M.G., Raychaudhuri P. and Madathil J., Application of an ultrathin LiF/Al bilayer in organic surface-emitting diodes, *Appl. Phys. Lett.*,78 (2001) 544.

J. Feng, Y. Liu, F. Li, Y. Wang, S. Liu, Opt. Quant. Electron. 35 (2003) 259.

Jiang X.Y., Zhang Z.L., Zhang B.X., Zhu W.Q., Xu S.H., Stable and current independent white-emitting organic diode *Synthetic Metals*, 129, 1, (June 2002), 9-13.

Jordan R.H., Dodabalapur A., Strukeji M., Miller T.M., White organic electro-luminescence devices, *Appl. Phys. Lett.* 68 (1996) 1192

Juang F S, Laih L H, Lin C J and Hsu Y J 2002 Japan. J. Appl. Phys. 41 2787

Kalinowski J., Cocchi M., Virgili D., Fattori V., and Gareth Williams J. A., Mixing of Excimer and Exciplex Emission: A New Way to Improve White Light Emitting Organic Electrophosphorescent Diodes, *Advanced Materials* 19, 22, (November, 2007), 4000-4005

Kamata, N. Ishi R., Tonsyo S., Terunuma D., Electroluminescence of mixed organic dyes via resonant energy transfer from polysilane molecules, *Appl. Phys. Lett.* 81, (October 2002), 4350, doi:10.1063/1.1527231

Kamtekar K. T., Monkman A. P. and Bryce M. R., Recent advances in white organic light-emiting materials and devices (WOLEDs), Advanced Materials,22 (2010), 572-582.

Kanno H., Sun Y. and Forrest S.R. "White organic light emitting device based on a compound fluorescent- phosphor-sensitized-fluorescent emission layer" *Appl. Phys. Lett.*, **89**, 5 Oct. 2006, 143516-3 (DOI: 10.1063/1.2357038)

Kawabe Y. and Abe J., "Electron mobility measurement using exciplex-type organic light-emitting diodes", *Appl. Phys. Lett.*, **81**, 2002, 493.

Kawamura Y., Yanagida S., Forrest S.R., Energy transfer in polymer electro-phosphorescent light emitting devices with single and multiple doped luminescent layers, *J. Appl. Phys.* 92, (March 2002), 87, doi:10.1063/1.1479751

Kido J, Shionoya H and Nagai K, "Single layer white light emitting organic electroluminescent devices based on dye dispersed poly(*N*'vinylcarbazole)", *Appl. Phys. Lett.* 67 1995 2281

Kido J., Hongawa H., Okuyama K., Nagai K., White light-emitting organic electroluminescent devices using the poly(N-vinylcarbazole) emitter layer doped with three fluorescent dyes *Appl. Phys. Lett.* 64, (November 1994) 815, doi:10.1063/1.111023

Kido J., Ikeda W., Kimura M., Nagai K., White-Light-Emitting Organic Electroluminescent Device Using Lanthanide Complexes, *Jpn, J.Appl.Phys.* 35 (1996) L394

Kido J., Matsumoto T., Nakada T., Endo J., Mori K., Kawamura N., Yokoi A., *SID 03 Digest*,34 (2003) 964 (×)

Kim J H, Herguth P, Kang M S, Jen A K Y, Tseng Y H and Shu T F, Bright white light electroluminescent devices based on a dye-dispersed polyfluorene derivative, *Appl. Phys. Lett.* 85 (2004) 1116

Kim J S, Ho P H, Greenham N C and Friend R H, Electroluminescence emission pattern of organic light-emitting diodes: Implications for device efficiency calculations, *J. Appl. Phys.* 88 (2000)1073

Kim J.S., Seo B.W., Gu H.B., "Exciplex emission and energy transfer in white light-emitting organic electroluminescent device", *Synth. Met.*, 132, 2003, 285.

Kim Y.-C. and Do Y.R., Nanohole-templated organic light-emitting diodes fabricated using laser-interfering lithography: moth-eye lighting, *Opt. Exp.* 13 (2005) 1598

Kitamura M., Iwamoto S. and Arakawa Y., Enhanced light emission from an organic photonic crystal with a nanocavity, *Appl. Phys. Lett.* 87 (2005) 151119

Ko C W and Tao Y T Bright white organic light-emitting diode, *Appl. Phys. Lett.* 79 (2001) 4234

Ko Y.W, , Chung C.H., Lee J. H., Kim Y.H., Sohn C.Y., Kim B.C., Hwang C.S., Song Y.H., Lim J, Ahn Y.J., Kang G.W., Lee N., Lee C., Efficient white organic light emission bysingle emitting layer *Thin Solid Films*, 426, 1-2, (February 2003) 246- 249

Kolosov D, English D S, Bulovic V, Barbara P F, Forrest S R, and Thompson M E, Direct observation of structural changes in organic light emitting devices during degradation, *J. Appl. Phys.*, 90 (2001) 3242–3247

Kondakov D Y, Sandifer J R, Tang C W, and Young R H, Nonradiative recombination centers and electrical aging of organic light-emitting diodes: direct connection between accumulation of trapped charge and luminance loss, *J. Appl. Phys.,* 93 (2003)1108–1119,

Koo Y.W., Chung C.H., Lee J.H., Kim Y.H., Sohn C.Y., Kim B.C., Hwang C.S., Song Y.H., Lim J., Ahn Y.J., Kang G.W., Lee N., Lee C., Efficient white organic light emission by single emitting layer, *Thin Solid Films*, 426,1-2, (February 2003), 246-249.

Kumar A., Srivastava R., Bawa S. S., Singh D., Singh K., Chauhan G., Singh I., Kamalasanan M. N. White organic light emitting diodes based on DCM dye sandwiched in 2-methyl-8-hydroxyquinolinolatolithium, *Journal of Luminescence*, In Press, (April 2010) Available online

Lakowicz J.R., Principles of Florescence spectroscopy, KA/PP, second ed., 1999.

Lee J I, Chu H Y, Kim S H, Do L M, Zyung T and Hwang D H, White light emitting diodes using polymer blends, *Opt. Mater.* (2002) 205-210,

Lee S T, Gao Z Q, and Hung L S, Metal diffusion from electrodes in organic light-emitting diodes, *Appl. Phys. Lett.,* 75 (1999) 1404–1406,.

Lee T. W., Park O. O., Cho H. N., and Kim Y. C.: Cascade energy transfer in dye-doped ternary polymer blend light-emitting diodes Synthetic Metals, Volume 131, Issues 1-3, 20 November 2002, Pages 129-133

Lee Y. J., Kim Y. H., Kim G. H., Lee Y. H., Cho S. H., Song Y. W., Kim Y. C., Do Y.R., Far-field radiation of photonic crystal organic light-emitting diode, *Opt. Exp.* 13 (2005) 5864

Lee Y.J., Kim S.H., Huh J., Kim G.H. and Lee Y.H., High-extraction-efficiency nanopatterned organic light-emitting diode, *Appl. Phys. Lett.* 82 (2003) 3779

Li G and Shinar J, Combinatorial fabrication and studies of bright white organic light-emitting devices based on emission from rubrene-doped 4,4′-bis(2,2′-diphenylvinyl)-1,1′-biphenyl, *Appl. Phys. Lett.* 83 (2003) 5359

Li, G.; Kim, C. H.; Zhou, Z.; Shinarb, J.; Okumoto, K.; Shirota, Y. , Combinatorial study of exciplex formation at the interface between two wide band gap organic semiconductors, *Appl. Phys. Lett.* 88, (June 2006) 253505,doi:10.1063/1.2202391

Li, M.; Li, W.; Chen, L.; Kong, Z.; Chu, B.; Li, B.; Hu, Z.; Zhang, Z. Tuning emission color of electroluminescence from two organic interfacial exciplexes by modulating the thickness of middle gadolinium complex layer, *Appl. Phys. Lett.* 88, (March 2006), 091108, doi:10.1063/1.2181194

Liang F.S., Cheng Y.X., Su G.P., Ma D.G., Wang L.X., Jing X.B., Wang F.S., "White organic electroluminescence based on a new boron complex", *Synthetic Metals*, 137, 1-3, 4 April 2003, (1109-1110).

Liang, C. J.; Choy, W. C. H., Color tunable organic light-emitting diodes by using europium organometallic complex *Appl. Phys. Lett.* 89, (December 2006), 251108, doi:10.1063/1.2405420

Liao L.S.,Klubek K.P., Tang C.W., High-efficiency tandem organic light-emitting diodes., *Appl. Phys. Lett.* 84, (2004),167 doi:10.1063/ 1.1 638624

Liew Y F, Aziz H, Hu N-X, Chan H S O, Xu G, and Popovic Z, Investigation of the sites of dark spots in organic light-emitting devices, *Appl. Phys. Lett.,* 77 (2000) 2650–2652

Lim J T, Lee M J, Lee N H, Ahn Y J, Lee C H and Hwang D H, "Efficient white light emitting devices with a blue emitting layer doped with a red dye", *Curr. Appl. Phys.* **4** (Apr 2004) 327

Lim J T, Lee N H, Ahn Y J, Kang G W and Lee C H, White-light-emitting devices based on organic multilayer structure *Current Applied Physics*, 2, 4, (August 2002), 295-298

Lim J., Oh S.S., Kim D.Y., Cho S.H., Kim I.T., Han S. H., Takezoe H., Choi E.H., Cho G. S., Seo Y.H., Kang S. O., Park B., Enhanced out-coupling factor of microcavity organic light-emitting devices with irregular microlens array, *Opt. Exp.* 14 (2006) 6564

Liu C., Kamaev V. and Vardeny Z.V., Efficiency enhancement of an organic light-emitting diode with a cathode forming two-dimensional periodic hole array, *Appl. Phys. Lett.* 86 (2005) 143501

Liu H. K., Hsueh H. T. and Chang C., Using holography to increase the light output efficiency of an organic light-emitting diode angular, wavelength, and near-field effects *Opt. Eng.* 44 (2005) 111306

Liu S, Huang J, Xie Z, Wang Y and Chen B, Organic white light electroluminescent devices, *Thin Solid Films*, 363, 1-2, (March 2000), 294-297.

Liu Y., Guo J.H., Zhang H.D., Wang Y., "Highly Efficient White Organic Electroluminescence from a Double-Layer Device Based on a Boron Hydroxyphenylpyridine Complex", *Angew. Chem. Int. Ed.*, 41, 2002, 182.

Lu M.-H., Madigan C. F. and Strurm J. C., Improved external coupling efficiency in organic light-emitting devices on high-index substrates, *International Electron Devices Meeting (IEDM)* 007 (2000) 607

Lu M.H. and Sturn J.C., Optimization of external coupling and light emission in organic light-emitting devices: modeling and experiment, *J. Appl. Phys.* 91, (2002) 595, doi:10.1063/1.1425448

Madigan C. F., Lu M. H. and Sturm J. C., Improvement of output coupling efficiency of organic light-emitting diodes by backside substrate modification, *Appl. Phys. Lett.* 76 (2000) 1650

Marks R N, The Optical and Electronic Response of Poly(p-Phenylene Vinylene) Thin Film Devices Ph.D. thesis, University of Cambridge, Cambridge, UK, 1993.

Mazzeo M, Pisignano D, Favaretto L, Sotgiu G, Barbarella G, Cingolani R and Gigli G, White emission from organic light emitting diodes based on energy down-convertion mechanisms, *Synthetic Metals*, 139, 3, (October 2003),675-677.

Mazzeo M, Vitale V, Sala F D, Anni M, Barbarella G, Favaretto L, Sotgiu G, Cingolani R and Gigli G, Bright White Organic Light-Emitting Devices from a Single Active Molecular Material, *Advanced Materials*, 17, 1, (January, 2005),34-39

Mazzeo M., Pisignano D., Della Sala F., Thompson J., Blyth R.I.R., Anni M., Gigli G., Cingolani R., Sotgiu G., Barbarella G., "Organic single-layer white light-emitting diodes by exciplex emission from spin-coated blends of blue-emitting molecules", *Appl. Phys. Lett.*, 82 2003 (334-336).

Mazzeo M., Thompson J., Blyth R.I.R., Anni M., Gigli G., Cingolani R., "White light from blue: white emitting organic LEDs based on spin coated blends of blue-emitting molecules", *Physica E: Low-dimensional Systems and Nanostructures*, 13, 2-4, March 2002, Pages 1243-1246.

Misra A., Kumar P., Kamalasanan M N and Chandra S., "White organic LEDs and their recent advancements", *Semicond. Sci. Technol.*, 21, 2006, R35–R47.

Misra A., Kumar P., Kamalasanan M. N. and Chandra S., White organic LEDs and their recent advancements, *Semicond. Sci. Technol.* 21 (2006) R35–R47

Moller S., Forrest S. R., Improved light out-coupling in organic light emitting diodes employing ordered microlens arrays, *J. Appl. Phys.* 91 (2002) 3324.

Movaghar B., Grunewald M., Ries B., Bassler H., Wurtz D., Diffusion and relaxation of energy in disordered organic and inorganic materials. *Phys. Rev. B* 33, (1986) 5545–5554,

Murano S et al, Highly efficient white organic light emitting diodes for lighting applications. *Proc. SPIE,* 5937 59370H (2005)

Nakamura T., Tsutsumi N., Juni N., Fujii H., Improvement of coupling-out efficiency in organic electroluminescent devices by addition of a diffusive layer, *J. Appl. Phys.* 96 (2004) 6016

Neyts K, "Microcavity effects and the outcoupling of light in displays and lighting applications based on thin emitting films", Applied Surface Science, 244, 1-4, 15 May 2005, 517-523

Neyts K. and Nieto A.U., Importance of scattering and absorption for the outcoupling efficiency in organic light-emitting devices, *J. Opt. Soc. Am. A* 23 (2006) 1201

Niu J, Li W, Wei H, Li M, Su W, Xin Q, Zhang Z and Hu Z, *"White organic light-emitting diodes based on bis(2-methyl-8-quinolinolato) (para-phenylphenolato)aluminium(III) doped with tiny red dopant"* J. Phys. D: Appl. Phys. *38 2005 1136.*

Noda T., Ogawa H., and Shirota Y., "A Blue-Emitting Organic Electroluminescent Device Using a Novel Emitting Amorphous Molecular Material, 5,5'-Bis(dimesitylboryl)-2,2'-bithiophene", *Adv. Mater. (Weinheim, Ger.),* 11, 1999, 283.

Okumoto K. and Shirota Y., "Exciplex formation at the organic solid/solid interface and tuning of the emission color in organic electroluminescent devices", *Journal of Luminescence,* 87-89, May 2000, 1171-1173.

Optoelectronics Industry Development Association (OIDA) 2002 Organic light emitting diodes (OLEDs) for general illumination update 2002 Report August

Paik K.L., Baek N.S., Kim H.K., Lee J.H., Lee Y., Macromolecules 35 (2002) 6782. White Light-Emitting Diodes from Novel Silicon-Based Copolymers Containing Both Electron-Transport Oxadiazole and Hole-Transport Carbazole Moieties in the Main Chain pp 6782–6791 Publication Date (Web): August 2, 2002 (Article) DOI: 10.1021/ma020406r

Park J H, Lee T W, Kim Y C, Park O O and Kim J K, "White polymer light-emitting devices from ternary-polymer blend with concentration gradient" *Chem. Phys. Lett.* **403** 2005 293

Peng H J, Wong M and Kwok H S, "Design and Characterization of Organic Light Emitting Diodes with Microcavity Structure" SID 03 DIGEST 516 78.

Peng H. J., Ho Y. L., Yu X. J. and Kwok H.S., *J. Appl. Phys.* 94 (2004) 1649. (×)

Peng H., Ho Y.L., Yu X.-J., Wong M. and Kwok H.-S., Coupling Efficiency Enhancement in Organic Light-Emitting Devices Using Microlens Array-Theory and Experiment, *J. Display Technol.* 1 (2005) 278

Pfeiffer M, Forrest S R, Leo K, and Thompson M E, Electro phosphorescent p-i-n organic light-emitting devices for very-high-efficiency flat-panel displays, *Adv. Mater.,* 14 (2002)1633–1636

Pope M., Kallmann H. P., and Magnante P., Electroluminescence in Organic Crystals, *J. Chem. Phys.* 38, (1963) 2042

Popovic Z D and Aziz H, Reliability and degradation of small molecule-based organic light emitting devices (OLEDs), *IEEE J. Sel. Top. Quant. Electron.*, 8 (2002) 362–371

Rai V. K., Sriva stava R., Chauhan G., Saxena K., Chand S., Kamalsanan M.N., Jpn. J. Appl. Phys. 47 (2008) May pp. 3408-3411, Trap Assisted Carrier Recombination in 4-(Dicyanomethylene)-2-methyl-6-(4-dimethylaminostyryl)-4H-pyran Doped Bis(2-(2-hydroxyphenyl)bezoxazolate) Zinc.

Rai V. K., Srivastava R., Kamalasanan M. N. White organic electroluminescence from fluorescent bis (2-(2-hydroxyphenyl) benzoxazolate) zinc doped with phosphorescent material *Journal of Luminescence*, 130, 2, (February 2010), 249-253

Rai V. K., Sriva stava R., Kamalasanan M.N. White organic light-emitting diodes based on blue fluorescent bis(2-(2hydroxyphenyl)benzoxazolate)zinc ($Zn(hpb)_2$) doped with DCM dye, *Synthetic Metals*, 159, 3-4, (February 2009), 234-237

Ries B. and Bassler H., Monte Carlo study of dispersive charge-carrier transport in spatially random systems with and without energetic disorder *Phys. Rev. B* ,35, (1987) 2295–2302, DOI:10.1103/

Ries B., Bassler H., Grunewald M., Movaghar B., Monte Carlo study of relaxation and diffusion in glassy systems *Phys. Rev. B*, 37, (1988) 5508–5517 DOI:10.1103/PhysRevB.37.5508

Romdhane S., Marai F., Hassine L., Fave J.L., Roussel J., Majdoub M., Bouchriha H., UV and white electroluminescence in AVB polymer Synthetic Metals, Volume 139, Issue 2, 5 September 2003, Pages 245-249

Sano T, Fijii H, Nishio Y, Hamada Y, Takahashi H, and Shibata K, Organic electroluminescent devices doped with condensed polycyclic aromatic compounds, *Synth. Met.*, 91 (1997) 27–30

Saxena K., Mehta D. S., Rai V. K., Srivastava R., Chauhan G. and Kamalasanan M. N., Implementation of anti-reflection coating to enhance light out-coupling in organic light-emitting devices, *J. of Luminescence* 128 (2008) 525–530

Shao Y and Yang Y, White organic light-emitting diodes prepared by a fused organic solid solution method.*Appl. Phys. Lett.* 86, 0(February 2005)73510, doi:10.1063/ 1.1866216

Shi J and Tang C W, Doped organic electroluminescent devices with improved stability, *Appl. Phys. Lett.*, 70 (1997) 1665–1667

Shoustikov A. A ., You Y., Thompson M.E., Electroluminescence color tuning by dye doping in organic light-emitting diodes, IEEE J. (Sel. Top.) Quantum Electron. 4(199 8) 3-13. DOI: 10.1109/2944.669454, ISSN: 1077-260X, Jan/Feb

Singha T., Fujikawa H., Taga Y., "Design of multiwavelength resonant cavities for white organic light-emitting diodes", J. Appl. Phys. **93**, 2003, 19.

Srivastava R., Chauhan G.,Saxena K., Bawa S. S., Srivastava P. C., Kamlasanan M.N., Fabrication of white organic light emitting diodes by co-doping of emissive layer, 47 (January 2009) 19-23

Strukeji M, Jordan R H and Dodabalapur A, Organic Multilayer White Light Emitting Diodes, *J. Am. C hem. Soc.* 118 (1996) 1213

Su, W. M.; Li, W. L.; Xin, Q.; Su, Z. S.; Chu, B.; Bi, D. F.; He H.; Niu, J. H., Appl. Phys. Lett. 91, 043508 (2007); July , doi:10.1063/1.2762298 (3 pages) Effect of acceptor on efficiencies of exciplex-type organic light emitting diodes

Sun J X, Zhu X L, Peng H J, Wong M and Kwok H S 2005 *EuroDisplay* p-55 397 (×)

Sun Y., Giebink N. C., Kanno H., Ma B., Thomson M. E. and Forrest S R., *Management of singlet and triplet excitons for efficient white organic light-emitting devices* Nature, *440 (2006) 908*

Szmytkowski J, Stampor W, Kalinowski J and Kafafi Z H, Electric field-assisted dissociation of singlet excitons in tris-(8-hydroxyquinolinato) aluminum (III), *Appl. Phys. Lett.* 80 (2002) 1465

Tang C W, VanSlyke S A, and Chen C H, Electroluminescence of doped organic thin films, *J. Appl. Phys.*, 65, 9 (1989) 3610–3616

Tang C. W. and VanSlyke S. A., Organic electroluminescent diodes, *Appl. Phys. Lett.*, 51, (1987) 913–915

Tasch S, List E J W, Ekstrom O, Graupner W and Leising G, Efficient white light-emitting diodes realized with new processable blends of conjugated polymers *Appl. Phys. Lett.* 71, (September 1997) 2883, doi:10.1063/1.120205

Thompson J, Blyt R I R, Mazzeo M, Anni M, Gigli G and Clinigolani R, "White light emission from blends of blue-emitting organic molecules: A general route to the white organic light-emitting diode?" Appl. *Phys. Lett.* 79 2001 560

Tokito S, Iijima T, Tsuzuki T and Sato F, High-efficiency white phosphorescent organic light-emitting devices with greenish-blue and red-emitting layers, *Appl. Phys. Lett.* 83 (2003) 2459

Tong, Q. X.; Lai, S. L.; Chan, M. Y.; Tang, J. X.; Kwong, H. L.; Lee, C. S.; Lee, S. T., High-efficiency nondoped white organic light-emitting devices, *Appl. Phys. Lett.* 91, (July 2007), 023503, doi:10.1063/1.2756137

Tsai M L, Liu C Y, Hsu M A and Chaw T J, White light emission from single component polymers fabricated by spin coating, *Appl. Phys. Lett.* 82 (2003) 550-552.

Tsutsui T., Tahiro M., Yokogawa H., Kawano K. and Yokoyama M., Doubling Coupling-Out Efficiency in Organic Light-Emitting Devices Using a Thin Silica Aerogel Layer, *Adv. Mater.* 13 (2001) 1149

Tu G, Mei C, Zhou Q, Cheng Y, Geng Y, Wang L, Ma D, Jing X and Wang F, Highly Efficient Pure-White-Light-Emitting Diodes from a Single Polymer: Polyfluorene with Naphthalimide Moieties, *Advanced Functional Materials*, 16, 1, (January, 2006), 101-106

Tu G, Zhou Q, Cheng Y, Wang L, Ma D, Jing X and Wang F, White electroluminescence from polyfluorene chemically doped with 1,8-napthalimide moieties *Appl. Phys. Lett.* 85, (July 2004), 2172, doi:10.1063/1.1793356

Turro N. J., Modern Molecular Photochemistry, University Science Books, Mill Valley, CA, 1991.

Tyagi P., Srivastava R., Kumar A., Rai V. K., Grover R. and Kamalasanan M. N., White electroluminescence from stacked organic light emitting diode *Synthetic Metals*, 160, 7-8, (April 2010), 756-761

Tyagi, P., Srivastava, R., Kumar A., Chauhan, G., Kumar A., Bawa, S.S., Kamalasanan M.N., Low voltage organic light emitting diode using p–i–n structure. *Synth. Metals*, 160 (9), (May 2010) 1126-1129

UV and white electroluminescence in AVB polymer Synthetic Metals, Volume 139, Issue 2, 5 September 2003, Pages 245-249

Wang J., Kawabe Y., Shaheen S.E., Morrell M.M., Jabbour G.E., Lee P.A., Anderson J., Armstrong N.R., Kippelen B., Mash E.A., and Peyghambarian N., "Exciplex Electroluminescence from Organic Bilayer Devices Composed of Triphenyldiamine and Quinoxaline Derivatives", *Advanced Materials*, 10, 3, February 1998, 230-233.

Wang L, Lei G and Qiu Y, "Bright white organic light-emitting diodes based on two blue emitters with similar molecular structures", *J. Appl. Phys.* 97 2005 114503

Wang Y M, Teng F, Xu Z, Hou Y B, Yang S Y, Qian L, Zhang T and Liu D A," White emission via electroplex formation at poly(*N*-vinylcarbazole)/2,9-dimethyl-4,7-diphenyl-1,10-phenanthroline interface" *Appl. Surf. Sci.* **236** 2004 251

Wong F.L., Fung M.K., Tao S.L., Lai S.L., Tsang W.M., Kong K.H., Choy W.M., Lee C.S. and Lee S.T. " Long-lifetime thin film encapsulated organic light-emitting diodes" J. Appl. Phys. 104, 014509(11 July 2008)DOI: 10.1063/1.2940727

Wu Y S, Hwang S W, Chen H H, Lee M T, Shen W J and Chen C H, Efficient white organic light emitting devices with dual emitting layers *Thin Solid Films*, 488, 1-2, (September 2005), 65-269

Wyszelki G. and Stiles W.S., (1982). *Color Science*, 2nd ed., Wiley, New York

Xie W, Wu Z, Lee N H, Lee M J, Song J H, Lee C and Hwang D H, Efficient white organic electroluminescent devices consisting of blue- and red-emitting layers *Materials Science and Engineering: C*, 24, 1-2,(January 2004), 233-235

Yamasaki T., Sumioka K. and Tsutsui T., Organic light-emitting device with an ordered monolayer of silica microspheres as a scattering medium, *Appl. Phys. Lett.* 76 (2000) 1243

Yang J P, Jin Y D, Heremans P L, Hoefnagels R, Dieltines P, Blockhuys F, Geise H J, Auweraer M V and Borghs C, White light emission from a single layer organic light emitting diode fabricated by spincoating, *Chemical Physics Letters*, 325, 1-3, (July 2000), 251-256

Yang K, Gao W, Zhao J, Sun J, Lu S and Liu S, An efficient and bright organic white-light-emitting device *Synth. Met.* 132 (2002) 43

Yang Y., Pei Q, Efficient blue-green and white light-emitting electrochemical cells based on poly(9,9-bis(3,6-dioxaheptyl)-fluorene-2,7-diyl), *J.Appl. Phys.* 81 (1997) 3294

Young R. H., Tang C. W. and Marchetti A. P., Current-induced fluorescence quenching in organic light-emitting diodes, *Appl. Phys. Lett.* 80 (2002) 874

Zhang X.H., Liu M.W., Wong O.Y., Lee C.S., Kwong H.L., Lee S.T., Wu S.K., Blue and white organic electroluminescent devices based on 9,10-bis(2' naphthyl) anthracene, *Chemical Physics Letters*, 369, 3-4, (February 2003), 478-482

Zhang Y, Cheng G, Zhao Y, Hou J and Liu S, White organic light-emitting devices based on 4,4'-bis(2,2'-diphenyl vinyl)-1,1'-biphenyl and phosphorescence sensitized 5,6,11,12-tetraphenylnaphthacene *Appl. Phys. Lett.* 86 (December 2005)11112, doi:10.1063/1.1845593

Zhang Z L, Jiang X Y, Zhu W Q, Zhang B X and Xu S H, A white organic light emitting diode with improved stability *J. Phys. D : A ppl. Phys.* 34 (2001) 3083.

Zhang Z., Jiang X.Y., Zhu W.Q., Zheng X.Y., Wu Y., Xu S.H., Blue and white emitting organic diodes based on anthracene derivative, *Synthetic Metals*, 137, 1-3, (April 2003), 1141-1142

Zheng X.Y., Zhu W.Q., Wu Y.Z., Jiang X.Y., Sun R.G., Zhang Z.L., Xu S.H., A white OLED based on DPVBi blue light emitting host and DCJTB red dopant *Displays*, 24, 3, (October 2003), 121-124